CW01190702

DREAM LAND

Foreword by **GEORGE KNAPP**

An autobiography by

BOB LAZAR

Dreamland: An Autobiography by Bob Lazar
Copyright © 2019 by Interstellar

All rights reserved. No portion of this book may be reproduced in any fashion, print, facsimile, or electronic, or by any method yet to be developed, without express written permission of the publisher.

Interstellar
2033 San Elijo Ave. #403
Cardiff By The Sea, CA 92007
InterstellarBooks.com

Book Design by Lamp Post

Manufactured in the United States of America

ISBN 978-0-578-43705-7 (Hardcover)
ISBN 978-0-578-52616-4 (eBook)

Distributed worldwide by Simon & Schuster

DREAM LAND

FOREWORD
by George Knapp

IN THE DECADES SINCE BOB LAZAR FIRST STEPPED FORward, Area 51 has morphed into an almost humorous contradiction. It is by far the best known "secret" base in the world. Whether this contradiction is good or bad, Lazar is the person most responsible for it. His tale changed an otherwise obscure outpost in the Nevada desert into a rock star of military facilities.

Since Lazar's story first exploded into the public consciousness in 1989, Area 51 has been catapulted to the forefront of UFO lore, along with the Roswell crash, Kenneth Arnold's sightings, the abductions of Betty and Barney and Travis Walton. It has inspired several movies, documentaries, books, thousands of magazine and newspaper articles, heated debates at UFO conferences and in online UFO social media forums, cartoons, and columns. It prompted the State of Nevada to create and christen the nation's only Extraterrestrial Highway; became the theme and namesake of a Triple A baseball team and a couple of roadhouse bars; and was adapted into rock songs, rock bands, poems, video games, posters, tee shirts, trinkets, shot glasses, key chains, snow globes, Christmas ornaments, viewer's guides, fireworks, jerky kiosks, and countless comedy routines. It is likely the only military base in the world to become the theme for a legal bordello and is also an interactive attraction in a Smithsonian-affiliated national museum. It is permanently perched in the pantheon of UFO holies, a god among lesser deities, much to the chagrin of UFO poobahs whose fervent protestations about Lazar have largely fallen on deaf ears.

A handful of revisionist writers have tried to defend some of the biggest, most self-serving lies told about Area 51, weakly arguing that it was never a secret base, was never all that classified, and isn't really a big deal. This is disingenuous hair-splitting at best, and deserves a spot in your mental filing cabinet under the letters B and S.

It is true that on May 18, 1955, a news release was issued by the Atomic Energy Commission announcing that "preliminary work" had begun "on a small satellite installation" within the Las Vegas Bombing and Gunnery Range. The construction will be "essentially temporary," the AEC claimed. That isn't how it worked out.

The statement was sent to eighteen news organizations, including my future employer KLAS-TV. In the months that followed, other details were shared in terse news releases, indicating the AEC's contractor was adding some "limited additional facilities and modifications" to a small installation near Groom Lake. In documents made public many years later, the AEC and its contractors referred to the location as Watertown, or the Watertown Strip. Government spokespersons said Watertown would be the hub of a program to obtain high altitude weather data for the Air Force, but the government wasn't being honest, and it wouldn't be the last lie that would be told about the base.

Watertown, the Watertown Strip, Paradise Ranch, the Box, Groom Lake, and Dreamland are names used over the ensuing decades to describe the enigmatic operation in the northeast corner of what was once known as the Nevada Test Site. The most common name for the once-tiny outpost though, the name that is now known all over the planet, is Area 51.

The cover stories issued by various authorities proved to be wildly inaccurate. The facility was built to house the U-2 spy plane, the most highly classified program of its day. U-2 flights did collect weather data, but their primary mission was to spy on America's adversaries. The small, temporary facility wasn't temporary at all. Today it is a sprawling complex that employs close to 2,000 scientists, technicians, pilots, radar specialists, contractors, security personnel, and assorted spooks from the world of military intelligence.

It is accurate to say the base's existence wasn't entirely a secret. I obtained a pair of phone books from the old AEC era, and inside the page are several now defunct numbers for personnel assigned to Area 51. Because the base was to the east of America's primary atomic testing ground, it was often bathed in radioactive fallout during the years when above-ground nuclear blasts were uncorked in the Nevada desert. The designation "Area 51" is found on several fallout maps from the '50s and '60s. But by the late '70s, Area 51 had all but disappeared from those maps. The quadrangle-shaped box on the maps was left empty. By the 1980s, reporters who asked questions about the base were told that no such base exists, even though, for a time, it was possible to see the base from the tops of nearby mountain tops. Soviet spy satellites released high-resolution photos of the two-mile-long runway and support structures. Something had changed. Like a military Brigadoon, the base had essentially disappeared.

As a journalist, I am aware that I will never be allowed to set foot on the base. Yet it has become a major part of my personal identity and professional life. It is as much a part of my life as it is for the men and women who worked out there, toiling in obscurity in order to protect the rest of us during the darkest days of the Cold War. When I kick the bucket, I'm sure my obituary will contain some reference to the Area 51 stories I have written throughout my career.

When I was first hired by KLAS-TV in 1981, my friend, mentor, and news director, Robert Stoldal, handed me a thin file he had collected over the years. It contained twenty-four or so pages of news clippings he had culled during his long career as a newsman. This was the sum total of what the public knew about Area 51 at the time. There were a few fuzzy references to the base in newspaper articles from the 1970s. Most of the pages were clippings from *Aviation Week & Space Technology*, an excellent publication whose readers sometimes referred to the magazine as "Aviation Leak." Stoldal had developed a passionate interest in the secret base and thought I might be the right guy to eventually find out what goes on out there.

I didn't write any Area 51 stories until a few years later. In 1984, the U. S. Air Force illegally seized control of 89,000 acres of public land adjacent to Groom Lake. One day, the land was accessible; the next, it was off limits. Armed guards patrolled the perimeter. Ranchers, rock hounds, even property owners who tried to traverse the new line in the sand were warned to stay out or else. Signs warned that the "use of deadly force" had been authorized. The Air Force acknowledged it seized the land without any legal permission. Months later, AF officials appeared before Congress to ask for retroactive authority to do what they had already done. Elected officials in Nevada griped about it, as did the public, but the Pentagon had no intention of re-opening the 89,000 acres to the public. Nor would they say why they needed all that acreage.

The extraordinarily brazen move by the Air Force generated headlines for a few months. A KLAS reporter named Richard Urey produced a multi-part series about the known history of Area 51, but the buzz died down.

DREAMLAND

For years, anyone who called Nellis Air Force base to ask about Area 51 was told the military could not confirm the existence of such a facility. Thus, it became a non-existent base, a preposterous position that was maintained for years. Taxpayers and voters funded all of the work that occurred at the base. Visitors could see it with their own eyes, or via telescopes. Foreign satellites photographed it and foreign planes had permission to fly over it via the Open Skies treaty, but American citizens were told it did not exist.

The Air Force position came to be even more extreme in the 1990s when it argued in court filings that it never used the name Area 51. The claims were made in connection with lawsuits filed by Groom Lake employees, who sued the government to find out what toxic chemicals they had been exposed to while working out there. Secrecy was so extreme that nothing could ever leave Area 51, not even trash, so the base's disposal system consisted of digging large trenches, filling them with piles of discarded equipment and composite materials used in the development of classified projects such as Stealth technology, then the entire pile would be doused in jet fuel and set on fire. Workers testified that huge clouds of treacherous black smoke would engulf the base and clog their lungs. After years of exposure to who-knows-what, some of the workers started getting sick. A few died. Still, their employer would not tell them what materials they had ingested. To do so might endanger national security, the Air Force lawyers argued. The case went all the way to the U. S. Supreme Court. (Attorneys for KLAS joined the case and argued for greater transparency.)

Eventually, the lawsuits became moot when an executive order from the White House exempted Area 51 from laws that govern every other piece of American soil. So much for transparency.

When I hear revisionists argue that Area 51 has never been a secret, that it isn't all that classified, it almost makes me laugh. The military and civilian contractors who operate out there have played deadly games with the public and with their own employees to maintain a level of extreme secrecy, so extreme that they deny the existence of the base, deny the name it has always used, and issue flimsy cover stories, false narratives, and flagrant denials in order to cloud and to protect one of the few places in our country where federal laws, even the Constitution, don't necessarily apply. Whenever someone tells me that Area 51 isn't all that secret, I suggest they try to drive across the invisible line outside the base and see what happens.

Security around Groom Lake was already tight when Bob Lazar entered the picture. In the years since his story exploded onto the international stage, tens of thousands of people have made the trek out into the desert to see for themselves what might be going on. Every major news organization in the world has beaten a path to Area 51's door. On any given night, dozens of aviation buffs and UFO hunters are camped around the perimeter, trying to catch glimpses of whatever might be flying around out there. In response to this ongoing invasion of the curious, base security measures have hardened and multiplied. There are sensors on public land, miles from the actual perimeter, which warn Area 51 personnel whenever vehicles are approaching. The primary entry points are ringed with cameras, some disguised as desert flora. There are other, more advanced sensors that can differentiate the smell of a human versus, say, a wild horse. There are layers of armed security on the perimeter, which is 12 miles outside the principal facilities at Groom Lake. Any plane that gets too close to the forbidden zone will be met by F-16s. Military pilots who accidentally stray into "The Box" are grounded, scolded, and debriefed when they land. Hikers who wander too close to the invisible line get buzzed by helicopter gunships or chased by the fearsome "Cammo Dudes." They take their security very seriously out there, especially for a facility that either doesn't exist or supposedly isn't all that classified. Many times over the years, UFO fans have contacted me with the suggestion that we should assemble an armada of Winnebagos and then coordinate a massive invasion of Area 51, in the mistaken belief that "they can't possibly stop all of us." I have strongly advised against this foolhardy plan because I am certain the base would have no problem repelling such an incursion.

A similar suggestion took on a life of its own starting in the spring of 2019. Bob Lazar and filmmaker Jeremy Corbell sat down for an interview with the world's most popular podcaster, comedian and television personality Joe Rogan. That interview was viewed millions of times within a matter of days. It also inspired one listener, a young man named Matty Roberts, to create a Facebook page which urged the public to "Storm Area 51." Roberts said the idea was a joke, pure satire, and he never meant for it to be taken seriously. But within a few days, the page he created became a viral sensation. Three weeks later, the page had been viewed 2.8 million times, and more than 1.5 million people had pledged to participate in the plan to perform a mass public invasion of the secret base. Another million or so people said they were thinking about participating. For one reason or another, Rogan's interview and Corbell's film helped renew interest in Lazar's tale, and the discussion touched a nerve with the public. Dozens of humorous memes popped up along with tweets from celebrities who say they planned to participate. At the time of this writing, it is unclear what will happen on the date designated for the event. The Air Force has issued a warning that it would be dangerous for anyone to try to enter the secure military facility, but businesses in and around the base are already bracing for a huge wave of visitors.

When Bob Lazar was first interviewed, anonymously and in silhouette, on KLAS-TV in May 1989, the story he told exploded. Las Vegas talk radio stations jumped on it, debating the merits almost nightly. A popular host named Billy Goodman organized bus trips to the outskirts of Groom Lake for UFO viewing parties. The innocuous but ominous-sounding "black mailbox" was transformed into a seemingly sacred gathering site. The public debated the identity of the mystery man who claimed to have worked at a secret facility near Papoose Dry Lake, a place he referred to as S-4. The tiny town of Rachel became a UFO mecca of sorts, and the friendly folks at the Rachel Bar and Grill were thrilled to welcome the large influx of UFO-curious customers.

That first burst of public interest was a blip on the Richter scale compared to what followed. For seven months between May and November of 1989, I was assigned to learn all that I could about Lazar, and about UFOs. It was a crash course, like cramming for a final exam. I read dozens of books, hundreds of newspaper and magazine articles, consulted with the most credible researchers in the field, and tried to get my head around the basics of what proved to be a very complicated subject. I was cocky and naïve enough to believe at the time that what the UFO topic needed was a thorough vetting by a journalist. I even boasted to co-workers that I should be able to figure the whole thing out in about six months. Bing, bam, boom—mystery solved. Suffice to say, that prediction proved to be wildly inaccurate. Three decades later, I still don't have solid answers about any of the biggest questions.

During those ensuing months, my photographer Bryant Blackburn and I traveled extensively in an effort to learn about Lazar and about the mysterious base where he said he worked. On dozens of nights, we parked our vehicle in various locations on the outskirts of Papoose Dry Lake and Area 5, foolishly thinking we were avoiding the scrutiny of security personnel. Our Wednesday night vigils never succeeded in obtaining any video of UF-type craft. I dug into Lazar's background, checked every reference on the unique resume he had provided to us, spoke to his friends, spent time getting to know his inner circle, and even traveled to Los Alamos National Lab with Lazar to find out if he really did know his way around the base, where he supposedly worked for a time in the early '80s. In the pre-Internet days, it was much more difficult to reconstruct someone's personal life than it is today, but Lazar proved to be in a category all his own. Much of what we tried to obtain proved to be unavailable or non-existent.

In the years since the Lazar story first surfaced, many researchers have conducted their own investigations into his background. Several well-known UFO authorities have claimed they managed to unearth definitive evidence to prove that Lazar is a liar. The late nuclear physicist Stanton Friedman, one of the best known and most respected UFO investigators, became the highest profile Lazar debunker, and although he often told his audiences that he was done with the case, he continued to write and speak about Lazar right up until his death in 2019. Bob Lazar became Friedman's white whale. Stan was a friend of mine and I greatly respected his decades of work on the UFO subject, but he had a blind spot when it came to Lazar. And contrary to his claims about how he exposed Lazar to be a fraud, the information he gleaned about Lazar's background did not constitute a "gotcha" moment. They were reported on KLAS-TV in the very first news stories we aired about Lazar.

DREAMLAND

In November 1989, we unleashed a massive, multi-part news series called *UFOs: The Best Evidence.* In Part 5 of that series, we revealed Lazar's identity, showed his face, and allowed him to tell the broad outlines of his astonishing tale. Our news series was—and still is—the highest rated, most viewed news project ever produced in Las Vegas. Transcripts of the nightly segments were sent all over the world via a fledgling service known as the Internet. UFO writers, researchers and luminaries all weighed in on the subject of Lazar's veracity. UFO tourism exploded to the point that the Rachel Bar and Grill changed its name to the Little A'Le'Inn.

The UFO critics who later claimed they had debunked Lazar's background conveniently overlooked the information that was conveyed in each of those initial news stories. The schools listed on Lazar's resume told me they had no records he ever attended courses there and no record of any advanced degrees. Los Alamos Lab told me they had no record of an employee named Robert Lazar. I included that information in the very first stories about Lazar, though—in their zeal to be the first debunker of the tale—different UFO bigwigs took turns pretending they had dug up inflammatory dirt on their own. I give credit to those who actually did some of the legwork in covering the same ground we covered months earlier, but the information about gaps in Lazar's story was part of the public record from day one.

During the early stages of our investigation of Lazar's claims, the absence of school records was a major concern.

If he would lie about where he went to college, might he also lie about working on flying saucers out at S-4? Lazar told us that someone was erasing the paper trail of his life to that point. We decided to focus on his claim about working as a physicist at Los Alamos National Lab. If he really was employed there, and if it could be proven that he worked on sensitive or classified projects, then it was at least plausible that he could have been hired to work on another sensitive, defense-related project in the Nevada desert. It would also make sense that he must have had some level of education to make it into the door of Los Alamos, given that no one is hired to work there without some sort of college degree. More importantly, it would mean he had been given a security clearance, which would have required a background check to verify his credentials.

After my first inquiry to LANL, I thought the story might be dead in the water. A spokesperson told me the lab had no record of anyone named Robert Lazar ever working there. Damn. In the years since I mailed that first letter to the lab, critics have made a big deal out of whether Lazar claimed to have worked "for" the lab or "at" the lab. He always told me he worked at the lab, including some time inside its Meson facility, a large structure that houses a massive particle accelerator. In real life, that might seem like a nitpicky point. In the bizarre parallel reality known as "UFO World," though, it can—and did—spark arguments that have persisted for decades.

The response from the lab took the air out of my lungs. I thought the story might have hit a dead end. If we could not prove Lazar was employed at LANL, if the official denials held up, we likely would have pulled the plug right there, and the rest of the world would never learn his name or hear the details about his experiences. But something was off. It didn't feel right. We kept poking around for more information and obtained a copy of the official LANL phone book, which listed the names of everyone who worked at the lab during the time Lazar said he was there. Sure enough, his name was listed in alphabetical order—Robert Scott Lazar. Lazar had also said in our first interview that he had been featured on the cover of the Los Alamos Monitor, a story about a jet engine he had installed in his Honda. The news report described Lazar as a physicist at Los Alamos Lab. That front-page story caught the eye of physicist Dr. Edward Teller, who was visiting LANL on the day the article appeared. I confirmed that Teller really was at the lab on that same date. The reason that point became important is that Lazar said a chance meeting with Teller on the grounds of LANL became a factor in his hiring years later to work in the Nevada desert.

Here's more minutiae that may or may not matter, depending on whether you reside in UFO World or not. I continued to communicate with LANL's public affairs team. I called. I sent letters. I sent copies of the newspaper article and the phone book entry. The staff was always courteous and professional, but it seemed like they grew weary of the back and forth. They eventually told me they found an employee number for Lazar, but added that he had been hired by a company called Kirk Mayer, a subcontractor at the lab. Kirk Mayer is essentially a headhunter firm. They hire technical and scientific personnel to fill positions at LANL. So my quest for employment records turned to Kirk Mayer.

My first inquiry was by telephone. I explained I was a reporter, seeking records about a former employee. I told them I had his signed permission to obtain the records. The person on the other end of the phone told me that yes, the company did hire Lazar and did have records and agreed to provide copies. Weeks went by, but no records arrived. I called, wrote, wrote again. Each time I communicated with Kirk Mayer, I got less information. I was told the records could not be located. Later I was told the records did not exist. Then they stopped responding to me altogether.

This went on for months. By that point, I knew Lazar was employed at the lab for a period. For one thing, we interviewed four former co-workers who verified it. In addition, Lazar told us he could take us on a tour of the lab itself. Bryant and I flew to New Mexico, met Lazar outside the lab, and carried our camera right through the gate. Bob waved to a man in the security hutch and we were allowed to drive in without anyone asking who we were or checking our identification. Lazar walked us through multiple buildings, said hello to several employees, took us on a thorough tour that culminated at the humongous particle accelerator in the meson facility. We recorded it on video. No one stopped us or asked who we were. Lazar seemed at home. (Snippets of the video from that tour were aired on KLAS months later.) In the years since the story broke, other lab employees have surfaced who remember Lazar, including a physicist who told Jeremy Corbell that he remembers being in classified briefings with Lazar. While the physicist says he never asked Lazar to prove that he had the credentials to be in the briefings, he figured Lazar belonged there.

As a young news reporter, I was surprised at how our Lazar stories were received by luminaries in the pro-UFO universe. I naively thought they might welcome a new source of information. But it quickly became obvious that in that particular slice of society, turf wars abound, backbiting and infighting are common, and everyone wants the truth about UFOs, so long as it is *their* version of truth. Mainstream media outlets at the time teed off on the story, poked fun at it, printed columns and cartoons lampooning the allegations, and generally had a good laugh—without ever doing any legwork to check out the basics of the story. Eventually, every news organization in the world would beat a path to Area 51's door, but for the first several years, the reaction among my media colleagues was pretty brutal.

The public, however, couldn't get enough. It's been thirty years since the story broke, but still, I get inquiries about Area 51 and Lazar every single day, whether I'm shopping in a grocery store or in a restaurant having dinner, even standing at a urinal in a men's room, random people approach me to ask what I really think about UFOs and Lazar. Most of them don't really care what I think. They mostly want to share their own UFO story or experience. The fact that this topic touches the public in deep and inexplicable ways is frustrating for media skeptics and professional debunkers. Too bad, I say.

No one is likely to ever prove for sure whether Lazar told the truth. I am not holding my breath for the government to suddenly open up the hangar doors and allow the rest of us to inspect, photograph, or essentially kick the tires on any recovered craft of uncertain origin. It isn't a black and white, true or false story. It is complicated, nuanced, difficult. In my experience, people tend to make up their minds about Lazar pretty quickly and then search for the set of facts that justify their point of view, while ignoring or excluding all the inconvenient parts. That is certainly the way it was with my late friend Stan Friedman. He and I had many pointed but friendly exchanges about Lazar over the years, including two face-to-face private conversations. Stan expressed the oft-repeated reasons why he doubted Lazar, and I told him the reasons why I felt differently. I am not sure he listened to any of them. He never addressed most of the counterpoints in any public forum.

My support for Lazar's claims is conditional, in that I recognize it will never be proven beyond reasonable doubt, but I have continued to report the story because of two factors. One is because of the things I personally witnessed during the months after he sat down for the anonymous "Dennis" interview. His house was broken into a few times. The intruders didn't burglarize the home. Instead, they played mind games. They moved around furniture, erased things that were written on his white board, opened windows, and made it clear that someone had been inside. They did the same to his car, leaving the doors open and the windows rolled down, his glovebox open, and a semi-automatic pistol lying on the seat. Multiple threats were made. Lazar took them seriously, and on at least two occasions, I witnessed how visibly rattled he was by the ominous phone calls he received. There also was a concerted effort to follow a small group of us that included Lazar, his friend Gene Huff, and—for a time—billionaire businessman Robert Bigelow during a period of several months when we would convene to chat about the latest developments. My phone at KLAS was tapped. I know this because six people who spoke to me by phone and offered to provide me additional information about unusual materials and technology being tested in the Nevada desert were visited by security personnel immediately after speaking to me. All six were warned to keep their mouths shut. That happened. I saw it. Today, we take for granted the notion that all of our phone calls and electronic communications are duly catalogued and likely stored in some gigantic NSA database somewhere, but back then, the knowledge that someone was listening to the phone of a journalist—without ever obtaining a warrant—was deeply unsettling.

The other aspect of Lazar's tale that is generally ignored by his critics is the list of things he told us that turned out to be true and verifiable. For one thing, he knew there was a place on the Nellis Air Force base range designated as S-4.

This had never been reported in any newspaper, but when I called Nellis to inquire about it, a spokesperson confirmed there was such a place on the Nevada range. In fact, there was more than one, according to the public affairs office at the base, though they would not tell me where the sites were or what activity takes place out there. Lazar also knew the name of the government agent who came to his home to conduct part of his background check for a security clearance. The name stood out—Mike Thigpen. Lazar thought Thigpen had said he worked for the FBI, but the FBI told me it has no such employee. A tipster confirmed to me that Thigpen was real, that he worked for an agency I had never heard of, something called the Office of Federal Investigation or OFI. OFI, I was to learn, was a little-known federal outfit that conducted background checks for persons seeking employment in classified programs. I even got a very surprised Thigpen on the phone but he would not comment. (Thirty years later, he spoke to Jeremy Corbell and confirmed that he remembered working on Lazar's security clearance.)

I filed multiple FOIA requests for Lazar's records but was always stymied. The FBI, the Navy, and other agencies denied having any records on Lazar. This became relevant in 1990 when Lazar got into some legal trouble and was facing possible prison time for his bad decisions. I reached out to U. S. Congressman James Bilbray of Nevada to see if he could help confirm Lazar's employment. Bilbray's office wrote to the court that it had made inquiries about Lazar to the FBI and had received an unusual response.

The FBI declined to provide any records and told Bilbray that the congressman "did not have a need to know" about Lazar's employment. The Bureau could have said it had no records on Lazar, but didn't. During that court proceeding, it would have been much easier for Lazar if he had fessed up to being a UFO con man, admitted he had made the whole thing up, and hoped for leniency. But he didn't do that. Instead, he told the court the same story he had told me about where he had been educated and where he had worked.

Most notably, Lazar somehow knew when and where the Navy was going to conduct test flights of its unusual craft. On three consecutive Wednesday nights in the spring of 1989, Lazar took groups of friends out into the desert to see the strange objects rise above the mountain ridge—not at Groom Lake—but rather in the area around the Papoose Dry Lake, 15 miles south of Groom and the location of what Lazar said was the S-4 installation built into the side of a mountain. Every communication I have received from the military as well as from former Area 51 employees who dismiss Lazar's claims have alleged there has never been any facility or base of any kind located near Papoose Dry Lake. I interviewed all of the people who made the treks into the desert with Lazar. All of them confirmed the glowing object they witnessed appeared to be a large, disc-shaped craft of some kind. It rose, performed simple maneuvers, then lowered back down below the mountain ridge. At least one of those sightings was recorded on video and the clip was included in my initial news reports. How did

Lazar know about the test flights? Was he, as some skeptics have claimed, merely a janitor at Area 51, or some low-level contractor who overheard the information about the test schedule? If so, then the spooks at Groom Lake had some serious security breaches back then to allow a mere janitor access to such sensitive information. Satellite photos have long confirmed that a road from Groom Lake to Papoose Lake exists, just as Lazar described it and contrary to government denials of ever having a presence at Papoose. How did Lazar know?

His most controversial claim might be the mystery material known as Element 115. Lazar is certainly not the first person to speculate about the existence of such a substance. Physicists have long theorized that a super heavy element might exist, an island of stability in the range of elements 115 and 116 on the periodic chart. When Lazar made the claim that he had seen pieces of Element 115 and that it was the miraculous power source for the craft that were being reverse-engineered at S-4, no such element existed on the period chart. It does now. Scientific teams have created miniscule amounts of Element 115 by using powerful particle accelerators. The isotope created in the lab isn't anything like the material described by Lazar, but those who are familiar with the full story know that Lazar told us in this first interview that it was unlikely humans could ever manufacture the isotope he encountered. The super heavy sample he somehow obtained was likely produced naturally, perhaps in a double star system where gravitational forces exceeded anything that could be reproduced in a human lab. By the way, for the record, the sample that Lazar possessed for a time is not something he stole from S-4. Instead, it came from Los Alamos, where, according to Lazar, there was a machining operation for a time in which employees used equipment that whittled and shaped pieces of 115 so they would fit perfectly into the advanced reactor in the Sport Model craft Lazar says he inspected. It's a lot to digest and I would not expect skeptics to believe it, though they could at least consider the full story that was related in 1989.

While many critics will never believe Lazar's tale about Element 115, someone takes it very seriously. During the production of Corbell's film about Lazar, the two men had a candid conversation about the capabilities of the mystery element and where a sample might have been stored for protection. The day after that conversation, a team of 24 law enforcement personnel showed up at Lazar's place of business in rural Michigan. FBI agents flashed a search warrant (but did not provide a copy) and said they were seeking sales receipts related to a suspected murder case involving someone who once bought chemicals from Lazar's company, United Nuclear. If the FBI wanted years-old sales receipts, all they had to do was ask. They certainly did not need to stake out Lazar's home and office for days, or conduct electronic surveillance, or send a team of 24 people to raid his business. To date, the information that was seized, including mirrored copies of Lazar's computers, have produced no criminal charges, for Lazar or anyone else. Is it possible the agents were there to find the missing piece of Element 115? I have my own suspicions, especially since there had been a similar raid and massive show of firepower at Lazar's home years earlier. Both times, agents left the scene without obtaining the clam-shell casing that held the precious sample I had seen at Lazar's Las Vegas home decades earlier.

═══

People are welcome to believe whatever they want about Lazar and his claims. It no longer matters to me, or to Lazar. He's told me many times since 1989 that he prefers it if the public does *not* believe him because his life is much simpler when the UFO World does not intrude on it. The fact is, his story is now bigger than it was thirty years ago. The documentary film about his claims has ignited all of the old arguments all over again. Joe Rogan's podcast kicked it into the stratosphere. At the time I am writing this, we have no idea how many people might show up for the "Storm Area 51" event planned for September 2019.

But we know for certain the story is not in danger of fading away. Much to the chagrin of Lazar's many critics, his tale is now permanently etched into the public consciousness. It is pop culture, modern mythology, carved into the collective psyche. Whether his story can ever be proven, one way or another, it has prevailed. The renewed interest in Lazar's claims comes at a time when the UFO subject is hotter than at any time in my professional life. The December 2017 revelations in the *New York Times* about a secret Pentagon study of UFOs and related phenomena, the blockbuster release of three UFO videos recorded by military pilots, and the testimony of credible witnesses, including Black Aces commander David Fravor about the capabilities of the craft he encountered, have only bolstered Lazar's story. Fravor, in particular, has remarked on the similarities between the technology Lazar says he tinkered with and the astonishing "Tic Tac" craft Fravor and other Navy personnel witnessed and documented in 2004 off the coast of Southern California. Fravor and Lazar had a chance to meet during the spring of this year and spent hours comparing notes. For what it's worth, Fravor thinks Lazar is legit. Members of Congress are, as of now, receiving closed-door briefings about multiple UFO encounters being reported by our best and brightest military aviators. The incidents prove that someone, somewhere, has technology that is far beyond the best aerospace platforms in the U. S. arsenal.

Lazar's story has come a long way since that afternoon in May 1989, when we first unleashed it on the world. Whether you believe it or not does not really matter to Lazar, but we can say with some confidence that it's not going away.

PREFACE

I sat in a windowless room in a building somewhere on the site of a military installation at Groom Lake in the Nevada desert. The only sound was the faint hum of the fluorescent lights and the occasional passing of my pen across the page. Seated at a small rectangular desk, for nearly an hour, I had been filling out forms for a job I was really looking forward to beginning. When it was to begin and what it was I would be doing was still a matter to be settled. At least I knew for whom I was going to be working—EG&G. Having previously worked at the Los Alamos National Laboratory, I was familiar with EG&G's history. Named after three of its founders—the scientists Harold Edgerton, his one-time graduate student Kenneth

Germeshausen, and, later, Herbert Grier—they had executed high-speed photographic images of implosions tests conducted as part of the Manhattan Project. The company continued as a defense contractor, working with the Atomic Energy Commission on a variety of projects at the Nevada Test Site to develop weapons. In Las Vegas most everyone either worked in the resort/gambling industry or as part of what hippies in the 1960s denigrated as the Military Industrial Complex. In that sense, despite all the glitter and glamour associated with one its industries, Las Vegas was kind of a company town all dressed up in either sequins or lab coats or military uniforms. Then there were those who worked in various states of undress. Simply put, I was aware of what went on at Groom Lake in the defense industry and of who some of the major players were.

If, in January of 1989, I was eager to get back to working in the scientific community after more than four years—instead of the fairly lucrative but ultimately not very mentally engaging photo development business I owned and operated out of my house—I was absolutely thrilled to be going to work for one of the biggest players in the defense industry. I was especially keen on working for them because I'd always had an interest in explosive reactions, the bigger the better. What could be better than working alongside people at the forefront of our defense industry who had a long and successful track record with the biggest of bangs—nuclear weapons?

Sign me up, as the expression goes, but the literal signing up was taking a lot more effort than I wanted it to.

Some of my excitement was tempered by the fact that I was sitting in that room trying to dredge up the names and addresses of at least five individuals who knew me when I was living with my adoptive parents back in Westbury, Long Island, New York, in the mid-sixties. Having worked at Los Alamos, I was used to the idea of having to provide lists of references for a job, and also a more comprehensive list of past residences, acquaintances, residences, in order to receive the necessary security clearances. For my work at EG&G, in what I was soon to learn was their Special Projects Division, I needed the highest clearance possible: Majestic. I was told that level was 22 levels higher than the highest civilian clearance, known as Q clearance. While I understood the need for that level of clearance, it was still a pain in the ass to be sitting there wracking my brain trying to recall the names and contact information of people I hadn't seen in years.

I shifted uncomfortably in my seat and used the silver push button of a Bic Clic pen to trace the wood grain pattern on the table. I smiled ruefully and shook my head. I knew that a retractable pen like the one I held in my hand worked as a result of the interaction between the thrust tube, the thrust device, and the ink chamber and spring. It could produce a reaction sufficient enough to launch the pen off the table's surface by as much as eight to eleven millimeters. I did not know, however, the contact information for my supervisor at Fairchild Electronics in Chatsworth, California, where I worked while attending classes at Pierce Junior College. I'd spent most of my life interested in taking things apart and putting them back together, such as retractable pens other devices, and I had made a jet-powered bicycle in my youth and a jet-powered car in my adulthood. But piecing together my employment history and personal life was proving to be a real evil.

I looked up at the security guard who stood sentinel to one side of the closed door and wondered if he found this trial by memory as tedious as I did. He was rooted there in what I'd jokingly come to think of over the years as penis protection mode—the manner in which many military men stood when in public. His crossed hands were anchored in front of his groin, his feet spread to just shy of shoulder width. His light blue uniform was military style, but bore no adornments—stripes, insignias, badges—to indicate rank or any other kind of identification. He stood square-jawed and staring straight ahead, as if there was something far beyond the walls of this twelve-by-twelve room that held his attention.

The third member of our trio—Dennis Mariani—was of similar military bearing and appearance. When I shifted my glance to him, he scratched at his close-cropped hair. It was the first he'd moved, other than to blink, shift his eyes, or slide, retrieve, or stack a form, since he'd sat down across from me. Unlike the security guard, Dennis wore a civilian uniform—a navy blue suit, something that could have come out of a man's closet in the 1950s or 1960s. The jacket had narrow lapels and a single chest pocket.

Normally, I wouldn't have noticed or commented on such things, but for the last three hours, since Dennis had met me at the EG&G offices on the grounds of McCarran Airport and accompanied me on the Janet Air Flight out to this remote location, we'd never been separated by more than a few feet. There hadn't been a whole lot else to look at but this sturdily built, heavy lidded man who seemed to be in his mid-thirties. In many respects Dennis reminded me of a boxer—not the athlete (although, come to think of it, he did have the air of a pugilist about him) but more the canine breed. His expression was seldom slack; he appeared highly attentive and yet his demeanor revealed nothing of what he was he thinking or feeling.

I'd met Dennis two days earlier at the second of my interviews with the team from EG&G. He sat on the periphery of our group interview, not asking any questions, in his dog-like manner, sitting at the edge of our firelight, observant and seemingly ready to pounce. Whether he was preventing any one from getting into or out of that circle I could not determine. He hadn't been present at the first interview, one that I felt had not gone at all well. I'd left, despairing of my chances. A few days lapsed and I grew more certain that I'd have to enlarge my circle of employment inquiry beyond the driving distance from our home, the one I shared with my second wife Tracy, in the west end of Las Vegas. When I was called in for a second interview, I was told that I was considered to be over qualified for the original position for which I'd interviewed. They were impressed with my knowledge and wanted to see how I might fit in as a senior physicist in the Special Projects Division. Obviously, they expected

I would accept, and for that reason I was now searching my memory for the data required to earn me that Majestic clearance.

I'd always had a bit of the arsonist in me, not so much as a prankster but as an agitator. I grew tired of dull routines, and particularly the display of self-importance that people who worked at places like Los Alamos (and, I presumed at this point, out at Groom Lake and at EG&G) wore like ID badges: *I do serious work; I am a serious person.* In the past, I loved to pop that bubble of arrogance with little jokes and witticisms. Mostly I did this to amuse myself rather than take down any of my colleagues. The atmosphere in the room that day was heavy with the elements of that kind of grandiosity, but I knew it was too soon to risk doing anything to relieve the tension.

I glanced at my watch. It was now 10:30 at night. I'd hoped that by noting the time, I'd set off some kind of normal human social interaction.

"Do you have somewhere else to be?"

"What time do you have?"

"Does that watch have a calculator function?"

"What kind of battery life do you get with that?"

"We won't be here much longer."

"Take your time we've got all night."

All of those floated through my mind as possibilities, but Dennis, and what I decided to now call his "Mariani-ette," remained statue-like, silent and still.

I decided to dip my toe into the water. "Time flies . . . "I muttered under my breath.

Dennis rose to the bait, "Do you have a question?" His voice sounded familiar, somewhat like the Long Islanders I knew back when, stretching the first two words out as "dew" and "yew."

"No. No question," I said, "Just thinking out loud. I just don't have all these names and numbers at my fingertips, you know?"

I paused waiting for him to respond. When he didn't, I added, "I know how important the security clearance is. I want to get this done as quickly as possible. Help make your job easier."

"Do what you can. We'll do our job."

"I wasn't insinuating—" I stopped when I saw something briefly pass across Dennis's eyes. A sign, maybe, of his irritation with me?

I thought again of that second interview I'd just had, how much more, if not jovial, at least collegial, it had all seemed. We were all there for the same reason, all men with technical/scientific backgrounds, and while we weren't cracking jokes, we were at least being human. Dennis had sat there then as unresponsive and blank as he was now. I couldn't tell you the names or the titles of the group of men I'd spoken with that day, but I did recall after the meeting ended, that I was most drawn to Dennis and wondered what his story was. Was he always in such a bad mood? As I sat there in that tiny office with him, I felt like my future was hanging in the balance. If this guy ended up hating me, which seemed to be the case, could he make the difference between me being hired full time and not?

As was made clear to me, I was being hired on a provisional basis. Until I received my Majestic clearance, I was only going to work on site on an on-call basis. I was pleased to hear that, and even more pleased that they agreed that I could continue to work during the day running the photo developing service out of my home. I needed to pay the bills and all that, I told them. They understood. Still, I'd have to be on call and when I received notification via phone that I was to go into the "office" that day, I'd have very little time to get to McCarran to get myself aboard that Janet flight out to the airfield at Groom Lake.

Earlier that evening, I'd received the call from a woman identifying herself as "Nancy from EG&G." I was to be at McCarran at 7:30 that evening. Having lived in the Las Vegas area for more than five years, I was already familiar with the sight of the so-called Janet aircraft taking off and landing from Las Vegas' major airport. McCarran was located only a few miles from the famous Las Vegas strip. Though Tracy and I lived in west Las Vegas, on the appropriately named James Lovell Street, our business took us all around the metro area and we both frequently saw the white Boeing 737 jets with the distinctive red markings along the fuselage at window height of the Janet jets. No other emblems were needed, especially since all the locals joked that Janet stood for Just Another Non-Existent Terminal. That was a reference to the fact that though we daily saw those planes ferrying employees to and from what was often referred to as Area 51, for a long time the military and the government disavowed the existence of any kind of operation going on around Groom Lake.

Many speculated about what went on "out there." Given that the Nevada Test Site was also in the region, and having worked at a government facility with a long history of secrecy, I was pretty sure that Area 51 was a place where experimental aircraft and weapons systems were tested. At the time, there were a few people on the fringe who were into UFOs and speculated about Roswell and the like. I wasn't into the whole extraterrestrial debate and mythos. As a kid I watched *Star Trek*, but I was more interested in how its warp drive functioned than I was in about the various life forms they encountered. Some kids might have wanted to be Captain James T. Kirk and engage in extraterrestrial relations, but I wanted to be Scotty and get my hands on dilythium crystals and not the body of a nubile female life form from the planet Zetar.

That said, when Tracy and I were house shopping in Las Vegas, we attached no special importance to finding a home on a street named after an astronaut. Jim Lovell was the commander of NASA's Apollo 13 mission. Whoever had developed our little area between Alta Drive to the north and Charleston Boulevard to the south must have had an interest in space exploration. In our little quadrilateral, there was Alan B. Shepherd Street, John Glenn Circle, Astronaut Avenue, Michael Collins Place, and Skytrail Place, mixed in among the usual suspects: Oakhurst, Larch, and Hemlock. You could be arboreal or patriotic or a combination of both, but to be honest, Tracy and I remarked on the name of our street once and then it just quickly became ingrained as one of the thousands and thousands of unremarkable facts of our lives.

So, once I got the call from Nancy to drive to the airport and EG&G for a flight I was thinking more about whether or not I should have brought a change of clothes or toiletries with me more than any kind of name coincidences. As a lover of explosive reactions, though, what was second to a nuclear blast in my mind was the thousands and thousands of pounds of thrust produced by those rockets used in the Apollo program. Little did I know how much of a role propulsion was going to play in my life in the months that followed that first call from Nancy.

As a scientist, I relied on rationality. I didn't do so to the extent that Mr. Spock might have, but certainly to a great enough degree that though I heard about things like the MJ-12 or Majestic 12 documents, I discounted them as a hoax. Friends and acquaintances knew about the conspiracy theory that existed around that alleged secret organization President Truman founded in 1947 to assist with the recovery and investigation of alien spacecraft. Those discussions were all a part of living in this company town, and the wild imaginings of those on the fringe were exactly that—imaginings: fun to ponder and speculate about, but, unlike the influence of organized crime on the development and continued operation of Las Vegas, without real substance to support them. For me, seeing, or having some other form of verifiable empirical evidence, was believing.

The issue of what to believe was on my mind the night of my first trip to Area 51 only because I was going to have to explain to my wife why I was leaving the house so late on a work night. Since Tracy's father had been employed at Los Alamos and the two of us met there before moving to Las Vegas, she understood why it was that I couldn't tell her much about my new job. I wasn't being evasive or secretive; as I set out from home that night I truly had no real idea what I would be expected to do that evening or on subsequent evenings. What did strike both of us as odd was that I had to jump immediately into action that night. Who, other than first responders and other emergency personnel, worked under such conditions?

"I need to report to EG&G," I said to Tracy. I rushed around the house trying to find my wallet and a lint-free cloth to clean my glasses. Oncoming headlights were bad enough, but when the light refracted off smudged lenses, the blurring and blinding was even worse. I was out the door in less than twenty minutes, minus the change of clothes and toiletries I previously mentioned.

My post–rush hour drive to the airport was uneventful, as was the flight out of Vegas. Only a handful of passengers was on the flight with Dennis and me. He had met me at the offices of EG&G and escorted me through a security gate and out onto the tarmac. We exchanged little more than a brief greeting and I took Dennis's command to "Follow me," to the most literal extent possible. If he wasn't going to speak, then neither was I. I was reminded of the times my father, Al, took me into New York City or

DREAMLAND

Queens on the Long Island Railroad. Most of the other riders on the train sat singly or, if forced into being seated next to someone else, regarded him or her as nothing more than a fixture of the train's cabin, something to sometimes bump up against but not to engage in a conversation with. Even the pilots on the flight were silent.

At the point when I felt the plane's speed slacken, Dennis turned to me and said, inclining his head toward the window, "That's the Groom Lake facility below."

"Yes. I've heard of it," I told him.

"That's where we'll be landing."

I didn't respond at all.

Dennis's brow raised in surprise. His reaction triggered a similar one in me. His statement seemed obvious, though when we departed McCarran, Dennis made no indication of where we were headed. I didn't tell him that I assumed that Area 51 was our destination. Maybe I'd spoiled the party for him a bit; maybe he liked to see others get excited at the mention of anything to do with Area 51. The fact that he was even speaking to me seemed like a spell that I shouldn't break with any unnecessary comment.

We covered the hundred or so miles quickly, and I imagined our flight path looking something like a steep spike on an oscilloscope. Once we disembarked from the plane, Dennis and I boarded a school bus. It was no longer in its traditional yellow and black livery but had been painted navy blue. All of the windows with the exception of the front windshield had been blacked out. We proceeded along a paved road to the first stop, a nondescript single-story office building. I followed Dennis out of the bus. The late December desert air was cool but refreshing. It was only then that I thought about how tired I was. I'd been unable to sleep, and had risen at six that morning.

Once inside the office complex, Dennis produced a plastic ID badge and held it out to a uniformed guard. The reception area had sets of doors immediately to the left and right of the desk and one directly behind it. I was led to the left. Dennis switched on the lights and it took a moment for the fluorescent bulb igniters to heat up the gas. In the dim half-light of the room, I made out a device in Dennis's hands—a camera.

"Stand there," Dennis said, pointing to a set of footprints embossed on the floor. He stepped behind a desk and I stood there for a few moments while he adjusted the equipment and then the flash stabbed at my eyes.

"No, 'Say *cheese*'?" I said, squinting hard to make out Dennis's expression. Didn't matter if I could see his face or not, I immediately thought. It was going to be same sullen yet alert look.

"Follow me."

As we passed by the reception desk, the man behind it lifted up a walkie-talkie and produced a manila folder that he handed to Dennis. Neither of them exchanged a single word, let alone anything approximating a pleasantry.

I thought that I might like working here, if in fact this was where I would be working. No idle chitchat, no feigning interest in relative strangers. Sign me up.

And that was exactly what Dennis had me doing for nearly the next hour and a half. As I was getting to the last lines of the final document, I could feel my eyelids scratching as I blinked. I tried to remember the last bit of liquid I had taken in. A coffee would have been nice or even just some water, but I saw no signs of the typical office refreshments anywhere in reception or the two rooms I'd been in so far.

"I think that should do it," I said at last, sliding the last of the documents across the table to Dennis, careful to turn it so that he received it in the proper orientation.

"Follow me," Dennis said after placing the last of the papers into the folder.

Outside Dennis said, "We'll take the bus back to the plane. That will take you back to Las Vegas. There you can pick up your car and go home."

"That's it?"

"For tonight."

I thought it a bit odd that I had to be on the premises to fill out that paperwork. I was considering asking Dennis why they thought it necessary for me to come all the way out there. Before I had a chance to ask, Dennis added, "It's late."

I shrugged and thrust my hands in my pockets. I couldn't refute the logic of that statement. I was more than a little amused by Dennis revealing the next steps we'd take—or the steps we'd retrace—and how unnecessary that was. It was also very much in keeping with how people in government jobs often conducted themselves. He stated very clearly the most obvious things but left unstated what I most wanted to know. When was I going to come back here? What exactly would I be working on? Who would be my boss?

My string of questions was interrupted by the squeak of the bus brakes. I bit my tongue when Dennis said the last words to me that he'd offer that night.

"The bus is here."

An hour and a half later, I slid beneath the sheets of our bed, hoping not to wake Tracy, but she snapped on the light on the nightstand.

"How was it?" she asked and tried to stifle a yawn.

"I couldn't agree more," I said.

She looked puzzled and I realized that my reference to her yawn wasn't clear.

"Hard to tell from a first day. A lot of paperwork. Boring stuff, really."

"I'm sure it will get better."

"It has to," I said.

I lay awake for a few moments thinking about the next day. I was going to have to drive out to a realtor on the north end of the city to pick up a few rolls of film. I couldn't remember if I'd filled the tank with gas.

Just as I was about to drift into sleep, Tracy said to me, "While you're out tomorrow, can you pick up a can of coffee and a pint of half and half?"

DREAMLAND

CHAPTER ONE

THE FOLLOWING MORNING, AFTER A QUICK CUP OF coffee, I headed out to pick up the photos from a regular client, a real estate appraiser, who worked out of an office near the Painted Desert Golf Course. The course had just opened and along with it a housing development had gone up, and many others as well, in that area far north of The Strip. Many of my clients worked in real estate. New construction was booming, current residents were taking advantage of rising prices to move up to new homes, the population rose from just shy of 275,000 people in 1970 to nearly 465,000 by 1980, and was destined to reach 740,000 by the start of the '90s. Simply put, business was good for anyone in real estate and housing, and I liked working with clients who were professionals rather than someone who came in griping that their vacation shots were over-exposed. Still, as I drove along with my window down, the scent of change was in the air as much as the exhaust fumes from the many diesel engines powering ready-mix cement trucks and semi-tractor trailers loaded with lumber.

I wasn't certain when I'd be called back to EG&G, and in some ways that was okay with me, at least temporarily. I figured it was just a matter of time before my security clearance came through. I knew that traffic along I-15 was going to be tight, so I chose Nevada 95 instead, and I felt a little sorry for those commuters sealed in their cars listening to one Morning Zoo Crew or another on their radios. The news was still full of stories about the horrific terrorist bombing of a Pan Am flight over Lockerbie, Scotland. I switched off the radio, preferring the company of my thoughts to how horrifying it must have been in those few instants before the blast obliterated part of the craft and the intact pieces hurtled to the ground. Maybe because we were nearing the end of one year and the beginning of another but I was in a more contemplative mood than usual.

My first day at EG&G didn't exactly begin with a blaze of glory, but at least I had made some headway in my desire to return to more meaningful work. I enjoyed the photo business that my first wife Carol and I had begun several years earlier when we were still living in Los Alamos, but it was hardly what I would call intellectually stimulating. For a while, that was fine with me. I had needed to clear my head. I adored Carol and the two of us had built a wonderful partnership together. As husband and wife, we shared a vision for what our lives should be like. In many ways, we were living the American Dream. I'd grown tired of working for someone else and had left my job working for Los Alamos National Laboratory (LANL) and set up Lazar Energy Systems, designing, developing and repairing alpha radiation detection systems. LANL used a lot of plutonium, which gives off alpha radiation, and for four years straight Lazar Energy Systems was granted a government contract to provide that form of radiation detection. As many had discovered before me, it was far better to work with a government agency that it was to work *for* one. It was certainly more lucrative, and I also got to be my own boss.

The photo business was Carol's to oversee, and that brought in more than a fair amount of revenue. We were doing well financially. We'd even invested in a legalized brothel in Las Vegas that was taking in $100,000 a month at the time we purchased it for a million dollars. The Honeysuckle Ranch had been in business for more than thirty years, had a solid management team in place, and with Carol doing careful oversight of the operation, we expected the wheels to keep on turning and the money to keep rolling in. Add to that the photo operation, and the two of us were incredibly busy, mostly working from home, socking away as much as we could. I did give in to one indulgence—the purchase of a 1984 Chevrolet Corvette—but we lived a quiet and unassuming life. That is with the exception of the jet car I built and tooled around town in.

That eventually got me some attention and figured prominently in my getting a job at EG&G.

I was too busy living my life to really reflect much on it and our reasons for doing the things we were doing. The general tenor of the '80s decade was examined in an Oliver Stone movie, *Wall Street*, that featured Michael Douglas as the financier Gordon Gecko. His lizard-like name was no accident, and what most people who saw the film will likely remember is a line from a speech Gecko delivered: "Greed... is good." I can't say that Carol and I were greedy, but we were definitely driven to succeed. Success, for us, and for most people, meant making money. If it took running three separate business, not taking vacations, and having work-related issues on your mind twenty-four hours a day and seven days a week, then that was a price we were willing to pay.

A price I wasn't willing to pay, and for a few months wasn't even aware I was paying, was the loss of Carol's life. For a time she was able to hide her pancreatic cancer diagnosis from me. Toward the end, her frequent absences from home, her weight loss, the sallowness of her complexion, and the anguished look of pain on her face when she thought I couldn't see her aroused my suspicion. Before I could get her to stop saying that it was nothing, that she was just feeling a little run down, she waited for me to leave the house to run a few errands. She sat behind the wheel of her car in the closed garage and ended her life.

I can't describe the feelings of devastation and loss that followed. I missed her and struggled with the kinds of questions that a man in his late thirties ought not to have taken on—mortality and the larger issues of why we're here. What do all our hard work and dreams mean when one day you can come home and discover that the woman you adore felt she had to face alone such an awful set of choices. To say that I was reeling is an understatement. I was overwhelmed by guilt and anger and sorrow. Responsibilities weighed heavily on me, and as much as I wanted to keep our businesses afloat as a way to honor what Carol and I had built, the feeling that none of it mattered eventually won out.

Whatever minor dissatisfactions I had felt about Los Alamos, the intellectual snobbery that many who worked at the LANL exhibited and lorded over the rest of the residents, paled in comparison to how much I associated that place with the loss of my wife. I wanted to get out of there, and eventually did. One saving grace in all of this, which I felt Carol had a hand in making, was that Tracy was in my life. Carol had hired her to work at one of our photo stores. She was a bit of a wild child to be sure, but the two of us grew close, drawn together by the belief that Carol had set in motion the events that led to Tracy and me getting married soon after Carol's death. Tracy and I had travelled to Las Vegas together to oversee some matter related to the photo business—what that matter was, is lost in a fog now. I don't know if either of us noted the irony of it, but we were married at the "We've Only Just Begun" chapel in Las Vegas in April of 1986.

DREAMLAND

When I arrived at the pickup point, I got out of the car and walked to the door. Taped to it was a nine-by-twelve mailer. Inside were a couple of rolls of Kodak Ektachrome film and a brief note of apology for not being there to meet me in person. I tucked the envelope under my arm and while walking back to the car, realized that I hadn't eaten a thing since lunch the previous day. I found a little mom and pop diner nearby and pulled in. The film would get developed and printed in plenty of time. Tracy would get her coffee and half and half. I settled into a booth, grateful that I could set my own schedule, and wondered how I might adjust to having to report and work under someone else's timeline.

In science, one of the ways that the word "entropy" is used is to talk about how any natural system tends to move from order to disorder. In many ways, that is a perfect description of that period of my life. Tracy helped keep things from falling apart completely, and there was something in my nature that also wouldn't allow me to just simply shrug my shoulders in resignation. It took me a few years of running the photo business in Las Vegas to regain my footing. I realized that when in doubt you should go back to the things you love, and for me, that was science.

It's funny to think of it now, but as I struggled that night in that office filling out paperwork and coming up with names and dates, the one that shouldn't have escaped me, and in fact didn't, was Alan Rothberg. Alan and I lived on the same block in Westbury and went to Bowling Green Elementary School together. His father was an instructor at Nassau Community College where he taught chemistry. He'd bring home various equipment, high-voltage demonstrator tubes, Geisler tubes and other things. Some kids might have had children's chemistry sets to play with, but with access to Alan's father's classroom and supplies we were conducting actual experiments with the kind of equipment you'd find in a research center's laboratory. My family moved around a bit at that time—back to Brooklyn, out to California, then back to Westbury—due to my dad's work. Alan was the one constant for me in that neighborhood, and I took great delight in attending the classes his father taught during the summer session. I was in junior high by that time and Alan and I both loved the idea that we would take the same test as the "real" students and could consistently outscore them.

In high school, I was an indifferent student in most of my other classes, but science was a big part of my academic life and success. Outside of school I devoured issues of *Popular Science*, *Popular Mechanics*, and *Science and Mechanics* as other kids my age lost themselves in the Marvel universe of Spiderman, Captain America, and other comic book heroes and villains. I still have many back issues of those science magazines in my possession today. What really enticed me about the magazines I read was that they weren't about pure, theoretical science. They provided me with plans that I could use to make things. The schematics for building a Wimshurst machine—an electrostatic generator—appeared in one issue. I was ready for the challenge and built one, essentially using two discs with metal sections attached to them spin in opposite directions producing high voltage through electrostatic induction. Instead of relying on friction, they produce the charges through influence. I also built a Tesla coil. I was stymied for a while on that project because I needed an automobile engine's ignition coil. Believe me, I was sorely tempted to raid my father's Oldsmobile, but I resisted. (Just prior to this I had installed in my room a box with a red light and a green light that I could switch between, indicating to my parents and my sister (like me, she was also adopted) when I'd allow them into my room. I did this because my mom had gone into my inner sanctum while I was away to do some cleaning. She stepped into the room, got tangled in some electrical wires setting off a shower of sparks and tripping a few breakers. I didn't need to add to their worries about what their mad scientist had done by creating car trouble for them.)

Fortunately for me, we were living in Brooklyn at the time and more than a few abandoned cars littered the neighborhood. While out riding my bike, I managed to scavenge an ignition coil from the dented, rusted, and windowless remains of an American Motors Pacer that other car carrion eaters had stripped. When the Tesla Coil was done, I could sit for hours watching it spew its colorful corona discharge, brush discharges and streamer arcs.

Almost as much as I got a rush of accomplishment when I built something from one of those magazines, I was also excited every time I looked at the Estes catalog. Estes Industries was based in Denver, Colorado, and they were producers of the finest functional scale model rockets at the time. Alan and I and a few other guys in Westbury were really into building them and launching them while in junior high and high school. The construction of the rocket wasn't that complex; in fact, quite the opposite. The engines were supplied either in the kit or could be purchased separately. There wasn't enough do-it-yourself activity to keep me interested in their flight. The engines were housed within a sturdy cardboard tube. Inside the tube were a ceramic nozzle, solid propellant, delay charge, ejection charge, and a clay retainer cap. We'd run a coated wire through the nozzle and make contact with the solid propellant. We then attached that wire to a battery and the electric current would heat the wire and ignite the solid propellant. An engine could also be ignited by the hot gases from the propellant of a booster engine.

I wanted to take on the challenge of building my own rocket engines from scratch. Small diameter cardboard tubing that would withstand the explosive forces wasn't difficult to source. Many wire coat hangers had cardboard wrapped around the bottom of the triangle. It was the solid fuel that we needed to make (cook is probably a better term), and I set about researching. I read a number of articles that made reference to something called "meal powder." At that point, I didn't understand what meal powder was—a finely grained gunpowder. To me, a meal was a meal, and the only meal I had access to was oatmeal. Oatmeal might be able to fuel muscles but it didn't do a thing for those early rocket engines I tried to create.

It wasn't that I didn't have the money for replacement engines. In my early teens I'd gotten a job at a local fireworks manufacturer. Talk about a kid in a candy store, but I didn't get involved in the manufacturing process at all. Mostly I just helped out by packaging the various items they produced. I didn't earn a whole lot, but enough. I also learned enough to solve the rocket propellant problem. That information, plus greater exposure to Alan's dad's chemistry classes, and a few of his chemistry supplies, had me launching rockets thousands of feet into the air at Salisbury Park.

I knew enough from chemistry to know that combining potassium nitrate (KNO_3) and sugar and heating them together would produce what those in the model rocket-know called "rocket candy" because of the distinctive smell of the sugar caramelizing as it was heated and absorbed the KNO_3. (Funny to think how much of my early rocket propulsion days were associated with food—oatmeal, Salisbury steak, rocket candy.) KNO_3 was the sole component of stump remover—a chemical application homeowners or contractors could use to dissolve tree stumps.

I also became our chief project engineer in the manufacture of pyrotechnic and other explosive devices. That was mostly due to the fact that my friends didn't seem as interested in blowing things up as I was. They were good with the rockets and enjoyed the pyrotechnic displays, but I was interested in seeing how chemicals could react with one another to produce small-scale explosions. That experimentation was a natural offshoot of my rocket engine making. Working with various compounds like the KNO_3 plus sugar formula, I was eventually able to produce my version of an M80 firecracker and other smaller firecrackers that produced a loud bang and little more. I tried to mass-produce them for the Fourth of July, along with cherry bombs, and managed to make quite a few, but as impressed as friends were with my efforts, their interest waned. The laws of supply and demand exerted themselves, but that only tempered my business interests and not my passion for the products and their explosive displays.

I don't know how much of the other guys' loss of interest was attributable to the one incident that resulted from faulty handling of the products. I'd taken potassium chlorate and red phosphorous and carefully wrapped them together in a tissue. This combination is often referred to in the pyrotechnic community as "death mix." The reason for that is because the chemicals are friction sensitive and thus very unstable. I'd take that mixture I'd placed into a tissue and then folded it around the shaft of a bolt, sandwiched between two nuts. I'd thread a nut onto the bolt to secure the little explosive package. After you tossed that bolt grenade and it hit the ground, it would erupt in a huge puff of smoke and a loud bang. Once, at Alan's house, he was standing on the sidewalk that led to their front door. He tried to toss the bolt underhanded toward the street, but it slipped out of his hand, went a few feet back toward the house and detonated near the porch. We spent a good hour cleaning up the front door and porch area in a complete panic as the hour neared for his parents to return home.

Fortunately, all the smoke stains came off and no one was the wiser—except us. We kept a safer distance from Alan's house from that point forward.

Looking back on it now, if that was the worst thing that happened, we were all pretty fortunate. I can't believe some of the stuff I was doing and how it might have all gone wrong if I wasn't as fortunate as I'd been.

As I sat in the booth of the diner, finishing my pancakes, I thought about those days from my youth because of the nature of the questions I was asked during my second interview at EG&G. I thought it a little odd that they kept asking me about my hobbies and other interests outside of work. I believed they might want to know more about my training, what I did at Los Alamos, that kind of thing. After all, as I was to find out toward the end of the interview, they were considering me for a senior staff physicist position. The reality was that my interests outside of work hadn't changed very much since I was a kid. I was still involved in pyrotechnics. I hosted the Desert Blast for a number of years, setting off a terrific display of fireworks each summer. I'd also created a jet car, installing one in a Honda Civic I owned, and later put a larger jet engine in a rail dragster. I took my jet cars to drag strips all around the southwest and beyond.

In a very real sense, if it wasn't for my interest in rocketry, pyrotechnics, and my love of science and powerful reactions, I might not ever have had an opportunity to interview with EG&G. Before all that, however, I had to get some additional technical experience. In 1976, my parents had moved to California again. I was determined to take a year off from college, but instead enrolled at Pierce Junior College on a part-time basis. I also got a job working for a company called Fairchild Electronics. They manufactured equipment to test something called "bubble memory" devices. Think of them as an early precursor of the hard drives that we have in today's personal computers. Instead of a semi-conductor, the main memory component of these devices was a slice of crystal. The people at Fairchild were absolutely certain that bubble memory was going to be the key to the future of computing excellence. Obviously, they were wrong, but at that time, bubble memory was a viable means to store data produced by mainframe computers. The other option was reel-to-reel tapes, the kind you'd see in movies and television encased in large metal cabinets. Just so you don't think that bubble memory devices were too out there, I can also recall a system of programmable memory that used ultraviolet light in order to erase the data on them. Given how sophisticated our devices are now, this seems impracticable at best and boneheaded at worst. But those were the times.

I originally worked at Fairchild as a technician repairing broken circuit boards, but eventually became a test engineer, and later an engineer designing circuit and logic boards. I loved electronics and I was earning money and going to school at Caltech by this time. I was studying electronics there mainly because the people at Fairchild thought that was the best use of my time. But in my off hours from school and Fairchild, I was working on my own projects at home—lasers had captured my interest and I was working with plasma containment and MHD (magnetohydrodynamics). But I knew that I had to pay my dues and if working with electronics was a way in, then that's what I was willing to do. But my desire was still to become a scientist and leverage my love of physics into a place in the defense industry working on weapons systems. At that period in our history, "weapons systems" meant nuclear weapons. We were still waging the Cold War, Ronald Reagan was in office, and his proposed Star Wars nuclear missile defense system was still making headlines.

By the summer of 1982, my feet had grown itchy and my desire to take the next step was too great to keep me at Fairchild. As much as the move to California from New York had initially been a bit of culture shock, I'd grown to really like the stereotypical, but true, laid-back surfer dude mentality of the people I worked with. They had treated me well, and there I was at the age of twenty-three, working as an electronics engineer even though I was still a few credits shy of actually having a college degree. I wanted more, so in the summer of 1982, I sent a cover letter and resume to Los Alamos National Laboratory. I spent an anxious few weeks waiting to hear back from them. I'd almost given up hope, but I received a call one day in September inviting me down for an interview.

I accepted, and as a part of the interview process, I was taken on a tour of the facility. I knew a lot about its history and the Manhattan project, but the Meson Physics Facility was all relatively new to me. I knew that the particle accelerator was there, but when you read about a device being a half-mile in length, it's hard to truly get the scale of the thing until you're on site. I was in awe. Seeing a Van De Graaff generator the size of small office complexes took my breath away. The fact that these machines were producing millions of volts was absolutely astounding! I was in power heaven. I saw bicycles propped up along various stretches of the runway (for lack of a better term) that ran alongside the accelerator and its associated buildings. The device was so large that technicians and scientists would pedal from one location inside the building to the next in order to save time. How cool was that? I will never forget looking at one meter and the lowest number on its measurement range was 110 kilovolts! And these guys were accelerating particles to such a speed that time itself was being affected.

I felt like I had stepped into the world of science fiction. I was moving among men and women with an intellectual capacity that was nearly as great as the power these machines were generating. I wanted to work there, even if it meant that I was going to be working in the electronic monitoring side of things and not conducting experiments. They also expressed a lot of interest in my work with lasers and during the interview and even after I was hired there was some talk of possibly utilizing that experience at the facility. The odd thing was, as was the case with EG&G, they wanted to know as much about my interests and activities outside of work, as they did my work experience or education.

Also, it's hard to imagine from the perspective of this post 9/11 world we inhabit that the level of security on the facility would seem completely lax today. I was free to roam around the facility on the day of my interview as well as after I was hired. I was reminded of those days with my dad that we spent at the Hall of Science in Flushing Meadows Queens and Corona Park, the site of the World's Fair in 1964. He'd take me to that museum and I could wander among the displays, knowing that I was walking along and seeing some of the great moments in human history, intellectual history, represented before my very eyes. I'm not a big sports fan, but having lived in New York, I know how revered the Yankees are. In many ways, those trips to the Hall of Science and those first hours and later months and years were like setting loose a Yankee fan on the field at Yankee Stadium. Not only was I allowed on the field, I got to play there.

Obviously then, when they made an offer that day for me to work there, I accepted. I returned to California, gave my notice, packed my belongings and within a month was in Los Alamos and working at Meson. I was in for a bit of culture shock there as well. The employees there, as I said, possessed some of the most powerful scientific minds in the world. I was used to working with lasers, but these were human beings with a similar kind of intensity and focus.

They weren't exactly like Dennis Mariani. The air of menace he gave off wasn't a part of these people's temperament, but they were similarly serious minded and watchful. Over time, I'd learn that the locals referred to the staff at Los Alamos as the Cone Heads, the alien family from Beldar who were brought to life on the TV show *Saturday Night Live*. Ironically, in Los Alamos, particularly among the staff at Los Alamos, it was generally "Saturday Night Dead." Make that "Every Night Dead," really. By 6:00 p.m., most people at the facility had their work day over. That part of town was so quiet that the traffic signals all flashed red lights, letting drivers roll to a stop and then proceed one at a time.

My newbie excitement wore off, not just because the place was boring, but because of the kind of arrogance I've mentioned before. Sure, in school I had to endure my fair share of taunting and teasing because of my interest in science and my disinterest in sports. I never thought myself superior to anyone or that I was in any way inferior. I just had different interests. That wasn't the case with a lot of people I worked with and the derision they felt for the rest of Los Alamos, the city, and people in general was more than a little off-putting.

I also didn't like working for the federal government. I hated to see the kind of waste of funds and supplies that working with other people's money produced. In fact, it was our money since we paid taxes like everyone else, but for as smart as these people were, they didn't always seem to understand that, or at least they didn't seem to care.

All of that culminated in me becoming a bit of a prankster, but I liked to think of myself as a prankster with a cause.

As its name implies a particle accelerator gets particles to move at a rapid rate—to the speed of light. Most of what were referred to as the targets—the particles accelerated—had to be cooled to close to absolute zero. That way the atoms and molecules inside them would be essentially inert—not moving. To do that cooling, you need vast amounts of a specific isotope of helium, helium 3. It is a rare and expensive isotope and we had it brought in by the truckload in specially equipped tanker trucks, which were essentially like scale vacuum thermoses designed to keep the incredibly cold gas as chilled as possible. Even in a vacuum-sealed environment, helium 3 is boiling off due to the tanks exposure to the ambient temperature. Generally, as soon as those trucks rolled onto the property and to where they were needed, the gas was offloaded.

Except on days when no experiments were being conducted. The trucks still rolled in, but they sat there, as the enormously expensive payload they were carrying dissipated into the atmosphere. The trucks would be parked on these radiation shield blocks with six-inch diameter hoses spewing off helium gas. I wasn't an accountant. I wasn't in the procurement department. But that kind of waste really irked me. I mentioned it to a few colleagues and they would shrug and say, "Hey, it's government work. They've got deep pockets so why worry about it?"

So I didn't. Except in my own little way.

One day when the accelerator wasn't in operation and I didn't have to monitor any of the sensors, I was little bored. I saw the trucks outside. I saw a few rolls of the tape we used to cordon off areas or to label any items that might have been exposed to even trace levels of radiation. It was kind of like crime scene tape except is was magenta and yellow and said, "DANGER RADIATION," all along it. I went to a custodian and asked for a few plastic trash bags—the kind that lined the large bins in the cafeteria and not the small office waste paper baskets. He gave me a box. I went outside and placed a bag over the pipe from which the now no longer super-cooled helium was being discharged. I tied off the bag with the "DANGER RADIATION" tape. I held onto it and I could feel it tugging at my arm, begging to be released. I let it go and watched it climb. That was fun. The second was fun as well.

I didn't want to keep all the enjoyment to myself, so I went inside and got my friend and colleague Joe to join me. We spent our lunch break filling and releasing helium balloons. Eventually, we stood there at the end of our lunch hour watching this dotted line of trash bags being carried south of our location on the prevailing winds. The facility sat on a high point and was separated from the main area of the town by a somewhat shallow valley. We could see the path they were taking, and, familiar with the area we knew generally where they were heading.

"What are those lights?" Joe asked me, squinting and pointing in the general vicinity of our flight path.

"I'm not sure," I said. "There's a quite a few of them. Looks like mars lights."

"You can't see Mars from here," Joe said. He couldn't keep a straight face and I knew that he understood that mars lights were the ones atop a police car or other emergency response vehicle.

"I think they're at the McDonald's," I said.

"Maybe it was held up," Joe said.

We headed back inside.

I've heard it said that smart people sometimes don't have common sense. In this case that was true of Joe and me. Not only did we not understand that imperfectly sealed bags would launch and fly but eventually come to Earth, we didn't understand the kind of panic that would result when those bags were found to have DANGER RADIATION emblazoned on them. We also didn't connect the dots between our location, the McDonalds, the flying bags, and the police lights. I can't say that our little stunt produced a reaction equal to Orson Well's broadcast of the *War of the Worlds*, but the facility did receive phone calls about the mysterious appearance of irradiated trash bags. The local authorities were genuinely concerned, knowing that there was likely only one place the items could have come from. They were not amused, nor was our supervisor, when Joe and I came forward and explained what we'd been up to.

During my interview with EG&G, I did not mention the gasbags. During my remaining time at Los Alamos, I did my best to curb my desire to make life there more interesting for us all. I had no desire to earn myself any more attention than I received. In a way, mine was a futile gesture. I suppose there were more appropriate ways to express my feelings about the wasteful practices. I also knew that doing so would have been shouting into a very strong and powerful wind that would have drowned my voice out and pushed me back to from where I'd emerged. Still, I didn't like feeling as if I was powerless to effect any kind of change—not about the bags and the helium 3, not about my growing sense that what I was doing at Los Alamos was only tangentially related to physics and the kind of work I truly wanted to do.

I'd taken what I thought was a step in the right direction, was grateful to the folks at Meson for sending me to MIT to further my education, but I felt as if I was one of those bags being carried along by the wind, unsure of how I could make any kind of course correction, and a little fearful that my scientific ambitions might result in me crash landing in the parking lot of a McDonald's, deflated and warning others of a danger that didn't exist. Lazar Energy Systems had been a step in the right direction, and if things had turned out differently, I may have kept that outfit running for many, many years.

In many ways, with Carol's death, the loss of various business interests and my interest in them, I had been drifting for a few years. Certainly not to the extent of those balloons, but as the last year of the 1980s approached, I had been wondering about waste—of my time, my knowledge, my passion. That was why I'd decided to send out those resumes and cover letters. As it turned out, though, my time at Los Alamos wasn't a waste by any stretch of the imagination. Sometimes it pays to go to class and to listen to lectures. The payoff may not come for years. Along with that, though, comes another kind of lesson—be careful what you wish for.

CHAPTER TWO

To give you some idea of how sleepy the city of Los Alamos can be, in June of 1982 a reporter from the local paper, the *Los Alamos Monitor*, contacted me. He identified himself as Terry England, told me that he was a staff writer for the paper, and he wanted to interview me. He'd gotten word that I had a jet-powered car. I agreed to talk to him, never figuring that the eventual story would wind up being on the front page of the Sunday paper on June 27th. I'd make some crack about it being a slow news day and he needed a fast story, but that would be too easy. The writer appeared to be as interested in the fact that I came from California to work there and still had vanity California plates on the Honda Civic that read,

"JET-U-BET." I know for a fact he was also extremely interested, or perhaps distressed is a better word, when Carol and I drove the car, using its conventional gasoline engine that still functioned in its usual place in the engine compartment at the front of the vehicle, to take it to the Pueblo High School parking lot. There, I fired up the jet engine for him and it howled like you'd expect a jet engine to. He stood there with his hands pressed to his ears, the pages of his notebook fluttering and threatening to take off. I explained to Mr. England that the sound waves coming out of the nozzle, which was exposed by pressing a switch that dropped the rear license plate out of the way, were synchronized with the intake, producing a fairly substantial amount of noise.

Starting it up for him a few times garnered us more attention that I would have otherwise wanted—a man drove over from another neighborhood a few miles away, and a police squad car from Los Alamos county showed up. The officer was nice and simply asked me to not do any more demonstrations due to the number of complaints.

The next week, the jet car and I got the attention of a man I never thought in my wildest imaginings that I would ever speak with. Edward Teller was back in Los Alamos, from the Lawrence Livermore National Laboratory, to deliver a lecture. I was excited by the opportunity to get to see and hear the man who was credited with being the father of the hydrogen bomb. Of course, given my interest in nuclear arms, I knew about him and was eager to hear him speak. Even if he hadn't been a part of the Manhattan

Project, he had an extraordinary career. Born in Budapest, he left his home country to be educated in Germany. He earned a PhD in physics under the renowned scientist Werner Heisenberg.

Heisenberg, of course, was awarded the Noble Prize in physics in 1932 for his creation of quantum mechanics. Even today, mention Heisenberg's name and people with only the most rudimentary understanding of the history of physics in the last hundred years will know at least about his "uncertainty principle." One of the biggest problems with quantum experiments is the seemingly unavoidable tendency of humans to influence the situation and velocity of small particles. This happens just by our observing the particles, and it has quantum physicists frustrated. To combat this, physicists have created enormous, elaborate machines like particle accelerators, such as the one I was hired to work with at Meson, to remove any physical human influence from the process of accelerating a particle's energy of motion.

Still, the mixed results quantum physicists find when examining the same particle indicate that we just can't help but affect the behavior of quanta, or quantum particles. Even the lights we use to help us better see the objects we're observing can influence the behavior of quanta. Photons, for example, the smallest measure of light, which have no mass or electrical charge, can still bounce a particle around, changing its velocity and speed.

To be able to hear a lecture by a man who studied with Heisenberg was a real privilege and one that I couldn't pass up. I also knew that Teller had studied under Neils Bohr while spending a year in Copenhagen. Bohr is another giant among scientists, in my mind, and in the mind of many others nearly as important a theoretical thinker as Albert Einstein. In 1935, Teller took a position at George Washington University where he worked as a theoretical physicist. Eventually Teller's attention turned to nuclear energy, both fission and fusion. In 1942, as the US involvement in World War II ratcheted up, Teller took part in Robert Oppenheimer's planning seminar at the University of California, Berkeley. From that summer meeting, the Manhattan Project got its start. Teller began his work on the development of a nuclear bomb at the University of Chicago, eventually heading a group at Los Alamos in the Theoretical Physics division. He was a big proponent of developing a hydrogen bomb (thus the nickname he claimed he didn't like) and that created some friction within the top level of scientists working on the Manhattan Project. Under pressure from the government and military to develop a nuclear bomb as soon as possible, Oppenheimer and others eventually created the two bombs that were dropped on Japan. Little Boy used enriched uranium. Fat Man used plutonium.

It wasn't until 1952 that the US successfully detonated a hydrogen bomb. One thousand times more powerful than what the Manhattan Project had developed, a thermonuclear as opposed to a conventional nuclear device, the hydrogen bomb was a truly frightening advancement in weapons technology. Many said that its development accelerated the arms race and heightened the Cold War tensions between the US and the USSR. That's true, but as a scientist, I could see why Teller had advocated for its use. When you're charged with producing the ultimate weapon, and you see how one can be made practicable, you advocate for it. By the time Teller was back at Los Alamos, he'd gained and lost notoriety and reputation among certain members of the scientific community. His advocacy on behalf of the Strategic Defense Initiative, what most people knew of as the Star Wars defense, a system combining ground-based units and orbital deployment platforms to destroy incoming nuclear missiles, lost him favor in some scientific circles. Some stated that the long-standing nuclear policy of "Mutually Assured Destruction" was working. What the SDI might do is be seen by the Soviets as an escalation of offensive practices and not as a defense.

In hindsight, given the collapse of the Soviet Union, it all seems to have not really mattered, but at the time, nearly the entire 1980s, the debate raged. In truth, we didn't have the technology to implement the proposal, but in my mind, that didn't mean that we couldn't develop it. Since Teller spoke on behalf of the plan, he was considered to be a kind of cock-eyed optimist about our scientific capability, more of a salesman of a plan, than someone who had really arrived at a workable solution. (In fact, a *Scientific American* article that several prominent scientists wrote said that any of our enemies could launch decoys for a fraction of the cost of our SDI system and wreak havoc on us with a follow up of actual missiles.) Teller didn't help himself by stirring up controversy throughout his career, even at one point, testifying against Oppenheimer when the man many considered a hero was being subject to a security clearance investigation in the 1950s.

All that taught me was that science and politics didn't really mix, huge egos often clash, and it would be better to toil in anonymity doing something you really enjoyed than to be mixing it up with the bigwigs in Washington, DC, and elsewhere.

I arrived at the site of Teller's lecture a few minutes early. A few people sat in the hall waiting. I decided to linger outside and enjoy some sunlight and fresh air. I enjoyed the nice break from the artificial light and the stale air inside the facility. We'd all heard rumors of how the Lawrence Livermore National Laboratory was run much more loosely. One article I read characterized Los Alamos as the Ivy League, all tweeds and ties, while Livermore was the counterculture, if not hippies than at least rule-breakers and risk-takers. In a way, I identified more with them than I did with Los Alamos and its conservatism.

I was hoping that Teller was going to speak about his Excalibur project, the precursor (or some said inspiration) for the Star Wars system. Teller had envisioned a device that would orbit in space, a spherical machine that would have elements sticking out of it, kind of like a sea urchin. Those elements would be fiber optic light guides to aim and focus immensely powerful X-rays that could evaporate the metal casings of Soviet ICBM missiles after they were launched. I would have loved to have been a part of the development team that worked on Excalibur, but even just hearing about it would have to do for the moment.

I took a seat on a bench outside a study area in the administration center. Above me, a series of portraits of Nobel Prize winners who'd worked at Los Alamos at one point or another in their career ran along one wall. In some ways, it was like being back in high school, though instead of varsity letter winners being commemorated, it was scientists. Across the way from me sat a man with an expansive broad nose and eyebrows that reminded me of a mink stole draped over a woman's shoulders. His heavy lidded eyes were deep set and focused on a newspaper he was reading. I recognized him immediately as Edward Teller.

I wasn't starstruck at all, but I did want to meet him. I stood and walked over to him and put out my hand, "Dr. Teller, my name is Bob Lazar, it's a real pleasure to meet you."

Teller looked at me over the edge of the paper, and said, his voice thick with an Eastern European accent, "Why, thank you."

He made no move to drop the paper and to take my offered hand. Feeling a bit awkward, I watched as his eyes focused on my face and then he turned his head slightly to the left as he squinted more forcefully. He looked down at the paper, closed it and then folded it so that the front page was visible.

"You're the man with the jet car," he said. He stabbed at the paper with my photograph on the front page.

"That's right, I'm Bob Lazar. That's my car."

"Lazar? I once studied with a man named Lazarev. A Russian. I believe the surnames are related. Might he have been a relation?"

I didn't tell him that I didn't know what my real surname was. I simply said, "I doubt that."

We spent a few more minutes talking. He seemed quite taken with the idea of me having installed a jet engine in my car. Though I wasn't applying rocket science in the ways that he was dealing with, he seemed both amused by and impressed with the application. He asked me what else I had planned on doing, but I wasn't able to answer. One of the staff members came and escorted him away so that Teller could begin his presentation. For whatever reason, Teller must have remembered me—whether it was our conversation, my jet car, my last name—I'm not certain. When it came time for me to send out query letters to get a new job, I wrote directly to him. He was still with Lawrence Livermore, but it seemed less and less likely that Star Wars was ever going to come to fruition. Various treaties limiting nuclear arsenals had been agreed upon. In a lot of ways, the nuclear weapons industry was contracting, but I hoped that my attempt at networking would at least get me some attention.

As it turned out, I got a job as a result of that meeting with Teller and, later, a whole lot more attention than I ever wanted or needed.

I was, of course, blissfully unaware of all that, when, two days after my first visit out to EG&G's remote site at Groom Lake, I got another call to report for work. The routine was much the same. Get the phone call, get to EG&G, get on the plane, and fly out to the base. This time though, there was a difference. When I arrived, I was the only one on the bus except for Dennis. This time, instead of remaining on the main base and proceeding to the administrative area, as I'd come to think of it, we headed south along a packed dirt road. Within minutes we were in the desert. With the blacked out windows, I couldn't see much to the sides of me, but it was likely what I was seeing in front of me through the frame of the front windows—a few rolling hills, sand, a smattering of creosote bushes, greasewood and cactus. We traveled for nearly forty minutes, covering approximately ten miles, the bus headlights sending out a flickering semaphore as we rattled over the corrugated sections of the road.

Eventually, we slowed and came up to a rather large hill. Above the loose pile of dirt and stone, a spire of rocks with the characteristic desert varnish darkening their surface sat like a broken jaw and teeth above it. As we moved left around the hill, it was as if someone had taken a giant shovel and squared the face of the hill at 90 degrees to the ground. To one side of that perpendicular rise stood a small entryway. Dennis led me inside the building and into a small 12 x 12-foot room that was completely unadorned and unfurnished. The only thing inside it was what I looked like a stand-alone ATM machine you might find in a convenience store. Dennis walked up to it, and put his hand on a glass platen. He splayed out his fingers and some kind of optical scanner or reader lit up from above.

Next, he typed in a few pieces of information on a small keypad. A moment later, his identification badge dropped into a slot at the machine's base. I repeated his actions and my badge popped out.

"That is yours. You'll need to have it on your person at all times. It's both your identification card and your key card." He nodded toward the door. Alongside it, a small stainless steel plate clung to the wall. "Without this, you won't be able to go in or out."

I immediately understood that with such a system all of my movements within and between rooms and buildings could be tracked. Not unusual for any kind of facility dealing with sensitive information.

We walked toward a second door. Dennis nodded toward the plate.

"Try it."

I swiped the key card and then heard a click as the latch was released. Dennis gestured for me to lead the way, and I stepped into a well-lit corridor. I immediately felt like I was back at Bowling Green Elementary School. I smiled to myself and wondered if the one of the colors on the walls or cinderblock was in, in fact, known as Bowling Green. The light green of the drywall and the darker green of the cinderblocks that rose to about waist height were the typical institutional green that I'd seen in countless hospitals and municipal buildings and schools. I tried to remember something I'd once read about the psychology of color, how green was supposed to react with our central nervous systems, but couldn't remember if it was calming or energizing and what any of that might have to do with being green with envy.

All I knew was that the corridor was very long, and it seemed as if it stretched along to the vanishing point in a long and narrowing row of doors on either side. The hallway was empty and the building nearly silent except for the clicking sound of our heels. Dennis stopped abruptly after just a few paces. He nodded his head to the right, "In there."

Unlike the other doors, this one didn't require me to use the keycard. I simply turned the knob and stepped in. Once again, I found myself in what seemed to be a waiting area of some kind. A few upholstered chairs sat along one wall and an office desk sat in roughly the center of the room. Alongside it, a two-drawer filing cabinet the color of unfinished brass sat with an English ivy plant and its tendrils draped along and down its flanks. A water cooler in the far corner of the room belched and bubbled before quieting. Before I could decide, or Dennis could indicate, that I should take a seat, another door opened at the back of the anteroom. A petite brunette wearing a bright red cardigan and black pants entered.

She smiled brightly and the room lost much of its cool indifference.

"Would you like to follow me?" she asked.

I knew she wasn't asking a question, but I appreciated her being far less curt than Dennis was.

"Absolutely," I said, not quite matching her enthusiasm but nearly.

"We'll need about thirty," the woman said, rising on her tiptoes and angling to the side to speak around me to Dennis. She held a clipboard folded in one arm and resting against her chest. Once I got into the room from which she'd come, I knew what was next. A small examination table, a series of glass fronted cabinets, a scale, and the other accessories you'd find in a typical doctor's office exam room clued me in.

"Would you like to take a seat?" she asked, and I wondered briefly if this was going to be a recurring pattern, if we were going to be playing an odd sort of game of twenty questions.

"Sure," I said. I sat on the examination table, feeling and hearing the rough paper cover shift as I did so.

I watched as the young woman moved past me. I noticed that in one corner of her eye, she had a faint mole that intensified the deep brown of her eyes. Once out of my line of sight, she must have opened a drawer. I heard the metallic clatter of things being moved about. When I saw her next, she had a stethoscope draped around her neck and she was tugging at the Velcro off a sphygmomanometer sleeve.

"Could I get you to take your shirt off?"

I was wearing a white oxford cloth dress shirt and a t-shirt. I undid the buttons and then shrugged out of the garment.

"May I?" she asked. She held out her hand and I gave her the shirt. She hung it from a hook on the back of the door. Sixteen more questions and we'd be done, I thought.

"Can you hold out your right arm for me?"

I did as instructed. She slid the blood pressure cuff on and began to inflate it.

"Any sign of rain out there?" she asked, "I thought that maybe we'd get lucky. I'm heading out to Death Valley this weekend with my boyfriend. I'm hoping the desert bloom will be beautiful. One of my favorite things."

"I hope I'm not heading to Death Valley soon," I said.

She laughed, "Not according to this," she said as she held the gauge out to me, "One twenty four over eighty two. Can I get your heart rate?"

We did a few more typical and cursory medical examination things to be sure my hearing and eyesight were okay. Without my glasses, I told her, I'd be lucky to see the chart let alone any of the letter on it. After that, we turned to what seemed to be the real reason for my being examined. After reading through a list of potential allergens—I'd lost track of the number of questions but was pretty sure we'd gone beyond twenty—she asked me if I was left-handed or right-handed. When I answered that I was right-handed, she said/asked, "Well, then let's have you do the left arm, okay?"

She retreated behind me again and returned with a small clear plastic case that was subdivided into compartments. She next produced a felt tip marker. Holding my wrist gently, she turned over my arm so that the bottom of my forearm was exposed. She swabbed the area with alcohol and then drew a neat grid on it producing a series of sixteen uniform sized squares. I was quite impressed with how symmetrical they were given that she'd done it freehand.

She shrugged off my compliment, "Done this dozens and dozens of times." Once she was done with the drawing, she retrieved a rolling stool and drew it alongside the table. She cranked a lever and rose up so that her head was nearly level with mine.

She looked at me and then shook her head, "Sorry." She retrieved the plastic case and placed it on a rolling table like the one dental hygienists use.

"Okay? Now. One the reasons you're here is so that we can do a baseline allergy test. What we've discovered is that some people working here get exposed to elements in the work environment that produce, in some cases, strong systemic responses. So, what I'm going to do is use one of these," she held up a small device that resembled a pushpin, "To place a small amount of different chemicals subcutaneously on your arm. I've got the grid there, and I'll know which, if any of these, produce any kind of reaction on your skin."

"What kinds of things are you testing for?"

"There could be redness. Swelling—"

I cut her off, "I'm sorry. I didn't quite ask the question properly. I know what the allergic reactions may be like. What are the allergens you're testing."

She frowned and let her shoulders slump, "I'd like to tell you but I can't. I really don't know. All I was told is that you, and a lot of other people here, will be working with," and here she used her fingers to form quotation marks, "'exotic materials.' What those are specifically, I haven't been told."

"Do you know if there have been any serious consequences of that exposure?"

"Can't say." This time her tone rose only slightly at the end as if she was questioning whether or not she was asking a question.

For my part, I was very concerned. This was a high tech lab, after all, so who knew what kinds of materials I might be working with? Better safe than sorry.

After she'd done the pricking of the various areas, she returned with a small plastic container, the kind that hospitals often use to place a patient's medications in. A faint yellowish liquid was inside, just a couple of ounces I estimated.

"Can you drink this?" She held it out to me while she made some notes on her chart.

I took the cup and brought it toward my mouth. I immediately thought of the household cleaner Pine-Sol. I drank it down and felt my eyes tear up briefly.

"What was that?" I asked.

"Not so bad, was it?" she asked. "Just another part of the testing."

She snapped the plastic case shut, and said, "We're all through here. Good luck with your work, and again, please be sure to monitor those test sites and if you notice anything unusual, let us know through your supervisor."

The nurse handed me my shirt, and I while I put it on, she went out into the anteroom. By the time I'd finished dressing and gone out there, she was gone. Dennis was nowhere to be seen either. I stood in the hallway for a few moments, thinking about the allergy test and what it might mean for my future employment. I was still waiting for the Majestic clearance needed to become a full time employee. I had no reason to doubt that I'd receive it, but I also knew that wheels often turned very slowly in these matters. I was accustomed to hard work, having multiple responsibilities, but still it would have been far more desirable to be able to shut down the photo business entirely—selling it off—and resume a more normal kind of nine-to-five existence.

A part of me really enjoyed the less structured schedule that my involvement in photo processing afforded me, but, of course, that came with certain trade-offs. Tracy and I were doing fine financially, but having to make myself available on an on-call basis for EG&G and in essence for the photo business could wreak havoc with a social life and our personal schedule. I knew that it was going to be temporary, but how "temporary" that was going to be was difficult to say. I knew that stress played a role in how our immune systems operated and responded to pathogens. Would my current mental and emotional state have an effect on the grid of substances that had been introduced into my forearm?

I didn't have a long time to reflect on the matter. I don't know how I missed seeing him at first, but Dennis was just down the hallway from me.

"This way, Bob." Dennis rarely used my name in our conversations and for as common as my name is and for as many times as I'd heard other people use it, it sounded strange coming out of his mouth.

I followed Dennis as he entered another unmarked door into another small room—this one even less spacious than the one where I'd been issued my ID keycard. Dennis gestured toward a small office desk, a metal framed affair with a cheap wood laminate finish. On the desk sat a stack of blue file folders. I took the seat behind the desk and noticed that the folders were blank, that is, there were no labels on the on the folders themselves to identify EG&G.

Dennis stood just inside the closed door and said, "It's important that we get you caught up as quickly as possible about where we've been and the direction the project is going in. Given your present clearance level, these briefings should give you the background information you need. Obviously, this is all confidential information. Once your clearance is upgraded, you'll learn more."

He paused and thrust back the cuff of his suit jacket and shirt to consult his watch, "I have a number of other tasks to attend to. Do your reading here, and when I'm done with those other matters, I'll come and get you. Any questions?"

He barely paused before saying, "Good. I'll be back."

The stack of folders was nearly a foot high. They were of various thicknesses, none containing any more than what seemed a dozen pages. I figured it best to begin at the top and work my way down. I opened the first folder that was labeled "Overview." The title page had nothing more on it than the words "Project Galileo." The prose was typical government plain speak, a kind of "just the facts" sparseness that suited me, especially given the stack of materials I was to read and having no indication at all from Dennis when he might return.

The short first paragraph would have pleased me except for the inclusion of one word—"extraterrestrial." It explained that the purpose of the program was to back engineer—the process by which you take knowledge or information from anything man-made and re-produce it or produce anything based on the extracted information—a propulsion system of an extraterrestrial craft. The power and propulsion systems of the craft were of the highest priority.

For a second, I stopped reading and considered just what was meant by "extraterrestrial." One thought I had was that EG&G, or whoever had authored this overview and whoever had approved it for distribution, was being either very precise or very loose with their use of language. Something that is considered to be extraterrestrial means that it is of or originates from outside the Earth or its atmosphere. Maybe, in the language employed at EG&G and out here at Groom Lake, the experimental craft they are presumably working on, get referred to that way since traditional nomenclature—a plane, a jet, or names for other land-based vehicles didn't quite fit with the scope of the craft's capabilities or intent. I knew how many scientists thought, and if this was something that a scientist or engineer had written, and it was intended as an in-house document, then the use of a potentially loaded word like that wouldn't be so odd.

On the other hand, I thought that maybe this whole document was another form of a test. I was still, in a sense, on trial with the organization. I'd been subjected to all those other questionnaires that dealt with personality and temperament, and maybe this was just a part of that larger assessment. Was I someone who would get frustrated or annoyed by references that were seemingly off topic? Would I be willing to wade through information or data that was seemingly irrelevant or confusing and not be able to plow ahead or keep an open mind.

I decided to do precisely what I thought they wanted me to do—to keep reading and not form any judgments until I had the bigger picture formed in my mind.

As I read on, a few other things became clear. Though no dates were given, it was clear to me that the project was not in its infancy. Some references were made to past attempts to understand the nature of the power and propulsion system. Several attempts had been made to reproduce the kind of system that had come into their hands, but with no success. Given what I knew of back engineering, it often involves disassembling something and analyzing its components and workings in detail. Any kind of propulsion system is likely to be highly complex and the disassembly and production or fabrication of new components could take a considerable amount of time.

I next read that a previous attempt to dismantle one of the existent propulsion systems had resulted in an accidental explosion. The document didn't go into much detail about that "incident," but I also knew that any time there was an accident like that, there would be a period of time when the work was halted, an investigation initiated, and only after a review process of the procedures—something that could have taken place in months or perhaps years—work would then commence. Reading a bit more made it clear to me that something like the process I'd imagined had taken place. The last bit of the overview dealt with the present reality. They were going to deemphasize the dismantling efforts and focus instead on other kinds of analyses to determine how it functioned. What those types of analyses were to be wasn't specified, but one other thing was made clear as a priority. We—it was a bit presumptuous of me to include my not-yet-fully-cleared self in this—were to duplicate this technology using existing materials and not utilizing what the briefing referred to as "exotic" materials.

None of what I read set off any alarm bells. I was pleased to read about the nature of the work. It would be a real challenge to take on the task that was set before me. What types of analyses could we devise to understand the nature and function of a propulsion system without having the ability to dismantle the existing, and presumably, functioning system? After all, in my mind, the word "analyze" means to break down into component parts to see how they interrelate. When you're working with a physical object, or a set of physical objects in the case of a propulsion system, not having the capability, or in this case the task, of seeing how those component parts all fit together, was like working with one hand tied behind your back.

We'd have to devise new approaches, and I wondered then if that was why I'd been hired.

Frequently, when a team is at a sticking point, it helps to bring in someone new who hasn't been through other parts of the process. I could look at things with a fresh set of eyes and none of the preconceptions, or at least a different set of them, from the other scientists already at work.

Still, as I returned the pages to the folder, the one question that kept coming back to my mind was what the hell these guys were up to. Was this just a test of my temperament and suitability to work in this environment, or was this reverse engineering of an extraterrestrial propulsion system a real work assignment?

The last sections of the Overview also cast some shadows of doubt on the whole enterprise and made me think that this had to be a kind of screening mechanism. If you could put up with this kind of disinformation, then let's see how you feel about working under these restrictive conditions. Project Galileo was comprised of a number of divisions—the power/propulsion team, and those that dealt with navigation, control, metallurgy and other component elements of a flying craft. These groups were not to work in concert, as you might expect, but were to be segregated. Not only wouldn't they work together, they were not to share information with one another. No communication between the work groups was allowed. If we had questions that we thought needed to be answered, we weren't to go to any other team members directly. A clear chain of command was in place, and it was a condition of our employment that we not share information with anyone other than our work partner. Failure to adhere to the security rules was grounds for immediate dismissal and potential prosecution.

Interestingly, some of the other briefings dealt with the other teams: Project Looking Glass dealt with the materials side of the craft, and Project Sidekick was exploring the possibilities of the craft's weaponization, but only in the most generic of ways. For example, I learned that the bulk of the craft was manufactured from the same material. Tests had shown that it was not metallic in origin nor was it ceramic. That intrigued me. Think of your car and how many different materials are used to produce its various components, aluminum, steel, glass, rubber, various plastics. And here was an object that was claimed to made of a single material?

I can't claim to be completely logical. I was a trained scientist and I was supposed to look at things with as much objectivity as possible. I found myself veering between two points—rejecting what I was reading out of hand as an attempt to use a fictional scenario to determine my suitability for some highly confidential project, and being caught up in the excitement of being exposed to new ideas that would test the limits of my understanding of the nature of things here on Earth. Or maybe the test was even simpler. They presented this scenario to me to see if I was going to share it with anyone. They could do all kinds of background checks, but the surest way to see if I was capable of keeping information confidential was to provide me with a story and see if I kept it to myself.

I briefly scanned the room, wondering if I was under surveillance in there, and if my eye rolling at some of what had been revealed to me might determine the fate of my employment.

For a few moments I considered the possibility that this craft, or maybe a better term was this "technology," came from somewhere other than the United States. We were still dealing with the effects of the Cold War, and perhaps we'd managed to gain control of some Soviet technology. Whether we had stolen it, had someone present it to us, or in some other way gained possession of it didn't really matter. What did matter was that we had it, apparently didn't really understand fully how it worked, and it was to be my job to determine how it functioned and how we could reproduce it. I'd wanted to be involved in working with nuclear weapons, but maybe this was an even more worthwhile task. Maybe this technology would prove to be the key to what seemed to be the dominant, and ominous, political philosophy of the day—Mutually Assured Destruction.

Even in hindsight, it is difficult to come up with a coherent explanation of my thoughts and feelings at the time I was reading those documents. I can say this with real clarity, however: I was thoroughly engaged in the reading— not by its language, but by what it offered as a possibility to utilize my interests and abilities. If this was all a game, a way to determine if I was cut out for the work, then how could I strategize and make the right moves to get to the end point, to be hired full time by EG&G? If what I was reading were true, that there was some extraterrestrial craft that we had in our possession, then I'd have to manage to accommodate a fundamental shift in my thinking, recalibrate, and proceed to the task at hand. In either case, it was all about the job in that moment. And for now, the task that was at hand in the immediate was to read more of the briefings.

I picked up a folder marked "Biology." As I scanned the first page, I was again torn in several different directions emotionally and mentally. The file identified the place from which the extraterrestrial craft originated. It stated that the craft came from the planet Zeta Reticuli star system. I wasn't familiar with the name, but the briefing went on to add that the constellation was located some 39 million light years from Earth. It was only visible from Earth from positions in the Southern Hemisphere.

The most obvious question was how could that craft have gotten here? Even if we could travel at the speed of light, approximately 186,000 miles per second or 671 million miles per hour that meant it would take 30 years to get there. That's a ridiculously long period of time even at that great speed and an unfathomably long period of time based on 1981 technology, and the fastest we could then travel through space. I couldn't even begin to imagine how much fuel would be needed to propel a vehicle that far, how life support systems could function and replenish, how food supplies could be sufficient, and all the rest. It just boggled my mind.

Those questions paled in comparison to the considerations that came to mind when I turned a few more pages in the file. I got to a pair of black-and-white photographs reproduced on regular copy or printer paper. One was a relative close-up of a humanoid organism. The photo encompassed what seemed to be the torso—no head was visible and nothing below what appeared to be the midline or waist. I could see just the stub of where a human arm would project from the torso. The rest of that limb was out of the frame. The same was true of the lower extremities. I could see just the suggestion of a pair of legs—a protrusion roughly where you'd find the hipbone of a human, more specifically the crest of the ilium, rising from the gray-white skin.

What most distinguished the photo was the classic t-shaped incision that ran nearly the entire vertical length of the trunk, with a far shorter horizontal line at the top of that major cut line. The skin had been pulled back revealing, if the subject had been human or an animal, a series of specific organs of respiration and digestion—lungs, heart, kidney, liver, bladder, etc. Instead, what I saw was an undifferentiated mass of tissue that filled the majority of that cavity. The second photo was much the same as the first, but it was smaller and in the white space that surrounded the photo on the page, it appeared as if several different people had added their notes. These annotations were in different handwriting, but they mostly expressed the same sentiment: how unusual it was to see this large central mass, and speculations about how it might function to perform the various specialized tasks that our human organs do.

For some reason, the inclusion of those notes from others really got my head spinning at a greater rate than before. Why would they—and to this point I had no clear understanding of who "they" were—include those notes in the briefing materials? What was the point of that?

I can see now that this seems like a minor question compared to the bigger issues these photographs and the information about Zeta Reticuli engendered. I was so wrapped up in what I was reading, so overcome with a kind of information overload that my mind fixated on a point independent of my own will. Maybe subconsciously my brain kicked into fight or flight mode. It sensed that this information could have done me some kind of harm, confused me to the point where I was in danger of losing touch with what I understood to be reality. I knew that from the time I first opened the pages of that first folder to viewing those photographs and reading the brief account of the planet from which the craft originated, I'd experienced something that had happened to me many times before. The passage of time hadn't been a part of my conscious awareness. I could have been reading for five minutes or five hours. I was so wrapped up in what I was reading that time had become irrelevant. I know that comparing the human brain to a computer and its operating system is a gross oversimplification, but the only way that I can think of to describe the state I was in is to say that information was being downloaded and I was in the process of formulating a vast number of questions. Because I was still acquiring more data, and wanted to, I didn't have the time to pose all of those questions to myself. They were being stored and queued up. Who I would ask, and what the answers might be wasn't something I knew as I sat there trying to assimilate all that I was reading into a coherent whole. I was failing miserably at that task, but was hopeful that at some point, I'd be able to put all those pieces together.

CHAPTER THREE

SOMETHING CAUGHT MY EYE AND I LOOKED UP FROM reading. Through the door's small rectangular window spider-webbed with wire, I caught a brief glimpse of someone out in the hallway. A few seconds later, I heard the electro-mechanical sounds of the door's latch being released.

"You can leave those there. There's someone I'd like you to meet," Dennis said from the doorway. He edged into the room and a uniformed security guard slipped past him. He bundled up the files and strode off. He didn't make any attempt to count them or inspect them and that made me wonder again if I had been observed the whole time I was in that room.

"Follow me," Dennis said. I did as instructed. We stopped in front of a door just a few spots down from where I'd been reading the briefing papers. Dennis carded us in and before we could enter, a security guard exited the room. Once inside I could see that we had entered a laboratory. A man was seated on a stool at a long workstation. When the saw us, he smiled broadly and walked toward us.

Dennis made the introductions, letting me know that my work partner was named Barry Castillo. He was an inch or so shorter than I, thin, and with an olive complexion and a shock of brown curly hair that he had the habit of running his fingers through whenever he spoke. Other than each of us saying, "Pleased to meet you," we were silent until Dennis left the room.

His parting remarks were simple and straightforward, "Barry will bring you up to speed. This is propulsion lab and you'll be spending most of your time here. He'll advise you on what your task assignment is, the goal that you're expected to achieve. I'll leave it to you gentlemen to get started. I'll see you both later on this morning."

As soon as Dennis left us alone, Barry eyed me, and said, his eyes sparkling, "You're not one of them are you?"

"I don't think so," I said.

"Military, I mean," he shook his head and waved both hands in front of him. "Forget that. I can tell. I mean, no disrespect, it's just that I can tell that you're—well, you're normal."

I laughed softly. "I was thinking the same thing about you. As soon as Dennis let me in here and I saw you, I was like, 'Thank, God. Somebody I can actually have a conversation with.'"

"I know what you mean."

Barry grabbed my biceps and said, "You've got to see this stuff."

I had that laundry list of questions I wanted to ask him about the briefings, but Barry's enthusiasm was so infectious that I let him lead me to his workstation. Sitting about fifteen feet from his high stool was a cylinder about the size and shape of a household garbage can. It was pewter in color and both from a distance and upon closer inspection it seemed to have no seams, no welds, no fasteners, no sharp edges, no marks on it at all. I saw nothing that hinted at how it had been manufactured. It didn't appear to have been cast, machined, molded, formed, or joined.

"This is one of the emitters," Barry said to me.

"What does it emit?" I asked, setting aside again the dozens and dozens of other questions about what I'd just read.

"I'll show you in a second," Barry said, sounding very much like an excited teen wanting to show off all the components of his stereo system before letting the music blast away.

On the desk surface of Barry's workstation sat a half-sphere of the same color and material. It was roughly the size of a basketball and sat on a one-inch plate, again of the same composition as the other two pieces—or so I surmised based on a quick look at it.

"This is the reactor."

Barry lifted the half-sphere and exposed a small tower-shaped object inside. It was approximately six inches tall and, as was true of everything Barry showed me, seemed to have been wholly formed through no process that I'd ever seen or read about. Sitting atop the tower was a cap—like all the other surfaces, it had a rounded or radius corner instead of a 90-degree edge. Barry retrieved a small triangle-shaped disc from a metal box and placed it in the top of the tower. He replaced the cap, put the half-sphere back on the base and stepped over to the emitter. He pushed it a few feet toward the reactor.

"Watch this," Barry said. He extended his arm toward the half-sphere but before his arm could reach its full extension, his hand moved back toward his upper body.

He looked at me expectantly, and I felt bad that I didn't share his enthusiasm in that moment. It simply looked like he just extended his arm as if to shake someone's hand and then withdrew it.

"Better yet," Barry said, stepping to the side, "feel this."

I stood in front of the reactor and did as Barry had done. When my hand got within approximately eight or nine inches of the devices, I felt myself being pushed back by some unseen force. If you've ever had the use of two fairly powerful magnets and put them so that two of the poles are placed near one another to produce a repelling effect, that was very much like what I felt.

"Antigravity," Barry said. "Can you believe it?"

"I read about that in the briefings," I said as I continued to test the field being produced. "I skimmed over that portion figuring it was part of some joke or something else. I don't know what. . . ." My voice trailed off. Like anybody who was interested in science and who grew up in the era that I did in the 1960s, antigravity was a concept with which I was familiar from watching science fiction shows on television or in films. And for the most part, that was where the notion of the existence of antigravity resided. Every schoolkid learned about Sir Isaac Newton and his discovery of gravity—one of the four fundamental forces at work in the universe. (The other three are electromagnetism and the strong and weak nuclear forces.) According to Newton, gravity was an external force transmitted by some unknown means. Antigravity, as its names implies, is the opposite of gravity: a hypothetical force by which a body of positive mass would repel a body of negative mass.

The key word there is, of course, "hypothetical." No one had been able to prove that it actually existed. When Einstein came along with his theory of general relativity, gravity was not considered to be a force, but a result of the geometry of space-time. Simply put, that theory offered no real hope that antigravity existed. Quantum physicists postulate that gravitons exist as a subatomic particle. These are massless elementary particles that transmit the force of gravity. How they could be created or destroyed is not yet clear.

We tend to think in terms of opposing binaries—light and dark, long and short, and so on. So it seems logical to us that if gravity exists, its opposite must as well. That's why going back as far as H. G. Wells and his book

The First Men in the Moon, published in 1901, we humans imagined a substance or a process by which gravity could be blocked or controlled. Wells called it Cavorite. He linked the idea to propulsion as well. He wasn't alone, and prior to his writing, scientists like Nikolas Tesla, Thomas Townsend Brown, John Worrell Keely, and others all worked on ways to control gravity, to harness the power of antigravity.

If that demonstration wasn't enough, Barry next went to a drawer and produced a golf ball. He wagged his eyebrows at me and stood several yards from the bench on which the devices sat. "Heads up," he said. He tossed the golf ball underhanded into the antigravity field being produced. The golf ball arced as it approached the field and then, once it entered into it, shot violently upward into the ceiling.

"Holy shit," I said. We were like two kids monkeying around while our parents were away. The ball had hit the acoustical tile in the suspended ceiling of the lab. Bits of the tile and a fine spray of dust fell and settled on the workstations and the floor.

"We better clean that up," I said.

"Isn't that so cool?" Barry said.

"It's amazing," I was still having trouble wrapping my head around what I'd just read and then seen demonstrated. As before, my brain focused on something more concrete, more of the present. "Do you have a dustpan? A broom?" This was really my first day on the job and the last thing I wanted was to get in trouble for breaking something or creating a mess in the lab. Funny how the mind works.

We spent the next couple of minutes with me sweeping up using a file folder as a broom and another as a dustpan. Barry took the half-sphere off the plate and removed the fuel disc.

It was showtime again. Barry reinstalled the fuel disc. He retrieved a candle from his desk drawer. He held it in the mouth of the emitter and using a striker, he lit the candle. At first it burned normally, then, as he pulled the emitter closer to the reactor and, presumably, focused the gravitational waves, the candle stopped flickering but the light remained.

"Bizarre," was all I could manage to say in response.

Next he took off his watch, a classic chronometer with a second hand, and placed it in the emitter. The second hand stopped.

"That doesn't mean that the device has slowed time or created frame dragging," I said. "It could just be electromagnetism affecting the watch's works."

Barry smiled and shrugged his shoulders exaggeratedly. "Maybe so. But check this out."

He removed the watch and then rotated the emitter 90 degrees. "Look inside," Barry told me.

I did, and there, about two feet from the bottom of the emitter, I saw that a black dot had formed.

"It's bending it. No doubt about it," I said just barely above a whisper. I was stunned. Anyone with a basic understanding of physics understands that light can be reflected or refracted, truly bending light is another matter entirely. We can use lenses, mirrors, prisms, and other means to change the direction that light is travelling—reflecting it or refracting it—but that doesn't truly bend it. That black dot, the absence of light in the visible spectrum, showed that the light was being bent.

The only force that can bend light is gravity. Having seen this latest of Barry's demonstrations, what I'd been puzzling over seemed to have only one possible conclusion. The device we were working with was able to produce a gravitational field. As far as I knew, as far as I'd say any terrestrial scientist knew, the only things that produced gravitational fields were enormously large objects—the Earth, for example. Here I was in a room with an emitter the size of a garbage can and a reactor that was smaller than a typical toaster oven, and the two of them worked in concert to produce a gravitational field. I was in awe at this technology.

Of course, I was dealing with a propulsion system, if I was to believe what I'd been told to read, that didn't originate on Earth. The only understanding I had of how the phenomenon of light and gravity operated was based on the work of scientists here on Earth. I had been standing there in that room with Barry for more than an hour, and in that time, I'd been processing what I was witnessing, analyzing based on my knowledge of, for lack of a better term, Earth physics.

That meant looking at this from both a Newtonian and Einsteinian perspective. Einstein's description of gravity, which is fundamental to his theory of general relativity, tells us that the effect of gravity is caused by distortions in space and time itself. Now, if you do something as fundamental as distorting space and time, and reshape it, anything that lives inside space and time will be affected. That includes waves, and so, waves can be bent and can follow different paths if you change the geometric properties of the space they live in. Gravity can effectively bend space and time. Meaning that anything in its field is also distorted and that includes light.

This thing was producing a controllable form of intense gravity. Once it was produced, they'd be able to do anything they wanted to do with it. It could push the craft or pull it. It would also produce some bizarre side effects as it bent light. Much like the thermal heat waves that rise up off a heated surface like a roadway in summer, it would distort our field of vision. A desert mirage is an even better example. You could see something that is not even there or the inverse of that—not see something that was there. Once you mess with the nature of light and optics all kinds of odd things could happen to our perceptions.

"It's the missing piece," Barry said. "If you can produce a gravitational field like that, then anything's possible."

"Force fields. Antigravity. Exotic propulsion. You can bend space and distort time."

"The thing is so simple. No moving pieces. How the hell did they do this?" Barry shook his head appreciatively.

"Elegant, really. I just—. I really don't know what to say. I just have so many questions. Do these devices focus the field? Do they produce their own artificial gravity field? Are they amplifying it somehow? Can that field be directed or controlled?"

"That's what we've been trying to figure out," he said.

"I mean, I get it that the reactor produces the field somehow, but what's really got my attention is the emitter. That's got to be one of the keys, doesn't it? If it were omnidirectional, I don't see how—"

"They're really interested in how the reactor operates," Barry went on. "That's to be our primary focus. I keep hearing that over and over. That makes sense in a way. Start with the power source. Move in a linear fashion from there."

"But thinking linear may not work."

"I know," Barry said. "But as much as I hate to approach things this way, we've got to do what they ask."

"I don't want to jump too far ahead here, but as far as we humans know, the only thing that produces gravity is an immensely large body of mass." I pointed at the cylinder and then at the reactor. "I've seen you move both of those things. That doesn't tell me we're dealing with a large amount of mass. Something fundamentally different from anything we know and have experienced is going on here."

Barry took a seat again on his stool. We let a silence linger for a minute or two.

"So? What do you think?" Barry asked.

I remained standing in front of the two devices. "Well, first of all, this thing works here, on Earth. In this atmosphere. Under all the conditions of our existence. That means that it operates for the most part in this physical realm and has to conform somehow to our understanding of our physics. Second, that makes no sense at all. Maybe it has to conform to some other physics. I mean, for it to produce a gravitational field as we understand it, it would have to produce terawatts, or even more, of power. A tremendous amount of energy has to be produced. All the power output possible on this planet focused into a small area. Unimaginable power, really. And the size of that reactor? Tiny."

I scratched the back of my neck and head. "How can I be standing here in front of this thing while it's still functioning? If it's producing enough energy to create a gravitational field, where's the waste energy going? The basic laws of thermodynamics are being violated here. We should be burned to a crisp. No reaction is one hundred percent efficient. None. Not nuclear, not chemical. None. It's not giving off any heat. It makes no noise. It makes no sense!"

"I know," Barry said. "These things have torn a huge hole in my understanding of science. A huge hole."

"None of this is supposed to be possible."

Barry pointed at the emitter and then at the reactor, "Yet there they are. Can you tell me that they're not there?"

"No. I can't." I paused for a few seconds. "Can you tell me how the hell this is possible?"

Barry laughed again and slapped the table. "What do you mean can *I* tell *you*? You tell me. That's why you're here isn't it?"

At least that night made one thing clear. Yes, indeed, that was why I was there.

I imagined that if I were a screenwriter or a Hollywood director, how I'd have been totally disappointed by what

Barry had shown me. More specifically I'd be disappointed by how little, for lack of a better term, "spectacle" there was to the whole demonstration that he'd produced. The reactor and emitter weren't that visually impressive at all. I knew that they were producing a reaction that was extraordinarily powerful, but it didn't *look* like much at all was happening. The room didn't shake, eye-searing beams of light didn't flash from it, it didn't produce any kind of discernible sound. I thought of the jet engines that I used in my vehicles, the rockets we used to launch and those that NASA used to launch spacecraft beyond the Earth's gravitational pull. Those things produced a kind of sound and fury, but what Barry and I witnessed produced nothing but a slight visual distortion.

If you've ever driven down a long stretch of highway and seen thermal heat waves rising from the heated pavement, then you've seen something similar to what I saw when an object was placed between the emitter and the reactor. The visual distortion I witnessed was milder than that, but still very much like it.

In the end, as much as I was always interested in powerful reactions, I was consumed by thoughts of how in the world, or in some other world, were these emitters and that reactor able to produce gravity? I now more clearly understood what our task was. This was my real "Aha!" moment. I was certain that the people in charge of this whole project had the same desire in mind that I did. We wanted to figure out how we could produce gravity in the same way this device did. That was the be all and end all of what we'd been assigned to do. In my mind, the whole alien origins of these objects, the whatever of their civilization, their purpose in coming to Earth, or whatever other far-flung questions about alien life forms and anything else, truly did take a distant backseat to the notion of how anyone could produce a force that was as fundamentally essential and powerful as gravity.

In a way, I was like a caveman who'd seen fire for the first time. I sensed that what I was seeing was so life altering in its possibilities that I had to do anything in my power to understand how I could produce that same phenomenon. As Barry later put it, what we were experiencing in witnessing what this device could do, was having a "science-gasm," the world's best.

The truth was, though, that what we were experiencing was a science-chasm. What that emitter and that reactor were able to do tore a gaping hole in our understanding of science. I was standing in a small room where an enormously powerful reaction was taking place, the freaking production of GRAVITY, and no residual heat was being given off. You can't have one hundred percent efficient energy transfer. That's just not possible. Nuclear reactions, chemical reactions, no kind of reaction in our world is one hundred percent efficient. And yet, this was one was. The entire output of power on our planet was essentially being focused in that room and I was still standing within a few feet of it and being completely unharmed. The nuclear bombs we dropped on Hiroshima and Nagasaki were like tiny sparks compared to the power needed to produce gravity. A nuclear power plant that provides energy to a small part of our planet has a veritable river of water, millions of gallons, flowing through it to keep it cool. But there I was standing and talking with Barry in a room so quiet that I could hear the institutional style clock on the wall click over to the next minute. In the quieter moments when Barry and I were both lost in concentration, I could hear the hum of the fluorescent lights above us.

I knew that I was repeating in my mind what I'd said to Barry minutes before, but it was like my mind was back to being an old-fashioned record and the needle was stuck in the groove. The question of how this thing worked spun around and around and around again in my mind. Somewhere I suppose the question of how could this even be happening was faintly playing, but far too softly to matter.

CHAPTER FOUR

AFTER THE FASCINATION CAME THE FEAR. I HAD NO idea how long I was supposed to work with Barry that first night. I considered it a kind of orientation session. We weren't going to get into any of the specifics of determining how the technology worked. That seemed obvious to me, especially since I was still reeling from the effects of what I'd witnessed. We'd had that science-gasm as Barry had put it, and in the lull afterward we sat at the workbench, each of us lost in thought. It was an awkward situation to say the least. Barry and I knew nothing about one another, our educations, our previous work experience. We'd been thrown into a room together and he'd proceeded to demonstrate to me how the technology worked. After that, it seemed as if there wasn't much else to talk about in terms of the devices. We had the *what*, or so we both believed, but now we had to turn our attention to the *how*.

Finally, I said to Barry, "So, how long have you been working at this?"

His expression darkened and he let out a noisy breath through his nose, "I don't know if I can tell you that." He screwed up his face in disgust at what he'd just said. "What I mean is, of course, I'm capable of telling you that. I *know* the answer to the question. I'm not stupid or brainwashed. I just don't know if I'm *allowed* to tell you that."

We both looked around the room. We were alone, but given what we know both knew, the enormity of what those devices represented, the kind of security clearance required, it made sense that we were being surveilled.

In fact, that was a part of the transition I was experiencing in going from being fascinated to being fearful. That's not completely accurate. I wasn't in a kind of binary situation where I was either fascinated or afraid. I was both simultaneously, but just in different proportions. That would prove to be the case the entire time I worked there. Part of that fear was a result of realizing my ambition. I always wanted to work with large and powerful forces. That was what I loved about jet cars and explosives and incendiary displays like fireworks. I knew enough about them that I felt I could control them. What I'd seen on the workbench with the emitter and reactor wasn't something that I felt I could control. I understood that it was "turned on" by placing the two objects in proximity to one another. (That kind of "powering up" wasn't that unusual by the way; a Tesla coil and some radio transmitters have a similar functionality.) Beyond that, I had no understanding of how it produced a gravitational field and controlled that field.

That lack of understanding was absolutely frightening. What went unsaid, and what I didn't really think of until after that first orientation session was over and I was on my way back to Las Vegas, were the larger implications of what this propulsion system represented. What if we were able to figure out how it operated? What if we were able to produce a similar kind of effect or the same effect? If we could help our government and military produce a gravitation field, what kind of weapons systems would be possible? What kind of propulsion systems? What kind of energy production? How would we be able to change the world as we knew it? And, as clichéd as it sounds, what if that technology fell into the wrong hands? To what lengths would nations or individuals go to get what seemed to me to be, as the expression goes, the keys to the kingdom? If you could produce and control gravity, if you could distort space and time as we commonly know them and accept them, what then?

I had to stop myself. I knew that those questions weren't going to get me any closer to doing what I had been hired to do. I'd learned while working at Los Alamos that the kind of institutional compartmentalization that went on there, and was also in place here, was the kind that I needed to put in place in my own mind. It's above my pay grade. I do my job as asked and leave the bigger questions for others. That was going to be more difficult here, but I knew that I had to make my best effort or else who knew what might happen to me. Fascination and fear were going to have to learn to get along.

Of course, though, I couldn't shut down every thought I had. I did wonder briefly why it was that what seemed to me to be a relatively small group of people were working on a project that had the kinds of implications that this one did. Los Alamos and the project to produce nuclear bombs involved an enormous number of people and resources. Still, very few people outside of the program actually knew what was being conducted out in the deserts of New Mexico. For all I knew, there were teams all over the country working on this same project as Barry and I. At least I wanted to believe that. I wasn't so egotistical as to believe that I was one of only a select few working on something so monumentally important as this was.

"I feel like I should tell you this, though," Barry said haltingly. "I have no idea how long this project has been in progress. I do know that I was brought on, and I assume you were, out of necessity. Maybe this is very obvious, but we need to be careful." He nodded toward the bench where the two pieces sat there looking as benign as some kind of Scandinavian sculptures that some Hollywood hipster might have had in his home.

He went on to explain that he'd been informed about an accident that had occurred that necessitated him being hired. He wasn't given any of the real details, but, as he put it, those involved were "unable to work again." By his tone and his expression, I knew that he meant that there had been a loss of life. I stood there trying to imagine what it was those individuals might have done. As Barry said, he wanted me to know because we were all under pressure to figure out the solution to this puzzle. There was also the temptation to take shortcuts, to take risks, and the consequences of a bad choice, being too eager to please or too frustrated to consider every possible effect when dealing with such an unknown could have enormous consequences. That was a part of the fear. On the surface, without all those science fiction-like displays of the propulsion system's power and capabilities, it appeared completely benign, a marvel of simplicity and elegant engineering of a type that was astoundingly sublime. Beneath that benign veneer was something more powerful than anything I had previously imagined. Yes, it was fascinating to consider. It had an allure that drew me in, but just as magnets have poles that attract and repel, I sensed in those first few hours in the lab that I was also going to feel respect for the intelligence behind its creation and a kind of nervous anxiety that I'd never experienced before in my personal or professional life. A whole lot was at stake here and to be incautious in any way seemed to be the least desirable approach to take, a last resort that I hoped that I would never even have to contemplate.

At some point, well into the early hours of the morning, Dennis came in to let me know that I was done for the night. He escorted me out of the room and sat with me briefly in a small office off the main corridor waiting area.

"This will be your routine for the foreseeable future," he told me. "Once you have your final clearance, you'll be housed on site."

"Barry?" I asked.

A flash of annoyance flickered across Dennis's face, but he answered, "Still working."

A faint ringing sounded in my ears. It was as if I'd been concentrating so hard for so long that the tension in every nerve and tendon in my body was as taut as a piano wire and was still vibrating long after the key had been depressed.

"You'll be partners," Dennis said. "You discuss your work with him, solely here, and with no one else."

"Understood." I was about to say that I understood that the first time that he'd mentioned that protocol to me but thought better of it. Above my pay grade to remind Dennis Mariani of anything.

We spent a few minutes going over my proposed schedule. Dennis wanted to be sure that I didn't have any conflicts with times and dates. Despite the tango of fear and fascination that I had, I assured him that I would clear my calendar to be certain that I'd be able to report as needed. He reminded me again that the security clearance could take some time.

"If you can think of anything that might get in the way of you being approved or even slowing the process, let me know now. It will be far easier and can speed things up."

"I can't think of a thing," I told him, and I spoke the truth as I did.

On the flight home, I thought again of the immense power the reactor and emitter produced and directed. I couldn't imagine what those individuals involved in the accident might have done to wind up dead. I imagined that they had to understand that what they were working with had the power equivalent to what our sun possesses. That is a mind bogglingly large amount of energy. I'd touched the emitter and the reactor, they were made of a material that could only have been cut into with something like a plasma torch. Who in their right mind, or with enough testosterone in them, would even consider applying that tool to something so powerful? Imagine someone taking the power of the sun and shrinking that reactor to the size of a small box and putting it into a room with you and saying, "Isn't that amazing?"

"Sure is."

"Let's cut it up to see what's inside."

"Okay! Sounds great."

As a result, I knew that it was unlikely these guys had done something really stupid. If they had, I'd feel bad for them, but essentially, they got what they deserved. Instead, they may have done something far less stupid, possibly something extremely intelligent or insightful and still they'd lost their lives. That scenario was far more frightening to me. I was going to engage in work with something unknown and I might even while being the most scrupulous and cautious as I could be, still end up doing great harm to myself and possibly others. Even when I got home and lay in bed next to Tracy, I contemplated that thought, and as the sun rose the next morning, I began to wonder just how many more of those beginnings of a new day I might end up seeing.

From seemingly out of nowhere, a thought occurred to me. What if the guys we had to replace had completely misunderstood the nature of the objects. What if they couldn't get their minds wrapped around the idea that something so small could be *producing* that kind of energy? Maybe they thought that it was retransmitting energy being produced elsewhere? That had to be it. It made sense then that since there were no outward signs—heat, noise, etc.,—that it was producing power, it made sense to them to dig beneath that surface to see how it functioned. The same with the emitter. It was essentially empty space, a kind of shell without any obvious other components, as we understood them, attached to its surface. The housing seemed inert because it was only when the fuel, the triangular bit that Barry introduced to the tower, was in place that the system functioned. Maybe those guys weren't doing something stupid, but only in hindsight could we see the error of their way.

Before I could go on any further, I felt Tracy stirring beside me.

"Good morning," I said.

When she didn't respond, I repeated my greeting, slightly louder. Still getting nothing in return, I slid out of bed and went to my desk. I had more pickups and deliveries to make, and despite only what seemed to be a few minutes of sleep, I knew that I had to get the day going. If I got the call, I was going to have to be ready.

By noon, bleary-eyed and better able to identify the cause of the ringing in my ears as sleeplessness, I'd decided that even after just a couple of trips out to Groom Lake, this schedule was unsustainable. I thought it was going to be just a question of a short period of time before I was on staff full time at the lab. I'd be living there, and I wouldn't be able to keep the photo business operation on my own. Tracy was familiar with the routine and could handle the workload on days I was called in or the mornings after. I'd have to show her how to use the machines. I knew she had the aptitude for it, since she'd shown some interest in the rocket car and other things mechanical. Besides, she was taking flying lessons, and the preflight checks and other routines associated with that were more complex than a photo processor.

The one difficulty was that she was working at a local airport doing administrative duties for the airport manager. She liked the work and had a great relationship with everyone there. I hoped that we could leverage that last fact to our advantage. I wasn't asking her to take on the whole photo business. She'd just have to fill in on an as-needed basis. The problem with that is we never got enough advance warning for her to ask ahead of time if she could be out for a few hours or a whole day. Maybe I was naïve, but it seemed to me that her job wasn't so essential that some accommodation couldn't be made.

And with that hope in mind, I returned home and we began the transition to a different way of life. Since I'd worked at Los Alamos before, and because Tracy's father had, I assumed that the secretive nature of the enterprise I was engaged in wasn't going to present any problems. Tracy knew that I couldn't talk about the exact nature of the work I was doing, and so I didn't really broach that part of the subject with her. I did address the fact that my schedule would be, for a while at least, irregular at best. I told her that I didn't like it, but if I wanted to do the kind of meaningful work that made me happy and provided a better living for us, then that was the price to pay.

She agreed. As she put it, "I'm not jumping for joy about it, but I get it. You have to do what you have to do."

So, with that assurance in mind, I brought her up to speed as best I could on what her new role would be. Her employers proved to be flexible, and so, on those occasions when I wasn't able to be working in the photo processing lab or needed a few hours of extra sleep, Tracy would cover for me. As it turned out, things didn't prove to run as smoothly as I'd hoped, and I was still being run ragged most days, adding to what would eventually be a heap of stress.

It was far easier for me to do that than it had been for Barry to do so for me, and I have to admit that I was frequently distracted while at home. I love problems and puzzles and the one that I encountered while working for EG&G was so engrossing that my mind would frequently take me back to the lab. I loved Tracy and the life we had together, but I did find myself hoping that every time the phone rang, it would the voice of the receptionist (as least that was the job that belonged to the pleasant but disembodied voice on the line belonged to) letting me know it was time to head to McCarran. I could never predict when the call would come, but when it did, I bid Tracy a quick goodbye and a "gotta go." A few times, dinner plans with friends had to be postponed, but Tracy and I were pretty much homebodies, so the disruption felt, to me at least, to be minimal.

My relationship with Barry was about as no-nonsense as a work relationship could be. I understood that I couldn't act out like I had in my Los Alamos days, but Barry kept himself on an incredibly tight leash. That first night we worked together was really the extent of him sharing any information that wasn't directly related to the work we were planning on doing that night or in the hours ahead. I suppose that part of that might have been that I was still considered, though no one referred to me this way, as a temp. The larger factor in Barry's tight-lipped approach was the presence of armed security personnel. Whether some member of the security detail was in the room with us the entire time, or dropped in unannounced and exited unannounced, they shadowed our existence within the confines of the lab, the building the lab was housed in, and the entire facility. Whether the security team members had pistols or rifles varied at times, but it was clear that they meant business. Whether they were there to protect us, or to keep us in line was never something I could decide firmly, but came to believe that it was likely both. Whenever we left the lab, to go to the restroom or the cafeteria, we were always escorted not just *to* those rooms, but *inside* them.

The second night I worked there I received a reminder of just how tightly monitored I was going to be. As was true the first time and, as it turned out, every time I was on the ground at Groome Lake, Dennis picked me up at the airfield. Unlike the first time though, when we entered through a side door into the administration/laboratory section of the facility, we pulled around to the front of the building. This time, the door to one of the hangars was open. Dennis indicated that I should go inside through the large open door and into the bay. Inside sat a cylindrical craft, typical of the "flying saucers" that I'd seen depicted in blurry photos and in TV shows and the movies. As I got nearer, I saw an American flag emblazoned on one its flanks. The flag was printed in reverse, so that if you were looking in a mirror the image would be correct. I thought that I understood better at that point. This was an experimental terrestrial aircraft. The story I'd been told in those reports was a kind of cover designed to keep me somewhat in the dark about the operational nature of the facility, etc.

As we walked past the craft, I did what I think anyone would do when walking past this type of vehicle. I ran my hand along the surface. Immediately, one the guards walking along with us snapped at me, "Hands off!" His menacing voice was so loud, I was startled. The sound of it echoed in the space, and I wondered if it was my imagination or did I really hear the sound of weapon being cocked?

In either case the message was clear: They had a very stringent set of protocols and I was being kept on a very tight leash. Not for the first time would I have to rein in some of my impulses if I wanted to get along with an employer.

With the institutional gray walls, the fluorescent lights, the armed guards, and no hint of collegiality much less joviality, even after the first week or so of working there, I very much began to feel like I was imprisoned in the confines of the facility that I would eventually hear being referred to as S4. I don't recall exactly the circumstances surrounding the discovery of that name. We didn't exist in complete monastic silence, though most conversations in the cafeteria were somewhat hushed and we were forced to eat at a table solely with our work partners. The tables we sat at were spread at a distance so that all I could generally hear were murmurs. Regardless, S4 was where I was now spending a few days a week at irregular intervals and for irregular spans of hours.

Though Barry and I had somewhat different temperaments (for example I seldom wore protective clothing while working with dangerous chemicals and was more willing to take chances with my personal safety than he was) and approaches to problem solving (I tended to trust my gut while Barry was more methodical and programmatic, i.e., by the book), we agreed on one point: The powers that be wanted to understand how this craft's propulsion system worked, so that meant we needed to concentrate our efforts on working with the reactor. As I saw in the first demonstrations Barry made, the reactor was comprised of three parts: the housing, the "tower," and the fuel. All three of those parts were relatively simple in design, seemingly manufactured from the same material based on appearance, yet they worked in concert to produce that astounding amount of power with one hundred percent efficiency. To put that into context, the most highly refined electric motors we have on Earth operate at about 85 percent efficiency. Some energy is lost through the production of heat. An internal combustion engine operates at about 35 percent efficiency; we lose some power again through the production of heat and gaseous exhaust. Electric motors have fewer moving parts than your car's motor does, so it made sense, in one way, that the fewer the parts the better. In the case of the alien technology, the parts within the reactor didn't seem to move at all, thus making for a highly efficient means of producing power with no waste.

Imagine having a fireplace that burned hot enough to heat your entire home, but the chimney or stovepipe remained completely cool to the touch. That's efficient. All the energy being produced is used to serve our purposes. If you've ever driven your car past electric power lines and had the radio reception interfered with, you've experienced the inefficiency of our systems. Some of the electricity being carried in those lines has "leaked" out and mixed itself up with the radio waves resulting in the production of static. We wondered whether or not there was any "leakage" in the propulsion system, so we measured all around the room to detect for stray energy waves and discovered none. Less scientifically, we had a radio playing in the lab, and its reception was as clear with the reactor and amplifier on as with it off.

Even back then, I realized that I was focusing, probably too much, on the concept of the system's efficiency. That it had allowed the craft to travel millions of light years meant that it had to be efficient, that its propulsion and other systems—life support, guidance, or whatever—had to remain operational over a long period and vast distances. As far as I could tell, based on the brief glimpse I had of the craft itself, it had no external fuel tanks or other obvious means of carrying enormous quantities of some substance to power the craft. That led me to think that maybe the entire craft itself was producing energy. Maybe what we thought of as the reactor was just a small part of a larger system that produced the energy needed for flight and guidance. Or maybe, what we considered the housing, the outer shell that contained the tower and the fuel disc, was more than just housing and was integrally involved in the production of the gravity wave we witnessed?

One of the first things we did was to consider how the device itself, its very structure, might contribute to the production of this enormous amount of energy. We decided to see if we could determine what kind of reaction was being produced inside the device that held the fuel cell even when that copper-colored triangle wasn't present. Did the housing itself produce some kind of measurable effect—electromagnetic, chemical, radioactive? Did the reactor's material composition contribute to the production of the gravitational field? Did it absorb some of the waste heat and, if so, did it then convert that energy captured into a different kind of energy that produced the field?

To determine this, we used a measurement device, a kind of miniature Geiger counter to detect the presence of radioactive emissions, beta particles or gamma rays. Instead of utilizing a Geiger tube, this much smaller device used a small chip transistor that was sensitive to radiation. We placed that on the tower in place of the small fuel disc. Almost instantaneously, the readout indicated that the space inside the reactor, even without the fuel being present, was being bombarded by particles, producing a relatively intense field of radiation within it.

"I'll be damned," Barry muttered.

"That's something," I added. "Have you never measured this before?"

Barry shook his head, "Thought about doing this for a while, but since we were told, well, I was told there was no radioactivity, I assumed . . ." There he sighed and then stopped speaking.

I didn't take the bait and talk about the dangers of assumption. Instead, I took what I saw as the next logical step in my thinking. "So, those particles are being accelerated somehow. And it can't be linear." I held out my hands like a fisherman indicating the size of the one that got away, my palms spanning nothing greater than the width of my shoulders. "Way too small to get them up to any kind of speed over that straight-line distance."

"They have to be moving around the reactor," Barry said, agreeing with my assessment.

"It's like a mini-cyclotron," I offered. Ernest O. Lawrence, in 1932, created the first cyclotron, a type of particle accelerator in which charged particles accelerate outwards from the center along a spiral path. The particles are held to a spiral trajectory by a static magnetic field and accelerated by a rapidly varying (radio frequency) electric field.

"Maybe we can detect those frequencies being produced by the electric field?" Barry frowned and shrugged. "I don't know if that would tell us anything."

"It might." I wasn't sure what that might tell us, but I was excited about the possibility that we might be heading down the right path at least. I didn't want to lose the momentum from having at least determined that a nuclear reaction of some kind, one that maybe didn't fit into what our current understanding of particle physics might allow for, was essential to how the space craft's propulsion system functioned. Even if our understanding was rudimentary, at least we had taken a first step forward. That didn't mean that a lot of other questions didn't remain unanswered. I was used to the idea that answers, even tentative ones, led to more questions. If the reactor's core, the structure that contained the site of the circling, spiraling particles, was central to how the whole system functioned and functioned so efficiently, it made sense to us to try to determine how it enabled this nuclear reaction to take place.

Barry and I spent the first few weeks of our partnership stuck on the idea of how this incredibly simple system could do the job. It frustrated the hell out of us that the device seemed to so effortlessly overcome the kinds of issues that would have plagued any kind of propulsion system we used on Earth—equipment placement and power loss over distance for example, and how it was that the signal being sent from the reactor to the emitter/amplifier travelled only to those components and didn't disperse more widely.

Our minds were boggled, and the only thing we could think to do that would be productive was to take a few steps back and ask a more fundamental question about the source of the energy, the fuel disc and the other parts as well. What material(s) was/were these things made of?

Here's where the idea of efficiency also played a role in guiding how I approached the problem of reverse engineering this system. The system under which we operated, how Naval Intelligence (or whatever agency truly guided these efforts) insisted upon compartmentalization and secrecy got in the way of all of us doing our jobs easily and productively. Given that there was a metallurgy group functioning somewhere on the base where Barry and I had worked, you'd think it would be an easy matter to get a company directory, dial that line (these were pre-Internet days so no email) and request the answer to our question. Because all our efforts were so segregated and knowledge not shared, that was impossible for us to do. I can cite a half dozen or more examples of how this inefficient system worked against our efforts to be productive but that one should suffice. We were in a classic Catch-22 situation. Get this job done quickly despite all the obstacles being placed in our way to move with anything approaching the kind of speed desired. If I hadn't worked at a government facility before and wasn't accustomed to this kind of quicksand-like environment, I would have either panicked or grown so frustrated I might have taken some action that was self-sabotaging in the long run.

Barry and I had to work within the confines of what we were given, and we did. I couldn't help but think that on some days, when we both sat there independent of one another silently pondering the problem that faced us, we were like shadetree mechanics staring at an inert engine, troubleshooting, hoping that somehow the device could speak to us to reveal what was ailing. In some ways, that comparison to amateur mechanics is apt. Given that this wasn't a terrestrially sourced propulsion system, we had limited knowledge about how it worked. When a car's engine failed to start, you could rely on some basic fundamentals. It would be silly to try to figure out what part or parts had failed that were causing the engine to not start. You could do that, of course, and start replacing parts until it finally did start, if it did at all, but that would be expensive and time-consuming. Instead, the better approach would be to think of the car's basic operating principles and the systems built to enable the three things needed for operation to function together—air, fuel, and spark. Those three elements had to be working together in proper proportion and timing in order for the engine to start. Looking at the problem of a nonstarting engine/propulsion system, you had a series of checks you could make to determine why the engine wasn't function. Was it getting air? Was it receiving fuel? Was there a spark present to ignite the air/fuel mixture?

At one of our early sessions together, when Barry and I were in that hands-in-pockets staring-at-the-thing mode familiar to anyone who has had been stranded on the roadside, I looked at him and said, "I wonder if there are any shadetrees on Zeta Reticuli?"

Barry looked at me with a startled expression and then shook his head like a dog trying to shed water from its coat.

"What are you talking about?"

"Nothing, really," I said, realizing that my alluding to amateur mechanics had eluded him. "Just that there has to be some kind of fundamentals to how this operates that we can focus in on."

"What does that have to do with trees?"

I went on to give him a quick version of the air, fuel, spark relationship of internal combustion engines.

It wasn't exactly an Archimedes-like Eureka moment and no apples fell from any trees here on Earth or on Zeta Reticuli, but we both thought about our stagnant situation for a minute.

"Seems to make sense then to focus on the basics. Go with what we know," Barry said after a few minutes.

"We don't know how the reaction takes place, but we do know that something has to be fueling it."

"So what? We're back to high school chemistry and talking about endothermic and exothermic reactions?" Barry sounded particularly glum.

"That's not the point," I said, straining to keep my frustration with his attitude at a minimum. "I know it seems like it's too much starting over, but if we begin at the beginning, maybe we'll see some things in a new light."

Barry smiled at my little pun regarding exothermic and the production of light. "The fuel's the thing, I agree."

"Then let's figure out just exactly what kind of fuel we're dealing with."

Barry walked over to the reactor. With it out of phase with the other part of the system it wasn't functioning. He took the top off the reactor, removed the cap, and lifted out the tower. He held the triangular-shaped, copper-colored fuel "piece" in his hand and squinted at it.

"I still can't get over how small it is," I said, glad also that it had been determined early on that it emitted no dangerous radiation. It was clear that it was part of a nuclear reaction, but it wouldn't be until years later that scientists here would be able to conceive and create devices that produced what would be called "low energy nuclear reactions." That meant that you could produce a nuclear reaction without nuclear and radioactive materials. We weren't there yet in our understanding of the capability to produce these kinds of reactions then.

Barry had taken the fuel disc and set it on a small piece of filter paper. With a very fine file, he scraped its surface, collecting a minute amount of the filings on the paper. He added them to a liquid solvent and prepared to inject them into a gas chromatograph where it would be vaporized and analyzed. I joined him at the machine and watched as he inserted the material into the sample port using a microsyringe to get it through a rubber septum and into the vacuum chamber. Helium (nitrogen is also often used) began to flow as the carrier gas. I checked the pressure regulator to be sure it was within parameters. It was and that gas joined the vaporized material we had injected in passing through a glass column packed with silica coated with a liquid. Since the material we placed in the solvent was insoluble—in other words it didn't dissolve into the liquid but was suspended within it—the whole process took a matter of seconds. We were using a thermal conductivity device (TCD) and helium provided a shorter analysis time due to its higher flow rates and low molecular weight.

The detection system in the device converted the property changes of the substances heated and vaporized into electrical impulses that a computer could analyze. Basically, it took an analog reading of the reactions taking place within the glass column and converted that reading to a digital one. The digital was less susceptible to interference and has a better signal-to-noise ratio. When we looked at the display the device provided, the chromatogram itself, literally a graph of the results, we noticed that there were no spikes, which would have indicated the presence of various elements. The x (horizontal axis) reflected the amount of time while the y axis measured the abundance or absorbance of the chemicals present. For example if you put a drop of water in the device, you'd find a spike indicating the presence of hydrogen and oxygen. What we saw was nothing at all except at the very far right of the graph indicating that something was in there, but it wasn't composed of any other known component elements. The obvious conclusion was that the material we were working with was an element itself—it wasn't a combination of other chemical substances but was one of the basic building blocks we call elements.

Barry and I stood looking at one another, shocked to realize that the fuel wasn't a compound, that it wasn't composed of multiple elements but was a single element itself.

"I didn't expect that," Barry said. "I was pretty certain it had to be some kind of alloy."

I'd thought that same thing. After all, we have a limited number of elements on Earth and most everything else is composed of combinations or variations on those elements. Carbon is an element but when those molecules get rearranged in a certain way, you end up with a diamond that bears almost no resemblance whatsoever to carbon. We figured this fuel had to be like that—something relatively common that was combined with other materials that, working together in some kind known process—fission, oxidation, or whatever—produced that enormous energy.

I also knew that terrestrial spacecraft used liquid oxygen and either liquid hydrogen or kerosene in their rockets as a propulsion system. Solid rocket fuel was made up of a combination of powdered aluminum and an oxidizer. Of course, we were dealing with an exotic fuel not of terrestrial origin, but the kinds of systems and processes we used on Earth were our only real frame of reference.

"We could try other kinds of spectography, but I'm pretty sure, given what this reveals and what we know of its origins, that it isn't going to be made up of anything we typically find here—carbon, for example—or any of the metals."

"That's probably the case, but we can't rule anything out," Barry said, stating what I knew to be obvious. On the one hand as a scientist I wasn't supposed to jump to any conclusions, but this was a special case. And, just because I made that statement hypothesizing the result, that didn't mean that I was not going to do the work needed to confirm what I suspected. The frustrating part of all of this chemical analysis was that we were both working outside our fields of expertise—physics. Eventually we subjected the material to mass spectral interpretation using a process called electron ionization mass spectrometry. It produced the same result—that same spike off the scale, more or less confirming what we suspected about the fuel being an element.

We also conducted something called a neutron activation analysis. Essentially we bombarded the fuel element sample with neutrons. As a result of that bombarding, we produced a radioactive isotope. Since we know the radioactive emissions and radioactive decay paths for each element currently on the periodic table, the spectra of the emissions will reveal what elements make up the sample. In this case, there was only one element that made up the sample, but it wasn't one that we could identify.

Every day that I was on site and Barry and I worked together, I could count on one thing. Dennis would show up and ask what we'd done that day, what progress we had made. We had to resist the temptation to oversell him on anything we'd done. We knew that they wanted us to get this job done and we felt the enormous pressure of those expectations. But as I've pointed out, we knew about the potential threat these materials posed. I also knew from my work in the scientific community, and Edward Teller's rise and fall within that circle was a clear example of this, if you hype something too strongly that is based on your speculation and highly reasoned and formulated opinions, you run the risk of being too optimistic. You had to produce results, verifiable and measurable results that could withstand intense scrutiny and analysis. We also knew that Dennis was a layman essentially. He didn't understand science to any great degree so until we had something that we could demonstrate to him as an unequivocal explanation of at least some part of the process that we could demonstrate, we had to be cautious about what we told him. It was a delicate balancing act trying to meet their expectations and ease their disappointment with our efforts. I was in a constant state of anxiety.

Worrying about my job safety, my personal safety, and the progress of my security clearance all became like a second job I had undertaken. The problems we were faced with solving occupied me day and night. Though I wasn't working every day at S4, and wasn't physically on-site there, my mind was definitely occupied by the work twenty-four hours a day and seven days a week. I was vaguely aware of Tracy and our life together. I wasn't checked out of the relationship completely, but I was preoccupied. She seemed to be in relatively good spirits. She was enjoying the flying lessons she was taking, the photo business had proven to be something she could handle with minimal stress, and even though it meant us being apart, I was looking forward to being hired full time. I wanted to be engaged in meaningful work and Tracy and I were both adaptable, and we'd find a way to accommodate the changes that loomed on the horizon for us. Life is all about change and adaptation, after all. I trusted in both our abilities and as frustrating as it was to work in that environment, the fact that I was working on such an exciting project with so many potential implications for me personally as well as for the world, from time to time pierced the protective cloud I'd inflated around me so that I could concentrate on the task at hand. I didn't spend any real time thinking about how my name might one day be associated with an incredible advance in our understanding of the nature of the universe, but I did wonder what kind of bonus might await those of us working on the forefront of an exciting scientific endeavor and what it might mean for our future.

CHAPTER FIVE

IN FEBRUARY, IF THE END OF THE PREVIOUS YEAR'S RAINS are plentiful enough, the Nevada desert blooms with wildflowers—various types of nightshades, geraniums and roses, miner's lettuce, and others put on a brief show. In 1989, when I made my way out to Groom Lake and area 51, I couldn't tell you whether or not a floral boom or bust was taking place out there. From my seat aboard the aircraft, or in the bus that took us to the S4 site, I had on figurative blinders—thinking only of the task at hand. My world was reduced to the colorless and sterile environment of the lab and its facilities. Funny that the word "facilities" is related to the word "facile"—which has as one its definitions, "easily achieved, effortless." Another of its meanings is, "appearing neat and comprehensive only by ignoring the true complexities of an issue; superficial." There was little about the job that was easy and effortless and we seemed to be expected to ignore the complexities of the bigger picture question of how the craft operated. We were to focus solely on the one piece of the puzzle we were assigned to, and even the environment in which we worked reflected that single-mindedness of purpose.

I'd never worked someplace that was so devoid of signs of human life. I never saw a houseplant or flowers. I never saw a photo of a loved one, a favorite vacation spot, a beloved pet on anyone's desk. No one had inspirational posters hanging on the walls of their workspace, no cats dangling from a branch urging us to "Hang in There." Oddly, or maybe I should say ironically, there was a poster hanging in one of the rooms depicting a saucer-shaped UFO with the caption, "They're here."

We were so segregated from others that even when we were in one another's presence, we were not allowed to speak. If, for example, we needed to borrow the salt from another table, we had to ask one of the security team's members to retrieve it for us. I would have thought that Barry and I would develop some kind of "us against them" mentality as fellow prisoners serving time together, but we didn't. I knew nothing of his personal life and he knew nothing of mine. I don't know if those who lived on-site during the week were more social, but given how oppressive the environment was where I worked, I highly doubted if there was a more lenient and relaxed and collegial atmosphere elsewhere in that part of the desert. In the world of animals and flora and fauna, some species manage to survive and even thrive in relative isolation, I didn't count myself among them. I was not then nor am I now a highly social animal needing to experience the gregarious pleasures of others to be happy, but given all that was going on in that world, I was greatly uncomfortable most of the time I was in it.

My irregular work hours also meant that I was sleeping irregularly and that contributed to my sense that I was living in a kind of fog. I've always appreciated an intellectual challenge and have always been able to dig into a deep reserve of energy to power my brain. When I was on site, I was fine and functioning well mentally. At home, that wasn't exactly the case. I was managing, but preoccupied, so when Tracy expressed some concern she had about seeing men parked in a car just a few hundred feet down the block from our house, I was a little confused at first.

"I don't know," I told her initially. "People do all kinds of things."

"They don't look like they belong here," she said.

"What does that mean?" My tone betrayed an annoyance greater than what I felt. I immediately picked up on it and apologized and rephrased the question, "What have you noticed that leads you to believe that?"

"This is Las Vegas, you don't wear a dress shirt and have your suit coat hanging in the rear window. And its two guys, always two guys. They just sit there and they don't even seem to talk to one another or even look around.

It's like they're at a drive-in movie or something." Tracy shrugged her shoulders. "I'm not paranoid, but something is up with those two."

It finally dawned on me, "They're most likely watching us. For the security clearance."

"Oh," Tracy said flatly. At the time, I didn't remark on how muted her response was, how a half-dozen states of mind or emotion could have circled around that tiny mass of sounds.

"It's all part of it," I added, needlessly as it turned out, since Tracy had already left the room. I went into the kitchen and let the water run for a long time before filling a glass. When I put it to my lips, I realized that I had turned the handle to the right and not to the left. I drank the too warm water anyway, as if punishing myself for some indiscretion I wasn't even aware that I had committed.

One night when I was on call, Dennis came into the lab. Barry and I had been working on an experiment to test how the location of the parts of the propulsion device affected its operation. We moved the emitter a few degrees from center, closer or farther to it. This, along with seeing how focused we get the gravity field to be, had been taking up much of our time.

"Gentleman," Dennis said. "I'd like to speak to you both."

Barry and I eyed one another warily.

"It's about the fuel."

We each sat on a stool and Dennis produced from his pocket one of the fuel pieces. It seemed to be an exact duplicate of the one we had been using. Dennis went on to explain that manufacturing an additional fuel element was necessary. He wanted our assistance with the process. I knew better than to ask, but I was immediately wondering, Why the two of us? Yes, the fuel element was an important component of the propulsion system, but we didn't have a clear sense of how it was made or what it was made of. Dennis mentioned that the metallurgy group had made some advances and there was some certainty—I was used to Dennis and his vague references and passive voice constructions that never clearly identified who did what or who directed or requested what—of its component materials.

He didn't tell us exactly what those components were, only that the fuel pieces needed to be machined.

"That's the kind of stuff that the guys at Los Alamos do a lot of the time," I said as much to myself as to Dennis.

"They do?"

"I knew a number of guys in the machine shop—well, one of the machine shops—and they were helpful. I know that from conversations I had with them that they were working on a lot of cutting-edge materials. High-precision materials, incredibly small tolerances. Weapons systems, I assumed. Highly classified work."

Dennis nodded. He produced another of the fuel triangles. He set it on top of the first one. Each was about the size of a half dollar and no more than a quarter-inch thick. I'd handled one before, obviously, but seeing the two of them stacked like that made me understand something I hadn't before. Truth be told, I didn't give that much thought to how they were made. But at that moment I got a better sense of how they were likely to have been machined. They weren't stamped or cut out of a single flat sheet of their material. Instead, it seemed to me, that a larger block of it would have to been shaped into a cone. From that cone, whatever cutting tool was needed could be guided to cut through that cone at various angles to produce the triangular-shaped pieces.

For a few minutes, Dennis, Barry, and I discussed how we thought the objective might be met. I offered my cone-cutting theory and Barry asserted that he was in complete agreement that my suggestion was the likeliest scenario.

"Los Alamos can do this," Dennis said while nodding.

He pushed himself away from the lab's countertop and walked out of the room. He came back in a few minutes with what appeared to be a cylindrical ingot of the material. I had no idea where he got it from—whether it was a part of one of the crafts or came from somewhere else—but he hefted it in his hand, according it no more regard than a deli counter clerk handling a bologna sausage.

"This will go on ahead of you," he said, nodding in my direction. "We'll get the schematics drawn up and then you'll be off to Los Alamos. As far as they're concerned, this material is known as LA1000. As far as they're concerned, and as you're concerned while with them, this is a new alloy used for armoring."

"Understood," I said, wondering how soon I would be going and if this little field trip was a good sign or a bad sign for my still pending clearance. Could the whole thing be a set up to test me? Ultimately, I decided it didn't matter. I was told to go to Los Alamos, and that's what I was going to do. The next time I got the call from EG&G, I was told to report to McCarran but that I'd be taking a commercial flight out of Las Vegas to Albuquerque. From there, I'd transfer to a different regional flight to the Los Alamos airport. I was only going to spend a few hours on site, delivering the instructions for the manufacture of the fuel disc/armor. The ingot was sent via courier to Los Alamos while Dennis met me at the airport and gave me a sealed 9 by 12-inch envelope that I assumed contained what I needed to deliver to the machinists at the New Mexico laboratory. Based on my experience working at LANL, I figured the ingot got sent out on what we referred to as a "dash flight." They took off daily from Mercury, Nevada, near the old AEC base camp at the nuclear test flight. Back in my days in New Mexico we often received materials from Area 51.

Landing at the airport in the city where I'd spent so much time felt both familiar and surreal simultaneously. I wasn't able to tell Tracy where I was going, only that I'd be gone for longer than I usually was. I had no idea how many hours that would be, but I told her not to worry.

"I don't worry," she had told me. "I just don't understand and I don't like that."

I had no way to reply, and just as I had landed in Los Alamos with no sense of how I was going to get to the location on the site's thousands and thousands of acres, I had no way to reassure or make Tracy understand that if I could tell her what I was doing I would. Trouble was, more often than not, I really didn't understand what I was doing. In my mind, that meant I wasn't withholding anything from her; I was the one from whom things were being withheld. I was being the good soldier, and had to trust that Dennis and those in charge had my best interests at heart. I knew that was never really going to be case. I was simply a moving part in a large and complicated machine, easily replaceable in some ways, of high value in others.

The answer to my wondering about how I was to proceed was answered by someone asking me a question. "Are you Bob?"

I turned to face a young blonde man in his mid-twenties. A wedge of hair angled across his forehead and over one bespectacled eye, like a curtain across a picture window.

I nodded and the young man said, cheerily, "Let's go."

A few minutes later we were in a white passenger van passing through a security checkpoint on the property of the LANL. I had never worked in that area before, and even the entrance was unfamiliar to me. My driver hadn't engaged me in any kind of conversation, and I wasn't feeling particularly chatty myself. I was very tired and may even have dozed off for a second or two on the drive over, just as I had, though for much longer, while on the flight in.

The van stopped in front of an administrative building, what looked like a typical office building. There, I gave my name to a man behind a reception desk and picked up my visitor badge. He made a quick call and then a woman came into the waiting area, and without introducing herself, asked me for the envelope I'd been carrying. She left immediately and I stood there wondering what I was supposed to do next. After a few moments of me standing there, the woman returned, without the folder, and said that I was to follow her. I went into an office and was introduced to the man who supervised this particular machine shop.

We went over the specs and how we wanted the cylinder to be sliced into discs, then stack those discs, fuse them, then machine that into a cone and then into the triangle-shaped bits we'd been using to help power the reactor.

None of what I asked registered on the face of the man I was speaking with. He spent most of his time looking at the specifications sheet and the schematic. In some ways, I envied him. He looked at the task as a kind of simple mechanical engineering fabrication problem. I knew that people like him worked with plutonium as part of the nuclear weapons making, so, given that none of what I was describing to him involved fissile materials—at least far as I knew—this was (pardon the pun) a relatively "run of the mill" operation for him and his people.

"Of course, we also need all of the material back," I said. There'd be a minute amount of residue or leavings produced in fabricating the fuel triangles, but we still needed even the tiniest fragment to be returned to us.

"Any idea what kind of phase changes or thermal expansion we might be dealing with?" he asked, ignoring my statement entirely.

"I don't believe there will an issue, but with the kind of tolerances we're talking about . . ." I shrugged but didn't go on. I was tempted to try to ingratiate myself to him and his people by letting him know that I understood about the work they did, how creating an alloy with plutonium and gallium and other things was part of what they were responsible for, materials that presumably were far more volatile than what had been shipped to them. Instead, I let him, the expert, work out for himself how they'd get the job done. I wasn't worried about that at all, and shortly before I left the office, the supervisor sat staring off into the upper corner of the room for a few seconds and then a satisfied smile spread across his face, "I think I've got a better idea of how to get this done."

He reached into a drawer and pulled out a small notebook and began scratching out a few notes. I didn't want to disturb him, so I simply said, "I'll let myself out."

Receiving no response, I walked out of the office and back into the waiting room. In what seemed like no time, I was driving back home from McCarran. I arrived home in the late evening, just around dinnertime, and noticed that the car that Tracy had talked about wasn't there. Tracy wasn't home either. She left no note, so I put a frozen meal in the microwave and sat there eating the overly crisp outside of a macaroni and cheese dish and its still only partly thawed innards. Fatigue overtook me while I sat in an easy chair, and before I knew it, sunlight on my face had awakened me. Stiff and sore from my cramped sleep, I stood in the hot shower hoping to ease the tension from my muscles. As soon as I had toweled myself off the phone rang and less than twenty minutes later, I was out the door and heading back to McCarran and on my way to S4, a cup of coffee and a fast-food egg sandwich of some kind jousting in my belly as I followed Dennis into an unmarked office.

"Take a seat," Dennis said.

Tired, and tired of Dennis's no-nonsense approach to things, I sat down in the chair and made no effort to engage him in any conversation at all.

"A new directive has been issued," Dennis rummaged through a desk drawer and withdrew some papers.

I thought that he'd tell me about the new emphasis we were to take in our research. Instead, he pulled out a small caliber revolver, what looked to be a .22 cal Smith & Wesson, and placed it on the desk.

"You're to carry this at all times when you you're off-site."

Whether it was exhaustion and discomfort or I'd bought into too much of the Iron Curtain propaganda that had permeated life in the US for a while, I immediately thought of the four Russian scientists who had once been seen regularly around the facility. They hadn't been seen for a few weeks, and even Barry, loathe as he was to engage in anything even approximating office gossip, had commented on their absence. I also speculated about what had transpired and was now putting two and two together and coming up with .22.

I didn't respond to Dennis at first but smiled ruefully to myself thinking that 2 + 2 = .44—the caliber of the long-barrel Magnum I owned. I wasn't exactly a gun aficionado, but given how much I liked explosives, I enjoyed owning a weapon and appreciated what one could do for me. I also owned an Uzi submachine gun that I appreciated for its design. I didn't let Dennis know about the Uzi, but I told him that I didn't need, or want, the .22 he was about to issue to me. I admired both the weapons I owned but also understood that, on a certain level—practicality—they were ridiculous to own. I had purchased a holster so that I could carry the .44 on my person, and that meant that I would have to wear a long coat to cover it—not ideal in the heat of Las Vegas and only a little more so when I lived in New Mexico.

At the mention of the Smith & Wesson Magnum, Dennis slipped a bit for the first time in the two months I'd known him. One eyebrow raised a fraction of an inch but then quickly settled.

"Have I made your day?" I asked him, referencing Clint Eastwood's .44 Magnum-carrying character Harry Caine and the famous line from the film *Dirty Harry*.

"Funny," Dennis said, showing no sign at all that he saw the humor in my remark. "At least I won't have to fill out these forms. That makes my day."

I'd expected him to put up more of a fight with me on this point, but was glad that he hadn't. I was familiar with my .44 and more comfortable with it. I wasn't comfortable with the idea that I *needed* to be comfortable with the weapon. I pushed as many of those thoughts outside my mind as I could.

"We're done," Dennis said.

I went back to the lab. Barry looked up at me but didn't ask a thing about my meeting with Dennis.

"There really is nothing unusual going on here," he said. He handed me a readout from the gas chromatograph. We'd decided to do an analysis of the copper-looking plates on the sides of the emitter. In appearance, they seemed to be the same material as the fuel triangles, but proved to be, based on this analysis, made of common elements.

"Not exotic at all," I said, handing the sheet back to Barry. "Not the most typical of alloys, but—" I let the sentence trail off.

Barry nodded distractedly.

"Have you given any thought to what I suggested about building that measurement device? With the right instrument we'd be able to produce more verifiable results."

"It makes sense. Time is the one concern I have."

"I agree, but with the kind of seat-of-the-pants tests we've been doing we don't have any way to quantify the effect of the reactor." I said thinking of the various demonstrations we'd been doing in the last few sessions. They were variations on the golf ball demonstration that Barry had done for me. We could see the results of the gravity wave, but we hadn't done a very good job of measuring the effect on an object placed in its path.

"I've got a pretty good idea of how to fabricate an instrument that will work. My first job out of college was with an electronics firm. I used to repair and refurbish pressure regulators. If they can measure minute pressure differentials, they should be able to measure the effects of the gravity wave. They'd also give us a sense of how the wave disperses."

"We'd have to build something fairly large to do that. What diameter would you estimate?"

"Sixty inches?"

"We'll have to get quite a few of those sensors and then we can get started. I'll talk to Dennis."

On the flight back to Las Vegas, I was thinking more about the conversation I had with Dennis than the one that Barry was going to have with him. I speculated about the presence of the men outside our house. I'd been fairly well convinced that they were part of the security firm who was investigating me so that I could receive the top-level clearance I needed. After hearing Dennis imply that I needed to be more concerned about my personal safety, I wondered if those guys had been assigned because of some imminent threat to Tracy or me. Worst-case scenario, what if those men hadn't been working for our side and instead were working against us. It seemed unlikely, but the whole "What happened to the Russians and why had we been cooperating with our sworn enemy" scenario was generating its own bit of paranoia. I wondered if there was some device that I could make to measure the influence of paranoia on me. Even without that device, I began to carry my .44 Magnum and the Uzi in my car.

We got the approval to build the device, and over the course of the next few weeks, I worked at wiring the sensors in sequence. We had to use dozens and dozens of them. The work was tedious but at least I could leave after each work session and see some progress. We also got some great news during that period of building the measurement instrument. The fuel triangles came back from LANL and they worked just as the ones that we'd first been supplied with. Everyone was very pleased, and even Dennis showed some enthusiasm with a "Good work," compliment. I hoped my involvement in the Los Alamos venture might offset a slight blunder that could have spoiled any chance I had of getting the job full time.

Before I began to carry my weapons with me, I decided that I'd better register them. The registration process was going to be easy, but I tripped up when I told a friend of mine that I could only spend a few minutes with him having lunch because I had an errand to do.

"What's that?" Gene asked, a perfectly normal conversational tactic.

"I have to register my guns," I told him, and then added, regretting the slip of the tongue as soon as it was out of my mouth, "I need to use them for work."

By this point in 1989, I had known Gene Huff for almost five years. He worked as a real estate appraiser and taking photos was part and parcel of his job. We'd met because of my photo processing company, and, since Gene worked very regularly and needed frequent photo service, we spoke on a number of occasions and eventually became friends. Actually, Gene dealt with Tracy originally when she was doing some of the drop-offs for me while I ran the equipment. She'd stop by his offices, chat a bit, and they talked a few times about living in Los Alamos. Gene assumed that I worked in photo processing there, and it was only later when we became friends and we'd talked about a wide variety of things that he finally said to me, "You know, you speak so knowledgably about so many things, you've got your rocket car, and everything else, you sound like a scientist and not just Bob The Photo Guy."

"Well, I am a scientist. I've studied and have degrees in electronics and physics."

"Why didn't you say anything?"

"What? Was I supposed to hand you your photos and say, 'By the way, I'm a scientist'?"

"Yeah. That's what I would have done."

"Well, I'm not like that," I told him.

Eventually, Gene and I grew closer and we'd talk about a lot of things, but what my new work was, and why I might need a gun, was not on the list of topics to be discussed. To his credit, Gene saw how I had blanched when I made the mistake of mentioning the gun and work, and he didn't press me for details. Still, that I had let my guard down that way bothered me a lot. I trusted Gene; after all, he trusted me enough to sit at my kitchen table with me while I mixed up a batch of nitroglycerin and went off into the desert with me to witness its explosive force. Gene shared an interest in explosives and pyrotechnics, and two years earlier, in 1987, we had initiated a gathering in the desert of fellow enthusiasts for what eventually became known as "Desert Blast." Through that and other shared interests, I also met John, the second son of the founder of the aircraft manufacturing firm, Bill Lear, he of the Lear Jet fame. John was an accomplished pilot himself, and later gained some notoriety due to his claims about extraterrestrial life. John and Gene were both very interesting men with active and inquiring minds and the kind of curiosity that I possessed as well.

I met John through Gene. Gene had watched a local TV show that George Knapp, a journalist at the local ABC television affiliate was doing called *On the Record*. John Lear was George's guest. He was an active believer in what was termed ufology. He believed in the existence of alien life, that extraterrestrial craft had come from other planets and solar systems. He appeared at panels on the subject, and he mentioned to me on several occasions that he believed that alien craft were being flown in the desert outside Las Vegas. Gene had some interest in the subject, as did a lot of people in the Las Vegas area, due to the frequent sightings and the "light show" out in the desert. He wasn't a proselytizer like John was, just, as I mentioned above, a really curious guy. John's views intrigued him, and he thought he'd call John up to talk more about the subject. That's just the kind of guy Gene was. When reached, John was a bit guarded at first, well at lot guarded probably characterizes it better. For whatever reason, Gene mentioned a couple of times that he worked as a real estate appraiser. The Lear family was wealthy and held several properties in the area. John lived in what I would eventually come to think of as a kind of compound—an enormous house and grounds. John seemed more interested in talking about having Gene come out to do an appraisal than he was in talking about UFOs. Gene thought it might be interesting to see the place where the pseudo-legendary, at this time, John Lear lived.

They made arrangements—in lieu of payment, John would share some copies of UFO videos and other material related to the subject. Gene called to ask me if I was interested in going. He said that I could pose as his assistant. Of course, I'd heard of John and his exploits as a pilot, so I agreed to go. He sounded like an interesting guy, and I was curious to see how the other half, the wealthier half, of Las Vegas lived. As part of the appraisal process Gene would have to take quite a few photos, so I became his "second shooter" and gear-toter. After a brief introduction, Gene and I went about our business. Gene always claimed that because I was just the assistant, John ignored me. Gene dropped names a few times, mentioning that I had once worked at LANL. The Los Alamos reference didn't sink in until that third mention. When John realized, belatedly, that I wasn't just a photo guy but had worked as a scientist, we started to talk about his experiences and mine.

The two of us hit it off. We had some similar interests in aviation, propulsion, pyrotechnics and other wide-ranging topics that caught our imaginations and intellects. This made us, if not kindred spirits, then at least individuals who could carry on a stimulating conversation with one another. I also knew this about John: He had some interesting connections in the government and the military. He'd done some work with the Central Intelligence Agency, and he could tell a great story. I never once, for a second, believed what he told me and others about alien spacecraft. I tolerated that part of him, mostly because the rest of his life and accomplishments were on the record and, at heart, he was a kind and fascinating guy.

I also eventually came to understand this about John: He had no bullshit detector. He'd never seen any of these objects himself, and several times expressed to me his disappointment that he hadn't, and, in my estimation, he indiscriminately took in what other people had to say on this subject. In that sense, he was what I would call a "true believer." As I looked at it, if someone said it; John thought it must be so. He gave the same credibility to something that someone at the CIA had told him as someone he met in the street for the first time. That's not to say that John was gullible or not a highly intelligent guy. He was an expert in the field of aviation, and his ability to recall information and to even reproduce on paper a diagram of the hydraulics that powered the landing gear of an L1011 aircraft—which I once asked him to do—was nothing short of remarkable. I took his belief in UFOs and aliens with a grain of salt. A lot of other people shared his belief, but that didn't make them bad people or uninteresting company.

I also knew that Gene was eager to learn as much as he could about what I was doing out at Area 51, but he didn't press me for details. The afternoon of my little slip up with him, Dennis showed up at my house and told me that I needed to go with him and that I had to bring my weapons with me.

We wound up at a Las Vegas Police Department substation at the corner of St. Louis and Atlantic.

After we got inside the building and past the desk clerk, and after Dennis had shown the man there his identification card, he instructed me to sit down in the waiting area. I didn't have to wait there for very long. Dennis and a uniformed officer came out a few minutes later. The officer pointed at me and shook his head, "This is the guy? Why would anybody let alone the Russians want to do anything to him?" I tried not to take the remark personally, but I knew I looked like what I was—a bespectacled scientist and not a James Bond-esque spy or whatever else this cop had imagined. I also didn't like that he'd brought up the subject of the Russians again. Dennis didn't look very pleased that the guy had said something in public about the nature of our visit and the need for me to have those guns registered.

That week I wasn't called in to work at all, and that troubled me a little. To offset that worry about how my security clearance was proceeding, two men from the Office of Federal Investigation showed up at the house. Tracy and I had a couple of people over, her sister Kristen and another couple we had become friendly with, Wayne and Robin. Wayne was a mechanic and serviced our vehicles. We became friendly due to my interest in vehicles generally and he was intrigued by the jet car that I used to take all around doing demonstrations.

At this point, Tracy had indicated that she was no longer interested in running the photo developing business. Wayne and his wife were looking for another business opportunity. They were over for the day discussing the possibility of acquiring our photo operation. Before finalizing any arrangements, I thought it best to get them acquainted with the machines and how they worked. Wayne was, obviously, mechanically inclined and used to working with various processes and systems, but this was unfamiliar territory for them both. They also weren't familiar with notion that some representatives of a federal agency could come to the house unannounced and begin looking around. On the one hand I was glad to know that I was still being considered, but the intrusiveness of it all didn't sit well with me. I was in a bind. If I reacted too strongly, then I'd amplify Tracy's irritation. One of the agents was named Mike Thigpen, and though he tried to be courteous and professional, something about him rubbed me the wrong way. Maybe it was because Wayne and wife Robin were there. Because they were, I had to reveal for the second time that I was doing work which required such high-level clearance. Prior to that, neither Wayne nor his wife were aware of what I was doing, presuming that I was operating the photo business.

Tracy knew about the need for security clearance, so I told her about the search. She wasn't pleased about the idea of the two men going through drawers, closets, cabinets, and nearly turning the place inside out. Unlike in TV shows or films where they leave the place a mess, these guys were very respectful.

Despite my assurances, Tracy was still shaken by the lengths the men went to. I told her that we'd done the right thing by allowing that search, though I was not really happy about it all. I kept from her my relief that their presence let me know that the investigation was ongoing. There had to be some reason why I wasn't called out to work at S4, but my being denied a security clearance wasn't the reason—at least not yet. Knowing that I was still in the running was far better than one of the other alternatives. I didn't want to press that point too much. Nor voice to Tracy my concerns that this was an odd way to go about this business. The less said the better. Let Tracy believe that this was standard operating procedure, and work to convince myself that this was true, and proceed with focusing on the job at hand.

I can't say for certain why it was that a brief period of inactivity went on, but there were a couple of occasions when the reactor and the emitter weren't present in the lab. Other units working on other systems must have needed to use them. Of course, we were never informed of this and it was mostly a surmise on my part and on Barry's part, but given what we eventually were allowed to do and to witness, that made the most sense.

I DON'T KNOW IF IT WAS BECAUSE I WAS RELIEVED TO LEARN that I was still being actively investigated for the security clearance I needed or if the home search indicated that I was under a new level of scrutiny, but I began to feel more comfortable talking with Dennis about our lack of real progress. Not only that, but Barry and I both began to let him know how much a hindrance it was for us to not be allowed to see and to inspect the other systems of the craft. We could study the reactor and the emitters in isolation, and we were getting closer to having a working instrument to measure the gravitational effects produced, but how that propulsion system functioned within the larger context of the craft was a gaping hole in our understanding.

"This isn't like a car engine of some unique design that's been dropped into a conventional automobile. We don't have a baseline knowledge of the drive train and the transmission and the steering and the suspension that are all common to most cars," I explained to Dennis in defending ourselves against his accusation that we weren't trying hard enough. "How this entire craft functions, how this reactor and emitter work in synch with the rest of the components and systems is something we have no idea of. We're working with an unknowing inside of a larger unknown. That's not a great situation to be in, especially if others have knowns that they can share with us."

My analogy seemed to work.

"Makes sense," Dennis said. "I'll see what I can do."

Barry and I took every opportunity after that to work at that tiny fissure in Dennis's armor. Every chance we got, we mentioned something about how seeing the entire craft was going to help speed the process.

"No one ever said that this was going to be easy," Dennis said to me one day.

"We're not asking you to make it easy," I told him, "We're asking you to let them know that they don't have to make it so difficult."

CHAPTER SIX

WE'VE ALL HEARD THE EXPRESSION, "BE CAREFUL what you wish for." I can't say for certain that thought passed through my mind during the extraordinary days and nights I spent on the grounds of the installation at Groom Lake. As is true of most experiences, when you are in the middle of one, deeply engrossed in the moment, you don't have a lot of time to consider anything but what is transpiring right then and right there. We've also all likely heard the importance of staying in the moment, staying present, and other variations on this idea.

After having completed the last of the work on the pressure sensors, I arrived at Groom Lake a few days after speaking openly with Dennis. I went to the lab building as usual. What wasn't usual was that Dennis was there to greet me. At first, I didn't think much about him saying to me, "You're not going to be working inside today." Instead of going inside, we walked to the hangar facility. Barry joined us.

I could see immediately that one of the hangar doors was open. With the light coming from it, the scene resembled a jagged jack o' lantern's mouth. I understood immediately that what Barry and I had been asking for was about to happen—this was the night we were going to see the actual craft, or crafts, themselves. My initial reaction was to feel enormously relieved. If they were allowing me this opportunity, then that boded well for my future with the team. I also felt vindicated—Barry and I had succeeded in convincing the powers that be that the best approach was a more open approach. Just how open that approach was going to be was still to be determined, but at least they'd acknowledged the importance of our input. In a work environment like the one we were subjected to, even the faintest of nods towards your needs seems like a monumental gesture. I was also thrilled by the opportunity to see such amazing technology up close. Even if what we'd said—that seeing the craft and the reactor, amplifier, and emitters in the craft could help speed the reverse engineering process—didn't prove true, the privilege of seeing something that presumably only a handful of humans had seen was going to be something I'd never forget.

Along with all those thoughts came another: How much access to the craft were we really going to get? The answer to that question came rather quickly. Once the vehicle came to a stop and we dismounted, Dennis said, "Take a good look. I can't guarantee you'll get another opportunity like this one."

I walked into the wedge of light coming out of the hangar and onto the concrete apron. I wanted to get the long view of the craft. The familiar saucer shape of the craft, like an inverted soup bowl resting atop a second one, sat on the paved floor of the hangar. It had no landing gear or other structure that might have supported its weight while on the ground. From what I could discern, it was approximately fifty feet or so in diameter and was roughly twenty feet tall. Where the two soup bowl discs met, the skin of the craft had a kind of rounded rim before the curves rose and fell to the narrower top and bottom. I moved closer and did a quick 360-degree turn around its perimeter, and where the two "halves" of the craft met, I could detect no seam. The same was true of the entire exterior of the vessel. I saw no panel lines, no welds, no rivets or other fasteners. As I had on first seeing the craft on that first day at S4, I ran my hand along the surface. I looked over my shoulder half-expecting to be reprimanded for touching it as I had been previously, but none of the security team members said anything to me. As before, the skin of the craft felt like a metal—cool to the touch and very, very smooth. It was dark aluminum in color, monochromatic across its entire surface except for the four black rectangles near to the top of the upper dome-like portion. (Later on I would speculate that they were sensor arrays of some kind, planar sensors that assisted in some type of celestial navigation. But at this stage I merely noted them and moved on.)

I was so engrossed in observing and noting as much as I could that I had nearly forgotten that Barry was also in the hangar with me. I entered the craft's interior through a small access hatch, just wide enough for me to put my shoulders through it with a fraction of an inch to spare. Once inside, I couldn't stand up straight. Work lights had been installed at various points inside the craft. I maneuvered on all fours, hunched and using my hands to steady me but not kneeling, along a honeycombed access way. Just as the exterior of the craft appeared to be seamless and was all rounded surfaces, so was the interior. It also appeared to be made of the same material of the skin of the craft. I was struck by the idea that it was almost as if the craft had been fabricated from melted wax and then cooled into this shape. Injection molding was the closest terrestrial machining or manufacturing process that I could compare it to.

On that lowest level, I saw three seats, similarly looking as if they had been part of the molding process and not manufactured separately and then affixed somehow to the rest of the structure. They reminded me of a Scandinavian chair, without legs, looking very much like a rounded flower petal, more cupped than a tulip's but nearly so. Just as the hangar had been lit so that we could see the craft's interior, so was the interior. The material only dimly reflected the lights, as if it had a kind of matte finish to it, but its color didn't appear to be layered on. Rather it was integral to the material itself. Integral and integrated were the two words that kept springing to mind. Whoever had designed and built this craft seemed to have no concern for aesthetics—at least not human aesthetics.

The three seats puzzled me. I saw no kinds of restraint systems, no indications of any life-support systems (vents, for example), and as I worked my way toward another access hatch, I was astounded by the fact that I saw not a single light, switch, dial, display or anything that I associated with a vessel that traveled through space. All that interfered with the open design of this level, besides the seats, was a length of pipe coming down from the ceiling and exiting through the floor—presumably, the parts of the propulsion system that carried the power/gravity wave from where it was generated to where it was emitted. I was in awe of the technology behind this elegantly simple and purposeful execution of a craft designed to travel enormous distances, and with what seemed to me, to be based on the craft's construction, relative ease.

It was on the second level that I saw the now very familiar elements of the propulsion system we'd had access to in the lab. The reactor sat on the floor, the wave guide piping ran from it to the amplifier, and then additional tubing ran down through the floor to where, I imagined though I couldn't see them, the emitters sat on the bottom of the craft. That platform section, or what is probably better described as a pedestal, was what the craft rested on. How, or by whom, the craft was piloted and navigated was something that I was hoping to be able to see for myself, but neither Barry nor I were allowed onto the third level where we presumed the pilot(s) would have been positioned. I had hoped that I could have been able to sit inside what is traditionally termed the cockpit, but that wasn't going to be possible.

Frequently, when I was faced with a design or some other kind of engineering or diagnosis and repair problem, I liked to sit with that problem. Often that meant walking away from the task at hand and just thinking about the issues or distracting myself with other work hoping that on some subconscious level I'd be able to come up with a solution. Other times sitting with it took on a more literal sense. I would sit near the device, or in the case of my jet cars and other vehicles—like the ones I converted to run on hydrogen—I sit in the "problem" and fully immersed in that environment experience a kind of osmotic absorption of the solution to the problem. As cramped as the interior of that craft was—I imagined that the life forms who could move comfortably about in that space had to be about the size of a six- or seven-year-old human—I wanted to linger inside for as long as possible. That possible wasn't too long since cramped conditions literally can produce cramps, but I did take a few extra moments to marvel at the intelligence behind this execution and implementation of the propulsion system within the larger context of the craft itself.

I was blown away by what I'd seen. I'd been convinced for a while that what we were working with at S4 was not of terrestrial origin. But seeing this entire craft really solidified what I'd come to believe. I knew with as high a degree of certainty that one can know anything, that we didn't have the capability, the Russians didn't have the capability, the US and the Russians working in concert didn't have the capability, all the world's greatest minds working together couldn't produce a working artifact like the one I was in. As I eased myself along to exit the craft, I kept shaking my head in wonder. My face had that half-flushed, half-swollen feeling of having smiled and laughed for too long. I briefly wished that I could share this experience with others.

At the same time, my attitude had evolved in the time I'd been working at S4. Initially, though I understood the government's and military's position about secrecy, I resented it. Now that I was literally and figuratively on the inside of the operation, a kind of elitism set in. I didn't think that many members of the American public could truly appreciate the magnitude of the achievement this craft represented purely on a technical, scientific, and production standpoint. I was completely blown away, and I'd already been privy to a lot of information about the craft and its origins. Whether this craft came from Zeta Reticuli or was a product of the drunken or febrile imaginings of the collective membership of the Theta Tau engineering fraternity, it represented such an enormous advance in moving a life form from place to place that I struggled to come up with an analogy that would make sense to explain what it meant and felt like to be in its presence.

Once outside the craft, I disengaged the fawning fan mode in my brain and resumed scientific inquiry mode. I wondered about the craft's structural integrity, given that the skin of the craft seemed so thin. I wondered how heavy the vehicle was, but I couldn't really imagine myself putting my hands on it and trying to lift even a corner of that pedestal off the ground. I looked around the hangar and saw that a small crane stood in a far corner. I walked over to it and saw that its load limit listed on a safety sticker indicated a maximum capacity of two tons. That at least gave me a ballpark sense of the craft's mass. I thought of the fuel discs again and wondered for how many hours or light-years they lasted and if the ship's interior was mostly used as a cargo hold for them. How long had the one we been using in our lab been discharging?

Barry and I stood again on the apron in front of the open hangar door, and I wondered if our mouths were hanging open. We didn't speak until we were back in the lab.

"Did you ever think you'd see anything like that?" Barry said. "Did anything in your life before this prepare you for that?"

"I've been trying to come up with a way to explain what this whole scenario is like, how to make sense of it for myself. The only thing I could come up with was this: Imagine that we were with the settlers on the Oregon Trail, riding along in our wagons back in the 1830s. We go to bed one night and in the morning we see this machine that has two wheels, a seat, handlebars, and some device cradled in its frame and a chain going back to the rear wheel. We'd know it as a motorcycle today, but they'd only seen crude early bicycles. Given enough time, they'd probably be able to figure out how to get that motorcycle started and even to ride it a bit. They wouldn't understand how the engine worked, and even if they did, once the thing ran out of gas, it was probably only good to them as a plow they could drag behind draft animals."

Barry nodded. "Something like that."

"More to the point," I said, realizing that maybe Barry didn't feel the need for that kind of explanation, "I didn't see any signs of crash damage."

"Listen," Barry said, "I was never a believer in the whole Roswell incident. I don't follow those kinds of stories. It seems strange to me that an intelligence that is advanced enough to produce what we just saw wouldn't be able to negotiate the Earth's atmosphere or have a landing system figured out or not be able to cope with our weather systems and phenomenon."

"I agree," I said. "Just doesn't seem like a plausible explanation."

Barry and I debriefed for a while longer, and we both postulated about how the propulsion system worked. We weren't any closer to figuring out how we could reproduce the devices to produce the kind of gravitational field we surmised was at work, but we had a better sense of how the reactor, the amplifier, and the emitters worked to produce a kind of antigravity effect that would allow it to render our usual concepts of space and distance immaterial. Without realizing it, we were on the verge of doing something that I had only presumed was possible, witnessing the craft in actual operation.

The next time I was called into report (I still kind of chuckled at the insistent regularity of the message, "Hello,

Mr. Lazar it is now [insert time]. We expect you to be at the installation at [insert time]."), Barry and I got busy working with the modified pressure sensor instruments we'd devised. We were about to use them to measure the intensity of the gravitational field when Dennis stepped into the lab.

"Follow me," he said.

We went outside and sitting in front of the open first hangar was the craft we'd recently examined. This time, and it struck me as odd, each of the interior doors connecting the individual bays, was open. In addition to the craft that we'd been allowed to inspect, I could see eight more saucer shapes through the procession of openings. I wondered briefly if we were going to be allowed to see all of them. Each of them looked to be of similar shape, and it wasn't just because of the distance I was from each of them, but they appeared to be of slightly different size. The one we'd inspected was smaller and sleeker than the others, as if it was the sports car in the line-up. In fact, from that point onward, I always thought of and referred to the craft I was able to enter as the "Sport Model."

"A low-performance test is going to be conducted. We thought the two of you should see this," Dennis said.

I looked over at him trying to see if he was pleased by his having made arrangements for this, but his face was impassive. I remembered the look of irritation on his face when I'd remarked that the military was still doing things then they way that they had nearly fifty years earlier out at Los Alamos and in Chicago and in other locations trying to create the atomic bomb. All that compartmentalization and secrecy had slowed progress. If they wanted us to be effective, they needed to let us share information. That was how most advances in science and technology—despite the belief in the romantic notion of the lone wolf working in isolation model—were made. Here we were getting some access, greater access to information than before, but still in a far more limited manner than I would have preferred.

I knew it was time to keep my mouth shut and just observe, and that's what I did. What frustrated me was that with the exception of Dennis, Barry, and I, there were few other S4 personnel around. The omnipresent security guys were there on the periphery, but there was one technician there sitting with a pair of headphones on while seated in front of what must have been a radio, and one other man who stood at a distance from us and kept his back to us the entire time. I stood there with my arms folded, rocking back slightly on my heels, and then the craft did something similar. At first I heard rather than saw any activity coming from the craft. A loud hiss, nothing painful but the kind of buzzing sound that an electric substation might produce, reached my ears. Then the craft lifted off the ground slightly, wobbling, the central axis tilting a few degrees from vertical. As it lifted off, I could see the blue glow of a corona discharge coming from the bottom of the craft.

That led me to believe that the air around the bottom of the craft, where we suspected the emitter was, was being broken down and photons were being emitted. The light was visible, just as lightning in the sky is, due to that incredible high-energy output. As the craft rose, the slight oscillations lessened and the hiss diminished. By the time it was thirty to forty feet in the air, lifting nearly perfectly straight up, the sound was completely gone. In all my years of working with jet engines and pyrotechnics, I was accustomed to hearing loud noises as objects were being propelled upward or forward. The silence was eerily exciting, and I felt a broad grin spreading across my face. I could hear the faint sound of an appreciative expulsion of air coming from Barry as I stood there wide-eyed and mind-boggled.

I could see that the craft had three emitters. Only one was creating the corona display and it was the one that was facing straight down at the ground. As the craft rose in the air, the display expanded. It began as a tight beam and became more diffused with each foot the craft rose. It was like watching someone twisting the lens of a flashlight, increasing the size of the circle of the light from its narrowest to its widest focal range. As that gravitational wave spread out, it helped to stabilize the craft. It hung there, completely still, as if it were suspended there by an invisible steel rod firmly anchored to the ground and some point above it.

With it hanging there above us, we could see more clearly the bottom of the craft and its triad of emitters. They were arranged in a pattern with the one that was functioning at the point of the equilateral triangle. The other two emitters pointed at a right angle to the downward pointing one. Given this orientation, because we were looking up at the larger spherical shape of the craft, I decided to think of the area to the left of the downward pointing emitter as the rear and the area in front of the two other emitters as the front or forward section. That was because the opening of the two forward emitters pointed in that direction, and from what we believed of how the gravitational wave was produced, the energy coming out of the open end of the emitter would allow the craft to move in the same direction as the opening. I knew this was a low performance test, but still I was hoping to see the craft maneuver in the air. It didn't. I looked over at the technician with the radio. He was far enough away that I couldn't hear him speaking into a headset, but I could see that he was talking, presumably to whoever was piloting the craft. I thought again of the cramped space inside and the childlike proportions. I also wondered how in the hell a radio wave—an electromagnetic wave—would be able to work in the presence of that gravitational wave. No kind of phase-locked loop should have been able to "survive" in that environment. The electronics guy in me engaged in a brief battle with the physicist in me over which phenomenal phenomenon to focus my attention on.

I didn't have long to contemplate them. After a few minutes of hovering in place, the craft descended and settled, reversing the process of moving from a stable orientation to a wobbling one before coming to rest.

"That's it," Dennis said, always the master of understatement. "Time for you guys to get back to work."

Barry and I did as instructed, both of us lost in thought for the next fifteen or so minutes. We each sat at on a stool in front of the lab bench. Barry scrawled a few lines in a notebook while I fiddled with a micrometer, measuring the thickness of the lead points of a few pencils and comparing them to the lead rods I used in a mechanical pencil. It didn't seem odd for us to be so quiet. While engaged in work, we both tended to go inward. We'd both witnessed something on the order of a miracle—if one accepted the definition that a miracle was the result of an agency from outside our terrestrial realm—and we both needed time to process what we'd seen. I suppose in some Hollywood version of these incidents, the two of us would have charged into the lab flush with excitement and gone to the blackboard and started furiously writing and erasing parts of a complex equation we'd left there while exclaiming loudly, "I've got it. I can't believe I didn't see this before!"

The truth was that since we'd started working, we had a sense of what was going on with the reactor, amplifier, and emitter. The flight demonstration simply confirmed that our theory was, for the most part, correct. We hadn't seen the craft move laterally nor had it covered a great distance in a short amount of time, but the fact that the device created a focused beam (for lack of a better term) of highly concentrated energy and a gravitational field, that it could also disperse more widely, would allow it to "cover" vast distances. I had suspected that the gravitational field was doing something that scientists and science fiction writers had been speculating about for many years. It was producing a kind of negative gravity, or antigravity that, in a sense, "removed" the gravitational force ahead of the direction in which the three emitters were pointing. In time, Barry and I came up the name "Omicron," to describe the action of the single downward emitter functioning to lift the craft above the ground. That Omicron action allowed it to initially escape the gravitational pull of the very large body it was near—the Earth. When all three emitters worked, the same antigravitational force was produced allowing the craft to move in multiple directions. This omnidirectional state of operation we termed "Delta."

We chose Omicron, the fifteenth letter of the Greek alphabet, because it means "small." In the Omicron state, the craft and the propulsion system configured in this single emitter operation, could only make relatively small moves. Delta, the fourth letter of the Greek alphabet, has its symbol used in science to indicate change. And we believed that when the three emitters were arranged in a kind of delta shape as they were, it allowed the craft to essentially move freely and change direction with stupendous rapidity that made our current capability pale in comparison. How the emitters worked was still in question, but orienting them in various directions was similar to how our spacecraft worked. It wasn't so much that the system defied our understanding of the laws of physics, its ability to move as quickly and as efficiently as it did was astounding. Essentially, when the propulsion system created that gravity wave or antigravity state, everything we knew about how objects could orient themselves in space, about the nature of flight, was altered. It was easy to get hung up on language and whether a gravity wave was being produced or if it was antigravity, but essentially, on a fundamental level, what the makers of this craft had done was create a device that could achieve something astounding—gravity control. It reduced or cancelled a gravitational field. No one on Earth had been able to do that to that point. It had been speculated about and hypothesized about, but it had not been achieved as we sat there in 1989.

Words fail me now, as they did then, to explain how great a paradigm shift in thinking was necessary to truly appreciate what we'd just witnessed. That said, as had been true for most of my adult life, I was able to come up with an analogy to explain in layman's terms how this process worked or at least to create a visual metaphor to illustrate it.

Imagine placing a bowling ball on a level mattress. It would sink into the cushion for a bit before settling and becoming still. If left undisturbed, the bowling ball would not move. You could push that bowling ball and get it to roll. In a sense man's early efforts at flight were like that. We used a propeller to push air behind the plane to get it moving, used the angle of the wings and their surface area to create lift, and the airplane would leave the ground. With a jet engine or a rocket, we expel hot gasses out of the rear of the engine to push the plane forward and create lift. What this alien craft did was like placing your hand on the mattress in any direction around the bowling ball and compressing the cushioned material. The ball would roll in the direction of where the resistance had been removed. A somewhat crude analogy but sufficient to understand the basics. What the emitters were able to do was to remove the force of gravity, not just in a tight beam enabling it to move forward, but widening the beam, and having the emitters pivot in various directions was akin to having a group of people standing around that mattress each of them pressing down on the cushion at various points, allowing the ball to move in multiple directions. Imagine those individuals moving with great speed and force, superhuman speed and force, and the visual becomes even better.

    I didn't share this analogy with Barry. He showed little interest in my previous motorcycle-dropped-in-among-the-settlers scenario. What was far less difficult to imagine was how this craft was able to cover vast distances. In a sense, it didn't cover any ground—as we so frequently describe an object moving from one point to another by going over ground from point A to point B. Instead, using that gravity control, this craft pulled distance objects toward it. Imagine a ball resting on a towel with a Frisbee at the opposite end of the towel. With the emitters in the Frisbee doing their thing, the fabric of the towel was pulled toward the ball that was anchored to it. It was as if the universe was being folded in an accordion-like fashion. As I noted when I first saw the propulsion system functioning, time and gravity are inextricably linked. If you controlled gravity, you also controlled time. That solved the dilemma we currently have with long-distance space travel. How could you possibly supply humans with enough oxygen and food, but even more so, how could you expect them to outlive our usual life span, in order to travel to distant galaxies. What this propulsion system did was render moot all those kinds of questions.

Normally, Barry would quickly shut down any kind of speculative questions about the origins of the device we were working on. He was far more disciplined in his thinking and with his emotions than I was. Following that demonstration, though, even Barry's mind was spinning.

"You know, I'm not sure if what they told us is true. But after seeing this, it opens up a whole universe of other possible explanations. I mean, I was sitting here thinking that this thing could have come from another dimension. It could have come from some point in the future. We just don't know. The whole Zeta Reticuli story could be a cover-up. I just don't think we've got the advanced technology to produce this ourselves in this time. Not even with the assistance of the Russians or teams of scientists from everywhere."

"I agree, but you know, Barry, like you always say, if we keep thinking about that part of it, we'll go crazy."

Barry laughed, "You're right. Kind of fun to think about, but we aren't here to have fun."

Frustratingly, as Barry and I talked about this gravity control device, we kept coming back to the same question over and over again. How was it able to produce such an enormous amount of power from such a simple system without the production of any kind of heat and sound and other indicators of the prodigious amount of energy being used to overcome what we on Earth considered to be one of, if not the, most fundamental forces in the universe—gravity. How could something seemingly so simple and fueled by such a simple and small object as the discs do that kind of work? In a way, I wondered if perhaps it functioned on the principles of some martial arts—using your opponent's energy against them, in opposition to them.

Ultimately as our discussion wound down and the frustration of feeling as if we'd returned to square one and were no closer than ever to being able to reverse-engineer the system, it felt pointless to construct ways in which to describe what was going on. I'd heard it said that language allows us to master and take control of our world, to give it shape and meaning. All I could think of as the hours went by and the lights above us hummed nearly as loudly as the craft had, was that a mechanical pencil produced a point far less sharp than a pencil sharpened mechanically. At that point, I wondered if maybe I was losing my sanity as more and more divergent and unproductive thoughts passed through my mind. After all, Einstein said, the definition of insanity is doing the same thing over and over again expecting a different result.

I was more than a little relieved to go home that night. I was tenacious when faced with a problem that needed to be solved, but now I was frustrated and intrigued, and the push-pull of those two competing forces would ultimately be won by another force, one that was far beyond my control.

CHAPTER SEVEN

ADDED INTO THE MIX OF ALL THE QUESTIONS I HAD regarding my situation came another and more frustrating question. Why was it, that after granting me access to the craft and allowing me to view a test flight, did the voice of the EG&G representative suddenly fall silent? I'd experienced one fallow period in my time with EG&G and been rewarded with those two opportunities. Maybe, I tried to console myself as week one of no calls turned into week two, I'd be similarly rewarded again. What form that reward might take occupied some of my time when I wasn't working, but not nearly enough. Tracy seemed very occupied with the business, her flying lessons, and her social life. She sensed that something was not right in my world, but when she asked me what was wrong, I told her that I was fine, that nothing was in any way out of sorts with me.

The devil's bargain that I signed in agreeing to a confidentiality clause required that I not divulge anything, good or bad, a potential state secret or a simmering dissatisfaction with a co-worker, the quality of the food in the cafeteria, the frustrations with having to respond on a moment's notice, the havoc the irregular schedule played with my sleep, were all equally off limits. I'd put up with the unannounced visits to my home, agreed to have my privacy violated by allowing our phones to be tapped, and wasn't even raising a fuss about the fact that I'd yet to receive a paycheck. All I wanted to do was work and to contribute to solving this problem, and I was being subjected to what I thought was a kind of tantalizing treatment. They'd tease me with privileged glimpses then back away again.

Also, for as much as I'd witnessed and believed, and now believed to a greater degree the truth of what I'd read in the briefings and later seen with my own eyes, there was still the possibility that I was somehow being set for a fall. What the nature of that fall was and why I'd been chosen wasn't clear, but I wasn't willing to dismiss anything as a possibility. After all, if my mind had been blown away and my perceptions opened by what I'd seen and learned while working at S4, then I would be foolish to believe that only the most righteous and above board and honest dealings with the employees at S4 were being conducted. Put another way, when your belief system in one part of your life is rocked, (that aliens existed in theory but not in actual fact) and now knowing that "visitors" from outer space had been on the planet, that seismic upset reverberates in other parts of your life as well. Maybe if I had someone with whom I could share what I'd experienced, another perspective to help me sort through all of this upheaval, then maybe I could have found a better way to respond to what happened to me as the Ides of March passed and I grew more and more anxious and uncertain about my future.

Even now, many, many years after those events in the early part of 1989, do I still wonder many of things I wondered back then. Hindsight generally affords us an opportunity to see things with greater clarity, growing older and having more experiences can make us wiser, but I still have trouble establishing a firm footing about whether or not the path I chose to take was reckless, wrong-headed, self-sabotaging, or right. Perhaps it was all of those in some combination. As a scientist, I was trained to look for and to find truths. But we all know that even in the scientific community and the undertakings made under that umbrella, there is much that appears initially to be black and white and absolute, but later is revealed to be more gray than white or black.

I write this as a way to preface what was to prove to be the unraveling of my life. I paid a heavy price, as did others, and I take full responsibility for my actions and regret deeply those who suffered collateral damage as a result of them. I won't go into much detail about the "civilians" who were affected; to do so would only open old wounds and expose people I care about to additional discomfort. In some ways, I wish that I had taken this notion of collateral damage into fuller account during those weeks in March of 1989 when I grew increasingly impatient and resorted to self-preservation as one of the central movers in my decision making.

GENERALLY, I'M NOT A VERY PARANOID INDIVIDUAL. I NEVER had a reason to be. But after learning what I had and seeing those demonstrations and crawling around inside that craft, and then not being called in for a week or so, just at the time when we seemed to be making progress and I was being given very exclusive access, I started to really wonder what the hell was going on. After all this transpired, someone told me that the writer William S. Burroughs once wrote, "Sometimes paranoia's just having all the facts." I wasn't aware of that quote back in 1989, but I had some thoughts along that line.

To contribute to my growing uneasiness, the house continued to be under surveillance. I did the best I could to ignore it, and I definitely didn't look at it as the positive sign I once had. I suppose that back when Mike Thigpen interrupted the meeting I was having with Wayne and his wife, when our deal for the photo business was still pending, I probably was engaged in a form of wishful thinking. Their search of the premises was very odd, mostly because as far as I knew, and Tracy had pointed this out, no one that we knew ever commented on the fact that they had been contacted by any agency. Not anyone in my family, no friends, no former employers or colleagues, no one let me know that they'd been questioned about me. That was very odd.

A few days after witnessing the low performance test, I was at home getting ready to go out on a delivery run. I stood at the kitchen sink rinsing my coffee cup and saw either a Buick Skyhawk or a Chevy Nova sitting across the street from our house. Whatever the make and model, it was one of those nondescript, bland kinds of cars that I seldom saw in our neighborhood. Inside the car sat two men in dark suits. They were looking straight ahead. I didn't see them using binoculars or a camera with a long telephoto lens, but I was pretty sure they were there watching our house. That suspicion was confirmed when I exited the house and then drove off to drop photos by at Gene's office. The car pulled out behind me, and unlike in thrillers or police procedurals, they made no effort to disguise the fact that they were following me. They didn't keep more than a car length or two behind me, accelerated to keep my pace, and went through a yellow light I purposely slowed down for and then sped through. When I got to Gene's office, the car parked a few spaces away. I was thrown off balance by it all, but managed to pull myself together and didn't let on to Gene that anything out of the ordinary was going on.

Gene and I transacted our business, and I went on with the rest of my day. I had to make a stop at Home Depot to get some supplies for a repair project on a leaky showerhead, and the mystery car was there again, trailing behind me. Like the car, the two men were bland and relatively nondescript. Caucasian, mid to late thirties or early forties, dark suits, white shirts, and sunglasses doing battle with the morning's low sun. I almost had to laugh when I thought of recounting that description to anyone, like I was some hard-boiled detective in a noir detective film. It wasn't so funny when they followed me home and then remained there, until another car and two men came to relieve them, early in the evening.

When they were still there in the morning, I decided to call the police. I recounted to them what I presented above, including the terse description. The officer who took the information assured me that she would pass it along and to expect someone to come to investigate the situation at some point. It must have been a slow morning in Las Vegas. No more than a half hour later, a squad car approached the vehicle and then parked behind it. A minute or so later, the officer got out of the car and approached the driver's side of the vehicle that had been sitting there. I wasn't able to hear or see much at all, but within minutes, the officer was back in his car and drove off. My "watchers" remained. While all this was going on, Tracy had come downstairs and was getting ready to leave for work.

"What so interesting out there?" she asked.

"Not sure. The police just pulled someone over briefly. Gone now."

"Are you turning into one of those kinds of guys," she said, teasing.

"What do you mean?"

Her eyes revealed her amusement. "Oh, you know. The 'stay off my lawn' kind of cranky guy who watches over everything in the neighborhood."

Not wanting to reveal my concern I played along. "No, not one of those guys. The kind of guy who buys a police scanner and monitors all the activity going on around the city. Your guy is way too low-tech for me."

She narrowed her gaze at me, unsure, it seemed, if I was kidding or not. "I'm surprised you haven't figured out how to make one, or maybe that's just one of your projects." She used her fingers to make air quotes around the last word.

"An idle mind," I said, reminding her of the famous Ben Franklin quote.

Tracy left a few minutes later, and my mind was not idling at all but racing. What had gone on between the police and the observers? Why hadn't they come to speak with me? I'd given my name an address and phone number to the person on the phone. Was I going to get some kind of response to my complaint? Was I going to be followed the rest of the day and for how far into the future? What did all of this have to do with what I most recently witnessed at S4, if anything?

I decided it was best to just go about my business. I could see how easy it would be to get too worked up about it all. I had planned to meet a friend named Mario at a local gym. He and I were workout partners. Every other evening, usually Tuesdays, Thursdays, and Saturday we met and did some cardio work or some weight lifting. Maybe a workout would help relieve some of the stress and help me put things in a better perspective. I pulled into the parking lot a few minutes after six o'clock and waited.

I thought about the previous time Mario and I had met there. We exercised and then left.

I had walked over to my Datsun 280Z and, while chatting with Mario, pulled out my car keys and went to unlock the door. I inserted it but noticed that the door was already unlocked.

"That's weird," I said.

"Yeah, it is but in Japan—" I cut Mario's explanation of sushi off. He'd been telling me about the new restaurant he'd just gone to.

"No, my car. The doors are unlocked. I never do that."

"No, you don't. That's your baby. You treat her right."

At the time I attributed it to just being tired.

That night, Mario was running a bit late. I got out of the car after Mario had walked up to mine. I got out, locked the doors, and then lifted the handle.

"They're locked, right?"

"Far as I can tell," Mario said.

We both wanted to be home at bit early, so we cut the workout short. About forty-five minutes later, we walked out of the gym. I took a few steps into the parking lot and then stopped in my tracks. I looked across the way and there was my car with both its doors wide open, looking like a fixed wing aircraft.

"Holy shit," Mario said. "What in the hell? You got broken into."

I stood there frozen in place, my mind ratcheting through several possibilities. It could have been a car burglary. The Z had a decent stereo in it. I had left my wallet and the Uzzi I was carrying around for protection inside it. It seemed strange to me that a thief would leave the doors open. Why leave a "Hey look what I've done!" calling card. Had they been interrupted in the middle of the job? None of that made sense.

I'd confided in Mario about my work situation and I think that though it took him a bit longer than it did me for it sink in, he got it.

"I don't know if we should walk over there. Who knows . . ." He let the thought drift on the night wind.

The car did have central locking but a malfunction of that system wouldn't have allowed the doors to open and the hatch to rise. Somebody had to open those doors and I had a strong suspicion the men who had been following me were sending a message. We're still here. I can't say that my blood ran cold or a shiver ran down my spine, but I did experience that gut-level twinge and burn of adrenaline kicking in. I had to fight the urge to jump in the car and take off after them, let them know what it was like to be pursued and followed. I knew it wasn't a good idea to enflame the situation. Look at what happened when I called the police.

Mario and I both jumped a bit when the doors behind us opened and rattled. A couple of other gym members walked past us.

"I guess I need to go over there," I said.

"Yeah, you do."

We walked over to the car, and I kept telling myself that I had nothing to worry about. I was definitely spooked, and I could tell that Mario was as well.

"If you don't mind," I said, "I'd like your help."

"With what?"

"We're going to have to start the car. That's your job. I'm going inside."

We both laughed.

They'd sent a clear message, a sign of the kind of power and control they held over me. They didn't like being messed with. I didn't like being messed with either.

I decided the best course of action was to try to de-escalate the situation. I needed to get better control of my state of mind as well. After all, I had no confirmation yet that I was no longer going to be working at S4 or anything like that. I had Dennis's phone number, so I decided the right course of action was to go to the source and not let my mind run away with me. After all, I also told myself, nothing had gone missing from the car. I got someone in Dennis's office and explained that I wanted to speak with him. After being told that wasn't possible at the moment and getting a vague to no response at all to my follow up questions about when he might be available, I left a message indicating who I was and that I very much would like to speak with him at his earliest possible convenience.

On and off for the next few days, I noticed that the car would be outside our home, but the following stopped and nothing bizarre happened while I was out and working on the last few of the photo jobs I was taking on. When forty-eight hours passed and I didn't hear back from Dennis or anyone at the number he'd given me, I made another inquiry and left another message. When the tailing resumed and my calls still went unanswered, I started to worry even more, and not just about whether or not I would continue to work for EG&G at S4, but about my own safety and Tracy's and other people I knew. I didn't believe then and don't believe now that my fears were the product of an irrational response to stress. Was I stressed?

Certainly.

Greatly.

Undeniably.

Was it plausible to believe that because of what I knew and what I had seen that some harm could come to me?

Yes.

Definitely.

For as much as I tried to focus just on the task at hand, the implications of what I'd experienced and knew had tremendous consequences. They wouldn't have been going to the lengths they were to do background checks on me and they wouldn't have required me to consent to my phones being tapped, and they wouldn't have asked me to sign a document that essentially stated that I agreed to waive every one of my constitutional rights (which I did sign) if there wasn't a serious need for them to keep this information in-house by whatever means necessary. That may seem like a large leap in logic, but given all that I've described about the need for me to carry a weapon, the heightened drama with the surveillance teams, I figured that I had to do something to protect my best interests. No one else was going to do that, so it was incumbent on me to think ahead and consider wisely what I should do to protect my professional reputation, my chances of getting gainful employment elsewhere, and the physical safety of me and others.

Only later, after I made the decision to share what I knew with a select few friends and later more widely, did I think more about the public's right to know this kind of information. I never set out to be a crusader, and though some called me a whistle-blower, I blew the whistle initially almost exclusively as an act of self-preservation, whistling to call attention to me to make it less likely that harm would come to me if I made clear what I knew, than it was to draw attention to the program at S4 and reveal the nation's deepest and darkest secrets.

All of those thoughts got generated in my head, were amplified, and were only emitted when I felt pressed into a corner and had to do so. I knew that Gene was the first person that I should tell. He was my closest friend at the time and he was about as level-headed a person that I knew. I felt that he'd believe me, and, because he didn't have as much at stake personally as Tracy did, he'd be better equipped to give me some guidance about how I should proceed. I had a few vague thoughts in mind. I was confident that everyone who knew me would believe me, but still thinking about the possibility that I might wind up somewhere in the Nevada desert with a bullet in my head and a fabricated suicide note left at home, I was leaning toward showing them evidence to prove the existence of those craft.

Sitting in Gene's car as we drove along Alta Avenue was both comfortably familiar and surreal. I prefaced my story with the kind of disclaimer you might expect, "This is going to sound really strange, but. . . ." When I was done, Gene said that coming from anyone else but me, he'd be incredibly suspicious. Then he added, "I knew that something was up when you told me about the gun registration and mentioned something about work. I couldn't have imagined that it was something like this. But, hell, who knows what kinds of things go on out there. I figured the UFO stuff was just a bunch of wahoos with nothing better to do with their time."

I didn't mention to Gene anything that I'd read in the briefings. For this first time speaking with someone openly about what I was doing, I wanted to stick to what I could definitively validate—what I'd seen, what I'd done, what had been done to me. As tempting as it was to go beyond those three categories of events, I refrained from doing so. As I pointed out before, Gene was an immensely curious man with wide interests, and his curiosity was reflected in the half-dazed, half-enthralled look that spread his face.

"Just so you're aware, Bob. I'm going to do a little digging around on the subject. Not to undermine you or anything, just because you've got me thinking."

"Gene, you're a free man in a free country, and I would never think of telling you what to do or how to go about it. Thanks for listening to me."

"No worries. Tracy know?"

"I haven't told her yet. Obviously, I want to keep as tight of control of all this as I can. At this point, I don't see the sense of pissing anybody off."

"You think you might have already?"

"I can't imagine how. Like I said, they let me on the craft, they let me see it low fly, and now nothing and some harassment. This is all just so nonsensical."

Gene nodded. "Hard to figure. Sometimes an irrational response is the only one you can make to another irrational act, but I don't think that's wise here."

"Agreed. I'm sure I'll need to talk to you about this again. I hope not, but even if I did get called back in, I'm not sure how I'd respond."

"That's understandable."

"I just don't know if it's all worth it." I produced from my pocket an envelope that contained the first and what would prove to be the only check I ever received. I slid it over to Gene.

"$958.11?" he said, "That's what the United States Department of Naval Intelligence pays a senior physicist?"

"That's right. You do it for the love of it and not to get rich. And to be honest, Gene, this isn't so much about the money as it is the headaches. If I'm getting this kind of a runaround and treated like this, the dollars-to-headache ratio is way out of balance."

Gene and I talked for a few more minutes and then he dropped me off at home. When Tracy got back from work, I inadvertently used the words that anyone in a relationship shouldn't use as a point of entry: "We need to talk." By the time I made it clear what the subject was, all the tension left her body. She sat on the couch, drew her legs up and heaved a big sigh while looking at the ceiling.

"This isn't some kind of goof is it? You're serious. I can always tell when you're trying to get me."

"No goof. No getting you."

"Holy shit," she said, drawing out the "Holy" for a few seconds. "This is nuts."

"Essentially, yes."

"I can't see you making up something like this." She drew a strand of her hair to her mouth and chewed on it, a habit she had when she was nervous. I hadn't told her everything about me being followed and the incident at the gym parking lot, hoping to spare her some worry. Later on, I'd come back to this moment and see it in a different light.

Of the three people I spoke to about my concerns, John Lear was the most receptive. That made sense given what he believed to be true about UFOs back at that time. I'd later come to learn more about the extent and the extremity of John's controversial views, but in those first few minutes, he, like Gene, seemed more curious than alarmed.

"Look," he said rubbing the stubble on his chin and neck, "We all know that something's out there."

"In the desert?"

John winked at me and said, looking from side to side as he did so, "Yeah, in the desert." He raised his eyes toward the ceiling indicating that it wasn't really just the desert he was talking about.

I looked around John's home office. The bookshelves were lined with various volumes of different types and sizes. In the corner, a telescope rested on its tripod and leaned against the wall. Along one wall were a series of framed certificates and awards John had received for his exploits as a pilot, among them some commemorated world records he had set. John was an accomplished pilot, more than an accomplished pilot actually, he's was fearless and also a bit reckless. Of course, once I got to know John, he shared with me the story of how he injured his feet and legs to such a degree that at times he hobbled around his home in Las Vegas. His father had created a manufactured updated version of a Lear Jet. (Over time the model designation has escaped my memory.) John was assigned a simple task. Do a flight demonstration for some clients considering acquiring the jet for their use. What John heard and what he listened to were very different things. Instead of an easy conventional demonstration of a take-off and landing and flight, John asked himself since this was the latest and most improved version of the jet, and highly capable, why not show just how capable it was?

He decided to do an outside loop, a very, very challenging and risky acrobatic maneuver. As John told me, once you committed to the move, you couldn't back out. Well, John committed, didn't back out even when he knew that he was in trouble and plowed the jet into the ground sustaining multiple leg injuries among others. John wasn't someone who was going to play by the rules, but I did trust him. In fact, I trusted him enough to take trips with him when he had to ferry commercial jets to one airport or another (he asked me along both for company and to wear a suit so that I could pass as an FAA inspector). Once aloft, John would light up his pipe, later to doze off completely with the autopilot of the L1011 functioning, and then wake up just in time to land the craft manually. I also trusted him with Tracy's life, and he gave her the first few flying lessons she had. He even allowed her, at one point, to take the controls of a commercial flight he was piloting.

To say John was a loose cannon is an understatement, and in looking back on it now, I can see to what degree how stressed and anxious I was to believe that John was someone who could help me put events into their proper perspective. By that I mean how I should approach the situation vis-a-vis my employment prospects. In terms of having someone who could assess the capabilities of the craft and determine whether they were anything in line with what terrestrial technology could produce, I knew of no one more qualified than he was. There was some risk inherent in sharing this information with John, but I also knew how much he believed in the existence of alien spacecraft and beings. My focus remained purely on the technology, and if John could offer any insight into that, then it was worth the price of having him possibly go off on tangents regarding the beings who produced that craft. I'd be able to tune that out.

That's not to say that he wasn't an interesting guy to be around, and he demonstrated his generosity toward me on multiple occasions. Still, John had his own theories about extraterrestrial life and government conspiracies that I can see now, in hindsight, predisposed him to believe what I was telling him. I also want to make it explicitly clear that he never urged me to do anything. He simply accepted an invitation to participate in acts that I instigated.

I was grateful that, for a few days at least, things seemed to settle down. No strange incidents occurred, and though the guys on over-watch were still around the house, I wasn't being followed—at least as far as I could tell. I supposed that in retrospect, I inadvisably let my guard down a bit. One afternoon, two days after letting Gene in on my secret, we were at his house. Over lunch, he told me that he'd been searching the Internet for information about UFOs. This was still early in the life of the Internet and information was relatively scarce compared to how it has proliferated. He told me a number of things, none of which really jibed with what I had witnessed, and even back then, we realized that you can't believe much of what you read on the Internet. We continued chatting, and I asked him about the source of the information he shared with me.

"There's two places in particular that seem to be gathering points. One is called MUFON—the Mutual UFO Network. The other is CUFON—the Canadian UFO Network."

"That's not very Canadian of them," I said, "You'd think the kindly Canadians would want to be part of the Mutual network."

We got to laughing and joking a bit more and it felt great to breathe easy for a bit. Eventually, Gene decided that since I had the real deal in terms of UFO information, I should be referred to as BUFON, for the Bob UFO Network. I countered with him transforming himself into GUFON, for obvious reasons. For the rest of the time we spent that afternoon, we continued to call each other by our new *noms de UFO*.

Gene's wife had given birth to a son on March 15th, and Gene was giddy with delight and sleeplessness. It was good to be around someone who was in such good spirits. He was grateful for the break from the routines of a young one and visitors. Somehow, all of this made me think that maybe my world had somehow been righted.

That feeling didn't last. I woke up before sunrise to see that the car was stationed in the same place it had been before. I wasn't physically tailed, but it was exhausting to feel as if my every move was being watched over. When I got home later in the day, Tracy was out. I felt like I needed to commiserate with someone. Deciding to not let on how down I was, I called Gene. When the answering machine responded, I said, 'GUFON this is BUFON. I have the baby pictures for you. I know you need them as soon as possible." Gene hadn't given me any photos and I hoped that he'd pick up on that right away. I sat there hoping he'd call back. In the interim, about forty-five minutes later, a car came into our driveway and braked hard. I heard a couple of car doors slam and then a rapid knock at the door. I answered it and two very official and distressed looking men in suits escorted me to the kitchen table. There, they began their interrogation.

Basically, they wanted to know who this BUFON character was or what it was code for. The same with GUFON. They then produced a printed copy of the message that I'd left for Gene not quite an hour earlier. I couldn't believe how quickly they'd gotten it and then gotten to the house. I was tempted to ask them about how they managed that task, but I could see they were in no mood for light-heartedness. I told them that they were just goofy names we'd made up a few weeks ago. We were in a silly mood at the time and that was it, nothing sinister at all. That seemed to appease them, but they produced a document that they made me fill out. I had to identify Gene by name, provide them with his mailing address, phone number, and other information that I was sure they already had. I also had to refer to Gene by his last name, Huff, and then add, Gene Huff, AKA GUFON. To this day, and even back then, Gene and I still laugh about that incident. Gene will still, on occasion, sign his name and then append it with AKA GUFON.

Even though I was amused by that incident, I still felt like lines had been crossed. I'd agreed to the wiretap, but that was to determine my suitability for a security clearance. It no longer seemed certain to me that they would conclude that I was a good candidate to work at S4, so I thought the investigations should cease. I'd also essentially signed away my rights to due process under the law. I'd been trying not to think of that while all of this nonsense was going on and had never mentioned that fact to Tracy. When I had told Gene about that, he said, "I'm not surprised you signed that document, but I sure would be worked up about it. You don't respond to things like that the way the rest of us do." Hearing him say that made me realize even more that I needed to bring this thing to some kind conclusion. If I wasn't able to keep my wits about me in the way that Gene had indicated I had in the past, it was time to make some changes.

In the initial interviews I'd been asked a lot of questions about how I managed stress, and I'm sure that my phone calls to Dennis's office—I'd made a few more with no success—calling the police, and now playing name games, all contributed to my sense that my days at S4 were over. I can see now that I did those things for another reason—I wanted the wait to end. I just wanted to be told yes or no and move on. I knew that security clearances could take time, but it was the frustrating stop and start of the work, being pressured to figure out the answer but not being given the proper time needed to solve it, that made me get to the point that I wanted something definitive to be said or done. It may be hard to believe, but co-existent with those feelings was a desire to see the job through. I can't say it was as strong as it had been before, but by still wanting to do the work, and not being able to do the work for no reason anyone could really offer, ratcheted up my anger at being put in that situation.

I decided to take action. As I pointed out, I wanted my wife and friends to have eyewitness certainty about the claims that I had made. That meant one thing—getting them out to the area to see a test flight. For years people in the area had reported strange lights and other activity in the sky. Because of my time in the lab at S4, I knew when one of those displays was going to occur. Wednesdays at eight in the evening was the usual time a high performance test flight was going to be conducted. I informed the rest of the others that I thought we should all drive out to the desert to witness one of those tests.

John Lear owned a Winnebago motorhome, and he volunteered to drive us all out to the perimeter of the base. In some ways, I thought that maybe including John wasn't the best idea. There was also some odd tension between Gene and him. But I knew that John's expertise and his 8-inch-diameter Celestron telescope would be valuable in assessing what we saw during the high performance test. As a bonus, we could travel in style in his motorhome. It was a 150-mile or so trip through some of the most boring desert landscape you can imagine. We'd be setting out in the late afternoon and getting there after dark when there'd be even less to see. Of the assembled group, Gene and John performed as the two ends of the poles between skepticism and belief. Gene wasn't rabidly anti-UFO belief, but given some of the strangeness between the pair, I figured they'd give each other crap. That would be good for some entertainment. We always seemed to be able to ride John pretty hard when he launched into some out-there discussion about whatever had caught his attention. John was kind of like a magpie, anything shiny and he'd take it and fly off with it. Yet, when you asked him a question about his area of expertise, it was like a switch flipped in his brain and he was the most articulate and focused expert.

Sure enough, after John picked up Tracy and me (Tracy liked John since he was the one who really introduced her to the new love of her life, flying) we headed to Gene's house. When Gene came out to the motorhome, I heard John snort.

"Where the hell did you get that shirt?" he asked.

I hadn't paid much attention to what Gene was wearing, but at John's cutting remark I saw that it was black and purple and looked like massive bruising had spread across my friend's upper body.

"Your wife got it for me," Gene snapped back.

It wasn't exactly a "your momma" joke, but close enough. John put the RV in gear and we headed out from Gene's place. The usual conversations about best routes began. Gene was in favor of taking the interstate until we got to Highway 93 to head north. John had to be contrarian, so, after making a few veiled references to President Eisenhower and his real reason for establishing the interstate system in this country, instead of the interstate, he took Power Line Road, until it intersected with the Great Basin Highway.

"How can you be in the middle of nowhere, if you're nowhere?" Gene asked as we bounced along the irregular pavement.

"That's for greater minds than yours to determine," John said. "I know this, though. I've flown over a few different nowheres, and I can tell you that the middle looks a lot like the rest of it."

Tracy intervened and asked about Gene's new baby and how he managed to get the night off. His in-laws were still in town and they looked at him askance when he told them he was going out with friends.

"I heard someone say once that it's easier to ask for forgiveness than it is for permission," John chimed in.

We all laughed at that. The conversation was on anything and everything but what we were all about to do and to see. I suspected that the worst thing that could possibly happen was the test flight to be cancelled for some reason. The weather wouldn't be an issue, but I knew that there was nothing I could do to control what the powers that be decided.

At one point, the RV began to stagger a bit on a climb.

"Damn it," John said, and pulled over to the shoulder.

Eventually the problem was diagnosed. The transmission was low on fluid.

"Call yourself a pilot? Have you ever heard of a pre-flight check?"

"And to think I let you take my wife up in a plane."

More good-natured insults and comments were hurled at John. Gene volunteered to hitch hike back to a gas station to get some fluid and then hitch a ride back.

When he came back he had a six-pack of beer in one hand and two quarts of automatic transmission fluid in the other.

"This," Gene said, hefting the beer, "goes in us." He then lifted the other containers. "These, go in the transmission. Got it, John?"

"Mind if I use that shirt to wipe off the dipstick, dipstick?" John looked very pleased with his comeback.

"Took you nearly a hundred miles to think of that one didn't it?" Gene offered, never one to allow himself to be topped.

Because of our delay, I was glad that we'd allowed enough time to make it to Groom Lake before night fell. The tests were generally conducted shortly after the sun went down. As we bounced along Groom Lake Road—a washboard dirt road at the far end of the site and outside the installation's boundaries—I had to laugh when John turned off the headlights.

"This is exactly the right vehicle for a stealth operation, isn't it?"

John nodded, "Thirty feet of bone- and machinery-rattling stealth at your service. Guaranteed to alert anyone of your presence within a mile and half."

"Let's just hope we don't have to make a fast getaway," Gene added.

I wasn't sure if he was truly concerned, but I reminded him, and more especially Tracy, "We're on a public road on public land. No one can tell us that we can't be here. Besides, with John with us, all we have to do is drop his name and his mom's and all will be forgiven."

Tracy looked at me and I watched a shadow of worry pass across her face before she smiled and squeezed my hand.

Along with some food and beverages, we'd also unpacked John's Celestron telescope, a couple of pairs of binoculars, and a video camera and tripod. We went about the business of setting up the devices. Once the equipment was set up, we stood, shifting nervously from foot to foot and waited. We didn't have to wait very long. Within minutes after settling in, John, who was looking through the telescope shouted, "I got something!"

He pointed into the night sky to our north and east toward Area 51 and S4. A light had suddenly appeared above the Papoose Mountains. A bright light, at the orange end of the spectrum, began to glow brightly. I stood focusing my attention on the bright object as it appeared approximately forty-five degrees above the horizon. A few instants later, it appeared at sixty degrees above the horizon. Then, seemingly in the time it took for me to blink again, it had climbed again and was then thirty degrees to the right of where I had just been focusing. I heard a commotion behind and a few muttered swear words and exclamations of surprise and delight. I kept my eyes on the lights.

Once again the light moved in what I could only describe as a staircase maneuver. Appearing at one height and the up, over, up, and over.

"That's no winged aircraft," John said, his voice firm with conviction. I recognized his authoritative tone mixed with a bit of awe. "No way in hell anything we produce could make those kinds of maneuvers. Flat out impossible."

"Did you see that?" Gene said, nearly shouting. "Zipping along like that and then just stopping. How in the heck can it do that?"

After I'd seen the low performance flight I was certain that the gravity control theory we'd arrived at was correct. Seeing this high performance flight only confirmed it. It's nearly impossible to convey in words how quickly the light moved from one location to another. I'd seen jets flying overhead in the night sky and I could easily scan their path of travel with my eyes while my head remained still. Not so with this. Even from that great a distance from the light source, at times the craft's maneuvers had the light moving out of our fixed line of sight. It was astounding and thrilling to see how that reactor and its other components actually worked in a real-time application. I had an even greater admiration for the engineering behind the emitters. How could they move so quickly to accommodate the kind of movements the craft was making? How could objects as large as they were and made from a material that was somewhere between a metal and a ceramic move so quickly to allow them to point in the direction that the craft was moving in? They appeared to be rigid structures, yet they moved with the fluidity and flexibility of a muscle. Were they, when under power, somehow transformed into a more supple, more elastic material?

I also wondered about the color of the discharge surrounding the craft. Why did it appear to be orange, not pumpkin orange but duller, with more brown mixed in. When I'd seen the low performance test, the corona discharge had been a bluish purple. Now, with the thing in flight and presumably expending more energy it was producing a discharge at the longer end of the wavelength spectrum. Orange wasn't in the spectra of atomic hydrogen that would produce bright reds like in the aurora. If oxygen molecules were excited, you'd see purples or blues from nitrogen being ionized. Just another piece of the puzzle. I thought of sodium vapor lights, the kind that cast an orange glow and are frequently used to light city streets. Was it possible that the craft was some sort of sodium alloy? Sodium is very light, it floats on water, and the parts we'd been working with were incredibly light. But sodium is also terribly reactive with moisture. Perhaps it overcame that characteristic somehow. Maybe this was a big anode, a positively charged electrode by which electrons leave a device. I wished that I had brought the compact spectrograph I owned. It would have shown the Fraunhofer lines in the light and therefore identify which elements were present.

Because we weren't in line of the direction the emitters were pointing, we couldn't see what form the distortion was taking, how it might have affected our visual sense of the area around the emitters specifically and the craft generally. If we had been standing right beneath it, right there at S4, we would have been able to see that and I could compare it to what I'd witnessed during the low performance test.

Eventually the light descended below the peaks and the test was over. We all stood silently for a minute. The sky filled with stars and the night air felt chilled again, crickets chirped and the wind rustled a creosote bush and I smelled its fragrance. The whole time the test flight had been going on, I'd been aware of very little going on around me, all of my other sensory apparatus shifting power to my eyes.

I heard a voice, as if I was coming up from being submerged in water, I gradually made sense of those sounds:

"Here we are at a super-secret government installation in the desert outside of Las Vegas near Groom Lake. What we've witness rising over the Papoose Mountains confirms what many have long suspected and the government has long denied."

The voice was John's. He stood in front of the video camera that he'd mounted on the tripod, and he was sounding very much like a news reporter doing one of those live-from-scene broadcasts. The camera's light cast him in a bluish glow, and just outside that light, I could see Gene standing, still scanning the sky, his eyes ablaze and his smile a wide gash across his face.

"Amazing," Gene said. "A-Maz-Zing!" He lightly tapped the side of his temple with the flat of his hand and kept his head wobbling for a few seconds.

John was still narrating his story, and Gene came up to me with his hand out, "Thank you. Thanks for this, Bob. I can't tell you what this means to me. That you shared this."

I was glad to have done it, and told Gene just that. Then he went on, "I know that maybe John won't say this. You know how he can sometimes be. I think I know why you invited him, too. Guy's spent a lot of time believing in this stuff, catching hell for it sometimes. You let him see what he's believed in and wanted to see. That's a nice thing, you did."

I had considered John's situation before, but Gene made me understand things on a deeper level. I could empathize with John. I knew some things and if events hadn't unfolded like they had, I might not have ever been able to share what I knew with anyone else. Little did I know just how far the parallels between John's plight and mine would extend.

I joined John and Gene in loading up the gear. Tracy was still standing in place, arms wrapped around herself, staring up at the sky.

"You think it helped?" Gene asked me.

"What do you mean?"

He nodded toward Tracy. "Had to be a little weird for the two of you. Leaving the house on short notice all the time things like that. Seeing is believing."

Maybe on some subconscious level I had wanted to prove to Tracy that my absences from home were for the reasons I'd told her. This might have made me a bad husband, but at the time, I really didn't have much of a sense that my work at S4 had put much of a strain on our relationship. I believed she understood what was necessary and believed her when she told me that she trusted me. Still, it seemed odd to me then, and remains so until today, that Tracy hadn't said a whole lot while watching the demonstration or after. She'd expressed some concern when I first told her of the plan and what it might mean for me and for us if I violated the security agreements I'd entered into. At least after the high performance test she could better appreciate what was at risk.

We had a long drive ahead of us, and we all were early risers, so we quickly got back into the RV for the drive home. I think that we'd all gotten a jolt of adrenaline through our systems, and were paying the price for it afterwards with lethargy.

A few days later, I let Gene know that I was planning to go out to the site again. This time I had invited another friend along. John couldn't make it. I was a little more concerned that time about being watched and tailed, so Gene and I worked out a plan to use a rental car to drive out there. We debated about meeting points and drop-off locations, working ourselves up into a bit of a state for what proved to be no good reason. Still, better to be safe than sorry. There was a kind of Three Stooges quality to our planning, as if we mixed up the two CIAs—the Culinary Institute of America and the Central Intelligence Agency. Tracy was going to join us again, and maybe it was the overlay of fear that clung to her that had Gene and me in a bit of a tizzy.

Some of that was due to the one discussion that the participants in the first trip did have on the way home. My reluctance to write about it is a reflection of what they discussed and my long-standing lack of any real focus on this issue. Gene had remarked that the entire time the flight was going on, I seemed lost in my own little world. That was true. I was fascinated to see the craft in flight and was trying to visualize how the systems I saw were capable of producing that kind of performance.

John reminded me of a remark that he had made at the time (while the craft was still in the air) and that I didn't even remember him making.

"You were still just standing there with your tunnel vision goggles on, and I said, 'At minimum that thing's got be going 700 mph. Then it's stopping on a dime. Then it's back up to 700 mph again. Can you imagine what that kind of lateral acceleration and deceleration would do the being inside it?'"

"You're right, I don't remember you saying that. And you're right. Essentially that would be the equivalent of hitting a wall at 700 mph. I don't know for sure if the human exoskeleton could sustain that kind of velocity, of having the organs sloshing around. I don't know if they'd come out, but they'd certainly be crushed."

John raised his finger in the air. "Precisely what I was trying to get you to conclude. A *human* body couldn't survive that speed. I'm not talking about the human body keeping its guts intact; I'm talking about there not being a human body in there at all. Some other being had to be piloting that thing, no doubt."

Gene said, "I don't know about that for sure, but it did get me thinking. I know that you said, Bob, that there were nine other of that kind of craft out in the hangar. I was just sitting here wondering what those beings' lives were like. Did some of them work at the factory or whatever that produced them? Did they go home to mate at the end of the day and be joined by their kids who went to school?"

I considered that for a minute. "I have to confess to you guys. To me that's the least interesting part of this. I didn't consider any of that at all while watching the test. The machine was the thing for me from the beginning and it still is. I don't know if I believe what I read in those briefings. I admit I didn't read them all, but still what fascinates me the most is the technology. The rest . . ." I let the words drift along the highway to scatter with the wind.

In the days after witnessing that first high-performance flight test, I'd been taken up with other ideas than the life of the beings that created those craft and how the craft arrived here. In some ways, that was immaterial. As I've said many times, the allure for me was the immensity of the power that propulsion system produced. What stymied me was the realization that it produced that power in a nearly reactionless fashion. As near as we could determine, apparently no nuclear reaction was going on—we could detect no change in the identity or characteristics of an atomic nucleus that resulted from it being bombarded with an energetic particle, and there were no fission or fusion products left over from a reaction. As near as we could determine, no chemical reaction was taking place—we detected no rearrangement of the molecular or ionic structure of a substance that produced a new one with a different chemical identity. No combustion. No decomposition. No synthesis.

We determined that the propulsion system bent light and distorted gravity, but that meant that we were dealing with tougher concepts than the kinds of reactions above. Most people could understand decomposition. They'd either seen it or smelled it. But when you were talking about the disruption of gravity, you couldn't visualize or smell a rotting piece of meat. Instead you had to play around with more abstract concepts like time and distance. For most us, those are firm parts of our reality. I look at a clock and see what time it is. I don't think about how that clock is, in a sense, an artificial construct. We as humans created the concept of time as existing in blocks of twenty-four hours. It's a good construct, but it is still artificial. The same with space and distance. I drove twelve miles to get from my house to my job. But how we broke up that distance into miles is completely arbitrary. That's why there's a metric system and an English system that give you different totals. Same "distance" but different numbers entirely. We've all agreed to use these systems of measurements but in trying to explain how this craft moved, how it bent gravity, you'd also have to understand that our usual, everyday conceptions of time and space and distance weren't going to serve us very well.

John and Gene had joked about the need for greater minds to solve the question of where is the middle of nowhere. The same was true of coming up with an explanation that really and truly and accurately described how those craft, and that one craft in particular, moved from one location to the next. While I spent the next week working, I anticipated that my fellow "spies" were going to look to me to help them grasp an explanation for what they saw.

For the 29th of March, 1989, the mean temperature was 70 degrees with a low of 57 and a high of 80. Those facts were noted and recorded. We set out to note and record what we believed would be evidence of the craft's capabilities and existence. Tracy and I took separate cars, Gene picked up the rental car, and we all met at the rendezvous point; Gene and I took more circuitous routes than necessary. Our fourth, a man I'll refer to as Jason, met us along the way just of off Interstate 15. We proceeded as before and set up our viewing area. As happened the first time, the light from the craft glowed in the distance, did its staircase maneuver and then others. I heard Gene ask, "Did you see that move it just did? It went *vroom, whoom!*"

"No I didn't, I said," I had just set up the video camera, capturing Gene's words and my response.

A few seconds later Gene said, "Look at how light it's getting."

Jason had joined me at the camera and I was still working on the focus. We exchanged a few words about whether or not we were getting anything on the view screen. Finally, satisfied that the camera was capturing the increasingly bright light, I stepped way and watched with a naked eye.

The craft hovered for a bit. I blinked and it had moved closer, the light increasing in intensity. The pattern repeated itself.

"I think it's coming at us," Gene said.

"It is, isn't it?" Jason said. "That's pretty cool."

By the time they were finished saying those things, the light had grown much larger and much, much closer. We all scrambled behind the trunk of the car, crouching and looking skyward.

Gene looked at me, startled. "What are you doing? You're never scared! You sit at home making nitro like it's pesto!"

I had to think about it for a second. I wasn't really frightened. I told him, "I'm a human being. Instincts kick in. Bright, glowing object in the sky moving my way fast, I go for cover."

"That's reassuring," Gene said. I wasn't sure if he meant that I was human and had natural reactions or something else. I didn't think any more about it.

"As the power output increases, the intensity of the light increases," I told them all. The light wasn't eye-searing, but it was somewhat painful to look at. We were able to speak in a normal conservational volume since the craft produced no discernible sound. It was still well above our position.

"It was like I couldn't see it move," Jason said. "One second it was there. The next second it was over there. Almost like a strobe effect or something."

"Looks like it would be a fun ride," Gene said, "Just be sure to keep your sphincter closed tight."

"Only you would say that, Gene," I told him. And then, cutting him off before he could ask, "No, I don't know if they have sphincters."

"Well, all I can say," Jason added as he laughed, "Mine is quivering right now."

"You weren't completely off track about the strobe effect. As the propulsion system produces more power, that glow is brighter. The gravitational effect also disturbs light, time, and space," I told Jason.

"Breathtaking," he said. "Stunning. My face feels frozen from having been smiling for so long."

By this time the lights had darted back over the mountain, still above them, before the sky above S4 went dark.

"Worth the drive out here?" Gene asked Jason.

"Hell, yes," he said.

"I've seen it twice now and this time was even better with the kind of fly over, or at least fly at us view that we got." Gene paused, "I don't know, Bob, is it right to use the word fly? I mean people call them flying saucers but how that thing moves doesn't really compare."

"I guess that 'fly' is the right word if you think about something being above the ground and moving."

"How can it possibly do what it did?" Jason asked.

We all stood in a tight semicircle, collapsing tripods and folding chairs. I'd been thinking about this moment for a while, how I might be able to provide a visual for everyone to answer as best I could, or at least provide a visual for the concept. I took a five dollar bill out of my wallet and held it up for everyone like a magician about to do a trick.

"So, imagine that this bill is the universe. If I take a pen and put a mark here at the far edge of the bill, that represents a starting point. When we traditionally think of time and space and travel, that dot would move incrementally across the bill from one edge to the other or in some other path. For the sake of argument, let's just say it is going to move directly across in a straight line. To do that, to make you be able to see that, I'd have to take my pen and, in essence, make a whole series of dots from one edge to the other. In other words, I'd draw a line across the bill. That's how we experience movement as a series of moves, one after another, across a surface."

"Got it," someone said.

"Well, with this craft's propulsion system capable of gravitational change, it's like it folds the bill in a series of moves bringing that far corner where my line eventually ended, closer and closer to it until it was 'across' the span of distance. Only it did it far faster than you or I could make the moves to fold that bill."

"I see. That makes sense." Gene said.

Jason nodded.

Tracy nodded as well.

"The thing is, we can all understand that on a certain level. I mean, that's a rudimentary explanation of it. Astrophysicists and others might have more technical, and, more elegant, even, ways to explain what I just did. But the most important thing about all of this is that here on Earth, we can conceptualize this, but we can't, or at least to this point, haven't been able to produce a machine capable of doing what that craft does. We simply haven't. Not by a long shot. And that's what I was being asked to help do."

"And if we could do what that thing did?" Gene asked.

"We'd be masters of the universe. I hope that we'd use it for good, but we could create unbelievably powerful weapons of mass destruction as well. We could become destroyers of worlds. It boggles my mind to think of what it would mean to be able to generate that kind of power over gravity."

"Antigravity?" Jason asked.

"I suppose you could call it that, but what's the opposite of gravity? What does that mean?"

"I see your point," he said, and then added, stating the thought for me, "At some point language just kind of falls short of all this, doesn't it?"

"Speaking of falling," Gene said, "I'm about to collapse. Let's get home."

As was true after the first test flight we all saw, we didn't speak much about what we'd seen on the drive home or in the days after. Part of that was because of the secretive nature of our visits, but mostly it was due to the simple fact that we all had lives and jobs and families to go back to. It was kind of funny to think of it terms of the capabilities that craft had over space and time, but the world didn't stop turning because of what we'd witnessed. We just had to keep on moving forward. To that end, I tried again to contact Dennis and never got through, never got a return call to my messages.

Strange to think that I was part of what seemed to be some unraveling of one of the great mysteries of the world, but I couldn't get a guy to return a simple phone call. I suppose that's what I deserved for choosing to do government work. No force of nature, terrestrial or extraterrestrial, could compete with that kind of stubbornness.

CHAPTER EIGHT

SOME SAY THAT THE THIRD TIME'S THE CHARM. IN this case that was true, but our case the charm was used to break the spell that had me in thrall to the work I'd been doing at Area 51's S4 site. In the intervening weeks since I'd been last called in to observe the test flights, I'd added to my list of frustrations the nature of how the work I was doing there was being conducted. It had troubled me from the beginning. In science, you should be able to take a linear approach to problem solving. You set up an experiment or your thinking with a clear beginning, middle, and end. You do as much reading as you can, familiarize yourself with the work that others have done, etc. We all learned about the scientific method in school, and it's a good model that has stood the test of time, but at S4 and working for whomever it was that I was really working for in the military and government, I'd been placed in a situation where the linear was out the window, where the scientific method didn't consist of a series of defined steps but was more scattershot. In a way, our approach to the problem of reverse-engineering the propulsion system was like what a diagram of that high performance test flight and the craft's appearance would have looked like: first here, then there, then up, then down. I'd really had it with every aspect of the job, if I could properly refer to it as a job.

And wouldn't you know it, just when I'd reached that point, they reached out to me and called me again to let me know that I should report the next day to work again at S4. I'd been so used to the regularity of the woman on the phone's patter, the it is now X time and you need to report at Y time, and that Y time being just an hour or two from X time, that the *next day* really threw me for a loop. Maybe it was an act of passive aggressiveness on my part, but I simply said, "Thanks," and hung up the phone with no intent whatsoever of going in as instructed. No one had bothered to respond to my messages, not even to say, "We can't tell you anything now, but we will be in touch when we have more information for you." Instead, they answered with silence. I was going to respond similarly, but with absence. I also wondered why the next day. Had they observed us somehow? The car outside the house and the following was irregular at that point, but then again, maybe they'd changed up their tactics? Who knew. If they *had* followed us, I thought, so what? We were on public land violating no laws; of course, I was violating all of the agreements I'd signed but desperate times call for desperate measures.

On April 2nd, another Wednesday and the day of the call to report back, Gene, Tracy, Jason, and Tracy's sister Kristen joined me on another trip up to Groom Lake. You might think that the novelty had worn off by then, but it hadn't. When the craft flew towards us that second time, and we'd all ducked behind the car, a new dimension had been added. Who knew what other elements to the high performance test might be added? What additional information could I glean from witnessing the craft in operation. At that point, obviously, I didn't care about how that knowledge might help me help them, but I wanted to know for myself the answer to the question of how the systems functioned.

We took more precautions on the trip out there. We made frequent stops and diverted from the highway a couple of times and looped around the interchange to see if anyone was following. That bit of trickery added to the excitement. We talked about a number of things, the subject getting serious only once when Gene mentioned that the oil spill near Valdez, Alaska, was spreading, and residents there on the Gulf of Alaska were up in arms. For my part, I was more interested in an item that appeared revealing that scientists at Brigham Young University had fused heavy forms of hydrogen into helium at room temperature. They hoped that the process, "piezoelectric fusion," in which a heavy form of hydrogen is electrically infused into titanium or palladium, might lead to a viable power source some day. They believed that they were still a long way off from it being a viable source of energy, but given what had happened with the oil spill, more and more concerns about greenhouse gasses being expressed, I was at least hopeful. Obviously, I also thought of it in terms of the propulsion system I'd been working on, how it gave off no real heat during its reactionless operation.

I saw Tracy roll her eyes a bit as she looked at her sister, they both nodded, and said, "Okay, Dad." I laughed, grateful that their scientist father had instilled in them a tolerance for that kind of talk.

Eventually, the sister's conversation turned to who might want to join them to see the Tom Hanks film *The 'Burbs*. I knew that they were teasing us, but the point was clear. Lighten up. They were both highly intelligent women, but this was an outing, an adventure and not a part of a seminar.

To avoid detection and vary our pattern, we drove down an even more isolated road off of Groom Lake, deeper into the desert on a track that the ranchers used to ferry their cattle. We were careful to still be on a public road and still have a good vantage point from which to view the test flight. Our growing anticipation had us all nearly giddy. We joked around with one another, and it was good to leave behind the burden of my decision to end my relationship with Dennis and the rest of the people who had anything to do with S4.

We returned to the theme of our ill-preparedness to be spies and to carry out our operation in a manner that even approached stealth.

"And you were the one who gave John so much crap the first time about not having enough transmission fluid," I said to Gene.

"Yeah, but, he's a pilot," he said again defensively, knowing what was coming next.

"But to not have enough gas in the car to even get us onto the Interstate?" Jason added, referring to part of our plan to stop shortly after meeting up to see if we were being tailed.

"All part of the plan. That quick pit stop was designed to suss out the situation."

"What's a suss?"

"Don't you mean 'assess?'"

"Or maybe that you're an ass," Jason said, eliciting an appreciative laugh from the rest of us for his word play. By this time, we had rolled to a stop and were setting up our viewing area.

Suddenly, we heard a soft thud and then saw a greenish round light rolling in front of us. A moment later, we all stood there with a deer-in-the-headlights look on our faces as a car parked no more than twenty feet from our position switched on its high-beams. I was at the rear of our formation, mostly hidden. I ducked down, knowing that of all of us gathered there, I was the one who was in the most vulnerable position if we were to be detained.

"What are you folks doing out here?" I heard someone say.

"Who's asking," Gene said.

"Installation security," a different voice replied, "You are on military property. I'm going to have to ask you to leave. If you don't comply, there's Nevada state troopers in the area that I'll radio."

At that point, I decided it was best for me to get out of there. I'd asked Gene a while ago to keep my gun in his car. I eased the door open, we'd switched off the dome light ahead of time to avoid detection, and eased the gun out of the glove box. Using the cover of the everyone else's voices telling the security personnel that we were on public roads and not violating any statutes, I hid out in the bush away from the car.

After a brief protest, our group said that they would do as asked and leave. I watched as the security detail, the two men, got back in their car and waited. Gene got everyone in his car and within a few seconds, they took off slowly in the opposite direction of the security vehicle. I waited for a bit until the security car began to move away. Once their taillights began to dim, I scrambled along in the underbrush and soon caught up with Gene and the rest of them and got inside the vehicle.

We all bemoaned our fate and grumbled about not being able to see the test flight, but we also all tacitly understood the precarious position I was in and to a lesser extent, the jeopardy I'd placed them in. We were still rolling along slowly, we crested an incline and Gene muttered, "Oh. Shit."

Ahead of us, a pair of headlights stared at us. A few seconds later, its police lights flashed.

"This might be funny, except it isn't," Jason said. "Like being back in high school and being hassled for trying to find a place to drink."

A uniformed officer stood in the middle of the road with a flashlight in one hand. As we neared him, he raised the other hand palm up. We stopped.

"Where are you going?" he asked. I knew something was up when he didn't ask to see license and registration immediately.

"Just saw some lights in the sky from the interstate and came out here to get a better look," Gene said. "We understood this to be public property. No signs for trespassing or anything like that."

The officer trained the flashlight all around the inside of the car.

"Shut off your ignition. Stay right here."

We did as instructed, speculating about what the cop was going to do. We saw him speaking into his radio's microphone. After a brief conversation he returned to the driver's side of Gene's car.

"There's five of you in this vehicle."

"Yes. That's right."

"I got a report from the security guys that they saw four people in a vehicle matching this description. You got an explanation for that?"

Jason spoke up from the back seat, "Well, it's dark. Those guys were just security men, you know. They came up on us and one of them dropped their night vision light. Not exactly the most competent guys in the world."

All of us in the car laughed at that insinuation. The police officer didn't.

"I'm sure they can count. Four when they stopped you. Five now. I'd like to see some identification from all of you. All *five* of you, please."

Each of us handed our driver's license to Gene who collected them and handed them over. The policeman looked at each one. Then the rifled through the deck, then walked around the car and used the flashlight to illuminate our faces to make sure that our faces matched the photos. I had gotten into the driver's side seat and had returned the gun to the glove box. I wasn't concerned about having it. I'd registered it and was legally allowed to transport it in a vehicle. I was concerned about what we had in the trunk. It was nothing illegal—John's telescope, the spectroscope I'd finally remembered to bring along, a Geiger counter, and a few other scientific instruments. Any one of those items would destroy our story about just happening by and seeing lights.

Of course, the police officer asked if he could search the car. This was just after he asked us all to exit the vehicle. We complied with that request, but not with the search.

"Well," he said, "I have to tell you, this doesn't look good for you and won't if you have to appear in court. What that tells me is that you've got something to hide."

Kristen spoke up. "Officer, you said that this was a voluntary search. Our not granting you that access isn't evidence of probable cause. I'm fairly conversant with the statues that apply here, so unless you can explain to us why it is that you've decided that this is no longer a case of us—"

The officer held up his hand, "Then I'm going to my car and I'm going to call a wrecker and have you towed out of here."

Kristen took a few steps forward, "No. You're not calling a wrecker. You can detain us for up to an hour, but that's it. You are not searching the car and you are not getting us towed. You have no probable cause. If you insist on doing what you say you're going to do, then you're going to have a lot of explaining to do to a judge, to your superiors, and I don't think you're going to like what they have to say."

The officer walked away. Tracy smiled animatedly, the first real genuine expression of pleasure I'd seen in a while. "Way to go, sis. Those law classes of yours are paying off."

"I may never become a paralegal, but at least I can sound like I know what I'm talking about when dealing with these guys. I can't believe he thought he'd scare us into letting him search us like that."

"Thanks, Kristen," I said. We could have stood there all night just saying 'no' to him," I said. In an odd way, being out in the desert, watching the police car's mars lights illuminate the landscape felt so much stranger to me than if we'd all been watching an alien spacecraft flying around. I'd been bored and unsatisfied with working as a photo processor. Life certainly wasn't boring anymore, but I wondered then if it was worth all this stress. There had to be some middle ground. I remembered my mother using the term, "happy medium," when I was a kid and telling her that maybe an "unhappy large" was more what I was going to be on the lookout for in my life.

Seeing the faces of everyone gathered around me heartened me. None of them were pointing fingers at me and blaming me for possibly getting them into trouble or for wasting their time. They knew what I'd put at risk in sharing the information I had with them. They were all doing their best to protect me. Despite the obvious potential complications, it felt good to be there with a group who were all working together.

After about five minutes spent in his cruiser, the police officer approached us. Checking to be sure he was returning the right licenses to the right individuals, he handed back everyone's identification but mine. He held onto the last one, looked at me, and then said with a tone that I had difficultly deciphering, something between sarcasm and admiration, "I guess they know who you are down there," he nodded toward the base. He handed me my license, stepped back and said, "You're all free to go. Drive safe and have a good night."

We all breathed a sigh of relief and got back into the car. On the ride back, we didn't talk much at all. Gene asked the question that was on my mind, and I assumed on everyone else's. "What next?"

"We'll see," I said. "I'm beyond trying to figure out what's going to happen next."

For a few minutes we tried to come up with a plausible explanation for our presence out there. It soon became clear that it was pointless to even try. It was so obvious what was going on that it was a waste of time and mental energy to go through an excuse-making exercise. A lot of thoughts passed through my mind as we headed toward Las Vegas, but I never considered that one of the group I trusted had broken that unspoken bond and said something that might have drawn the security guards' attention to that area. I was sure of that and little else. I remembered a conversation I had with Dennis after learning about the scientists and engineers who'd been killed while working with the reactor. I would have thought that an incident of that type would make the news. When I asked Dennis why it hadn't, he simply said, "We took care of that."

I had understood before that these men had control over every bit of information. If something was going to happen to me, no one would ever know the truth of how I came to lose my life. Oddly, that didn't trouble my sleep that night, and I woke up feeling more rested than I had in a while.

By sunset the next day, I knew what was next. I answered the phone and Dennis Mariani was on the line.

"I'll be at your home in twenty-five minutes. We need to meet."

I knew it was useless to make any kind of excuse, so I said, "I'll be here."

When I answered the door to find Dennis standing on the porch, I was surprised to see that there was no vehicle other than my car in the driveway.

"You're going to be driving," he said while looking at his watch. "Let's go."

After I got through buckling in, I asked him, "Where to?"

"Indian Springs."

Though he didn't elaborate, I understood that he meant Indian Springs Air Force Base. It was located just north of Las Vegas. I had thought it was deactivated and that troubled me for a few beats. Why was I going to an abandoned military base alone and at night? It all seemed too sinister, then I remembered seeing some craft taking off from there every now and then. Maybe there'd be a full staff on site, people who would be able to account for my presence there.

We drove along in silence, the tension thick. I was trying to stay calm and not let on that I was thinking too much about what was going on, not let him see that I was on edge.

"Didn't they once use Indian Springs back in the day to aid nuclear testing? I thought I read something about sniffer planes being launched from there."

Dennis grunted an indistinct response.

For the remainder of the hour drive, neither of us spoke. I squirmed in my seat, glancing nervously into the rearview mirror, dreading what was to come. As we all had at the start of the drive back home after being detained by the police, I frantically tried to come up with a reason why I'd been out there with my friends and family. Eventually, I gave up. By the time I saw the first signs for the base, I was resigned to doing this—all I could do was fall back on the truth.

Once we were through the gates and then into an office, Dennis was joined by two other men, both armed. They looked like they were part of the security team out at S4. Dennis made it apparent that he was really pissed off. He sat down in a chair and pulled it close to mine. With his face inches from mine he said, "When we told you this was a highly classified project, how did that get twisted in your mind to think that you could tell your friends about it?"

Something about the way he phrased it and the absurdity of the premise made me smile and chuckle a little bit. I truly thought at first that Dennis was injecting a note of levity into the situation. I was wrong.

The man seated next to me reached for his weapon and said, "That's probably not funny."

Dennis continued. Though he didn't raise his voice much, his accusatory tone was sharp, and he seemed as if he was biting off each of the sounds he made. I sat there eyeing a black box on the table in front of me. Was it a microphone—part of some intercom system—or a recording device?

Do you not understand the nature of the agreements you signed?

Are you not aware of the consequences of violating those agreements?

After a few minutes of simply sitting there and essentially being pummeled by the comments and questions, I was asked the most obvious questions, the first one that seemed to be one Dennis wanted me to answer: What was I doing out there and why was I there with those people?

"I wasn't sure what you guys were going to do with me. Once you stopped calling me in to work in the lab."

Even before I could finish that second statement, I sensed that this was the wrong tack to take. Each of the men displayed body language that this was not the avenue to go down—arms folded across a chest, a lean back in the chair, a lean forward and a slow, nearly imperceptible shake of the head, an eyebrow quivering and then lifting.

Thinking quickly on the go, I then smiled and said, "Look. Everybody knows that something to do with experimental aircraft and testing goes on out there. If you live in Vegas, you know that. So, when they heard I had a job out there, they assumed. All we did was go out there and watch the lights in the sky. I figured that would answer any of the questions they had, alleviate their curiosity. Do a little disinformation. No big deal. No breach of security. I didn't tell them anything."

Once I got started talking a kind of mental momentum took over. By the time I was halfway into it, I'd almost convinced myself that what I was saying was true. At the same time, I wondered about my assumption that these guys were part of the security group. Would they have been allowed to hear what Dennis was talking about with me?

"Don't you understand that this project is far more important than any one single person?" Dennis asked. "More important than your life. More important than mine."

"Like I said, they knew that things were going on with aircraft," then, uncertain of what the other two men and careful not to say anything about the alien nature of the propulsion system and craft, I added, "These were lay people. The general public. They wouldn't have been able to figure out anything but what I told them. And I told them nothing."

"Nothing?" Dennis jumped on that word immediately.

"Essentially nothing. Just that there were flights——"

"What do mean 'essentially nothing?'" The guy to my left said.

"If you let me finish——"

"I think you're finished," he said.

Dennis intervened, "And just what is it that you think we're going to do with you and your friends."

"Nothing with them. I figured I would go back to work on the project."

"You smiling again? You think this is funny? Don't you understand how serious this is?" The interrupter was at it again. Dennis sat back as if to get a better view of the show.

I wasn't aware that I had been smiling. Maybe under the stress of the situation I had. I knew that I had to fight the urge to say that I understood the gravity of the situation.

"This isn't funny. This is no joke," the Interrupter went on. "Well, maybe Mr. Lazar doesn't understand what the rules are around here."

"They didn't see anything. A few lights in the sky. Oh, what a big deal. A few lights in the sky."

The interrupter must not have like my tone, he nudged the gun against me, and I lost my composure for a bit, "This is ridiculous. It's not like I told them that what they were seeing was aliens flying around in their saucers. This is—" I stopped myself. Dennis looked like I had kicked him in the balls. He recovered quickly and said to the Interrupter, "Please leave the room."

Once the man left, Dennis switched off the device on the desk. He tore a sheet off the legal pad that had been sitting there, presumably the whole time though it hadn't registered with me until then, I was so shaken.

"Write down their names and their contact information."

I hated the idea of having to do that. It was one thing for the police to have gotten it from their driver's licenses, but this was far more Judas-like an action. I felt my throat tighten, "They have that information already."

"I want you to write it," Dennis said, his voice reaching a level of menace that I hadn't heard before. I knew then that this writing had some kind of significance beyond the simple gathering of information. My writing those names and figures down would be an act of betrayal, a demonstration that I was capable of violating any kind of trust.

"I don't know all that information. I don't have people's addresses and phone numbers committed to memory." Dennis continued to glare at me. "Look, I'll admit it. I'm scared. I can't think straight right now."

Dennis tapped the sheet of paper, "Write it. Now."

"I'm telling you," I began, my voice sounding even to me like the bleating of a sheep. I stopped and looked at the third who was in the room. The entire time he hadn't spoken and he had displayed none of the gestures of disapproval or dismay that the others had and that I'd mentally catalogued earlier. He simply stared ahead impassively. Our eyes met and he continued to regard me as if I was no consequence to him whatsoever. That indicated to me that he was the one I should be most concerned about, he was the one most capable of doing me harm.

I was in such a quandary. They would know if I was giving them false information. That would lead them to believe that I had something to hide, that I was capable of other kinds of duplicity, that I had no regard for what they demanded of me, that I was loyal only to myself. If I didn't, I had something to hide, that I'd already committed some acts of duplicity, that I had no regard for what they demanded of me, that I was loyal to someone else. The last was true if I gave them the full information they asked for. Seeing no other way around it, I gave them some information, substituted Gene's office number for his home number, left many things blank and gave them an empty promise that I would get them what they needed.

I could neither be the stand-up guy who openly defied them or the man who was willing to lie down and be run over by their threats. Instead, I sat there in that middle ground, finding nothing to be happy about and wondering if the man's indifference was somehow a reflection of how I truly felt about myself. Was I too indifferent even to feel self-loathing?

"Next, I want you to tell us who you're working for," Dennis said, his lips pursing as he scanned the paper I'd returned to him.

"For you. For whoever you work for. For EG&G. I don't know. Naval Intelligence was on the check, so I guess them."

"You know that's not what I mean. I know that you wanted to sabotage the project. That's clear."

"No," I said recoiling in my chair. "That's not true at all. Not one bit of it."

This went on for a while until Dennis looked at his watch.

"We're done here. For now. Go back home. Wait for my call. We'll probably have you come back in."

His last statement was ambiguous—to work or to be further interrogated? I didn't dare ask.

I stood up, and as I did, the Interrogator came back in the room. He handed Dennis a stack of papers.

Dennis's eyes lit up. "That's right. That's right. I almost forgot." He looked at the papers for a few moments. "Sit down. Sit down." He waved the papers at me and they fluttered briefly before coming to rest.

"You do know you're wife's having an affair," Dennis stated flatly.

I sat there not willing to take the bait.

"Been going on for a while. Her flight instructor, a guy name Tony. Started back in February. A Valentine's Day thing, I guess."

"Why are you doing this?" I asked. "You could make anything up."

"Even these?" he said, holding up the papers. "These are the transcripts to every call between them. Funny how nearly all of them took place when you weren't around. Seemed like your schedule was ideal for them to meet up."

I tried to tell myself that what I'd said to Dennis was true: they could make anything up. But something deep in my gut told me that wasn't the case. I sat there feeling scooped out, like someone had taken a melon baller and eviscerated me from neck to nuts. When I told Dennis that I didn't want to see the transcripts, my voice sounded tinny, as if it was echoing inside that hollow space inside me. He slid them over anyway.

I sat for a few seconds not looking at them. In that brief span, I knew that the empty space inside me was all of the denial I'd kept stored in there leaching away. I'd noticed that Tracy had been distracted and distant. I'd noticed that it seemed like every time I left the house, Tracy did too. At one point early on in my time at S4, Jason had come over to the house. We were going to test some new rocket motors for the fireworks display we put on. We left in my car, and I wasn't sure if I had all the motors, couldn't remember if I'd left a box on a shelf in the garage. Rather than go all the way out into the desert and then discover they weren't there, I pulled over and got out of the car to check what was in the hatch. Everything was there. I climbed back in and checked the wing mirror before I pulled out. There was Tracy's car. She had to have seen me, but she just drove right past. I dismissed it then, but after that noted but tried not to focus on the dozens of other similar things that happened—phone calls ending abruptly, her changing plans at the last minute. I stopped myself from thinking anymore about them.

I wanted to look, but I didn't. I decided that I didn't need to look through a catalog of my despair and devastation selecting which item to punish myself with to suit my mood at the time.

"Well, thanks, Dennis," I said as I slid the papers back to him.

"No, keep them," he said, sounding more jovial than I'd ever heard him before. "Least we can do."

"Yeah," I said, channeling Gene at that moment. "Your tax dollars at work."

To say I walked out there feeling completely devastated would be an understatement. Tracy's betrayal gutted me. I didn't want to linger on the Indian Springs site, thinking that there were likely to be surveillance cameras. I imagined Dennis sitting there and laughing at me, a deep satisfaction animating his face, if he saw me showing any signs of the anguish I was feeling. I drove away, and on the way home, my sorrow and disillusionment weighed heavily on me. I felt as if I was back in the Puritan days, and someone had attached stones to my body and submersed me in water. I had no energy to struggle against those forces dragging me down. Tracy had once rescued me from the despair that I experienced following my first wife's death, had given me hope that life could be good again. Now that hope had diminished. As the days went on after Dennis's revelation, I became resigned. I truly didn't care whether I lived or died.

In addition to feeling betrayed and the sickening thoughts of her having a sexual relationship with another man and how clichéd the whole thing was—with a co-worker, with a man whose contact I'd encouraged and helped pay for, a flight instructor—I came to another gut-wrenching conclusion. One of the reasons why my security clearance was being delayed was due to Tracy's affair. Whoever was administrating the project and was in charge of evaluating my suitability for the security clearance knew this. A man whose wife is cheating on him is likely to find out about her infidelity at some point. He's likely to be emotionally unstable as a result. As a result, he wasn't the best candidate to be entrusted with the kinds of information that I was going to be, and had been, privy to. Not only had Tracy's affair destroyed me emotionally, it had severely damaged my chances of doing what I considered some of the most meaningful work I might ever do. As frustrated as I was with the processes in place at S4, the eventual product of our work could have proved to be life-altering for me and for millions of others. Just because I had sabotaged any chance I had of working there didn't mean that I wouldn't have liked to have continued to do the work. Complicated and conflicted seemed to be the buzz words for everything I experienced at S4.

As I neared our neighborhood, I remembered something else that Dennis had said. Maybe I could reapply in six to nine months. I don't know if he was simply throwing me crumbs or a lifeline. At the time, I envisioned that in the weeks and months to come I might have to rely on either of those two to help sustain me and keep me moving forward. I came home from the "visit" with Dennis to an empty house, and in a way that was fine with me. After everything else that had gone on, I didn't know if I had the energy to confront Tracy immediately with that I'd learned. I knew that I wanted to be on my game, so to speak, but this was anything but a game.

I sat in an easy chair in the living room with a drink I'd made, hoping I could be distracted by something mindless on the television. When that didn't work, I prowled around in my home "lab." I'd recently bought some electronics equipment I'd had my eye on for a while and I distractedly read through the manuals while sipping a second drink. Though alcohol should have numbed some of the pain I was feeling, it didn't. It eased my nerves a bit and after a while what was going on with Dennis receded, but the Tracy betrayal came rushing in. Simply put, I was pissed off. We'd meant a lot to each other and she'd been there for me during one of the darkest moments in my life. I didn't deserve this. Nothing I'd ever done or said could justify the pain she was inflicting on me. I would never do anything like that to her. If I thought that things weren't right between us, I would have said something, tried to work it out, and if that failed, we would go our separate ways. That would hurt, but nothing compared to this. I must have drifted off. I went upstairs to bed and Tracy was already there, sound asleep.

I got out of bed first the next morning and brewed a pot of coffee. Tracy came down while I was finishing the first cup. She filled her own and then mine. She sat down and looked at me, widened her eyes in expectation. I didn't say anything. She frowned and said sarcastically, "Good morning."

"No it isn't," I said. "Not by a longshot."

"What's going on? What happened?"

I told her what I knew and how I'd found out. She sat there looking at me, her expression steeling itself as she fiddled with a spoon.

"I don't know what to say," she said after I'd concluded. She shifted in her chair and brought one leg up underneath, turning away from me slightly.

"How about I'm sorry. I'm treating you like shit and I'm sorry?"

"I'm not treating you like shit. Things happen. I wasn't trying to—"

I immediately cut her off. "Don't even . . ."

"What?"

"Say that you weren't trying to hurt me. How did you expect me to feel? I've been doing nothing but trying to do the best for you. Busting my ass to make a life for us, make things as easy as possible for you."

"Running around at all hours of the night? Leaving me here all the time. What did you expect?"

That really got to me. "What did I expect?" I was nearly shouting at this point. "I expected you to understand that I was working on a highly sensitive project. I expected you to understand that the hours were strange. I didn't expect you to start fucking your flight instructor as thanks for me taking on a second job and trying to better things for us."

"It isn't just the sex."

"I don't want to hear that. I needed you, especially with everything else going on."

"Well, if you would have listened to me then maybe all this Dennis shit wouldn't be hitting the fan. I told you that you were playing with fire, but you love that. You love taking risks and flying in the face of authority."

I pounded my hand on the table. The utensil jumped and a tiny pool of cream shuddered. "Don't turn this around on me. I'm not the one who did anything wrong."

"I'm not turning this around. I'm saying that things are really complicated right now. I heard what you told me about Dennis hauling your ass in. I knew this was going to turn out bad. I knew it."

"Now's not the time."

"The hell it isn't? Yeah, we have to talk about marriage. We'll do that. But for now, don't you think we've got a few more pressing items on our agenda to deal with?"

The phone rang.

I locked eyes with Tracy and laughed ruefully. "This is how fucked up things are. I'm wondering about the lesser of two evils. Is that your boyfriend calling and wondering where you are or it's that Dennis luring me out somewhere so he can finish this whole thing off?"

Tracy and I went around and around for a while until neither of us was sure who was the monkey and who was the weasel and who was chasing whom. We at last agreed to table for the rest of the day any discussions about next steps—her calling it off, us calling it quits. Let things settle down. Let me focus on what I needed to do. After the reminder she made that I'd put us in jeopardy, that I'd, as she put it, "potentially screwed the pooch professionally at the very least," I seethed. Two wrongs don't make a right. In my mind, they raise the number of wrongs to another power, keep raising the number of sins of omission and commission until you can't see the forest for the trees, or the clichés.

A FEW DAYS PASSED WITHOUT INCIDENT. I GOT BACK INTO the regular routine of picking up rolls of film, processing them, and delivering the finished pictures. Wayne and his wife were still not certain if the business was right for them. That was fine with me. Now that it looked as if the S4 work was truly over, I needed some income. Having the familiar rhythms of work also soothed my admittedly raw nerves. The work also gave me an excuse to get out of the house, both to be away from Tracy and to allow me to meet face-to-face with Gene, John, and Jason, each separately. I told them about the meeting I'd had with Dennis. Our conversations were brief. We commiserated for a while. I told them that I had things under control. It was worrisome, of course, but I thought that having gone out there with each of them had provided me with a kind of insurance policy. Gratefully, they were each okay with that, and didn't feel I'd betrayed them or put them in jeopardy. John's opinion held a little more weight with me because of the kind of work he sometimes did for the government and the people he knew in the CIA. He suggested that I get the hell out of there right away. He'd asked if the surveillance teams were back, and I told him that they were.

"You can hunker down here in the bunker if you need to. In fact you should." He unlocked one of the gun cabinets in his study to show me a part of his collection, unnecessary since I already knew about them. "They'd be fools to try to come storming in here. A. It would be really hard for them to get in here, and B. It would be too high profile of a scene. That's not how they like to operate."

I thanked him and told him I'd consider it. And in fact, I did stay there for a few nights over the next couple of weeks. With me barely noticing it, we turned the page on April and moved into May.

I still had in mind the fact that Dennis was going to call me, and he'd instructed me to be around for that. At that stage, I didn't want to piss him off any more than I already had. By the start of the second week of May, nearly a month after the meeting with Dennis, I was driving back from a photo drop-off at Gene's office. I was on the east side of Las Vegas on Eastern Avenue, heading for home via Charleston Boulevard and then the freeway. A half-mile or so from the on-ramp, I noticed a car coming up from behind, closing somewhat quickly. I was driving the sports car that I'd taken out to Indian Springs, the 280-Z. Maybe it was a remnant of my jet-car days and being around people at drag strips, but my first assumption, confirmed when the car didn't overtake me but pulled alongside, was that this was someone looking for a race or otherwise messing around. Instinct took over, and I accelerated thinking I'd beat them to the on-ramp. I cut in front of them and they pulled alongside again. A quick sidelong glance allowed me to see two head shapes, nothing more distinct than that, in the car's front seat. We did this catch up, speed up thing one more time before we got to the on-ramp. I had to slow at that point, in order to make the turn. I was ahead of them and figured that was that.

It wasn't.

The on-ramp was a single lane, and the other car drove with two wheels on the shoulder and two in the dirt and grass alongside it. Adrenaline shot through me, as much because I wondered if I was going to witness a crash, and also from fear of who these guys were and what they might try to do to me. A shot rang out, and I felt the rear of my car slewing sideways. Instinctively, I turned the wheel hard left to correct and that took me off the ramp and into the grass. I came to a stop. I was paralyzed with fear at that moment. I sat with my arms, elbows locked, head pointed straight ahead, just waiting, certain that someone was going to come up to the window and fire another shot at me. I don't know how long I sat in that position, but by the time I looked around, all the dust had settled, the other car was nowhere in sight.

I eased the driver's side door open and got out. Traffic thumped by on the Boulevard and I could hear the faint *whish* and *whoosh* of interstate traffic in the distance, keeping time with thumps was my thudding heart. I walked around the back of the car and saw that my passenger side rear tire was flat and that a round hole was visible in the sidewall. I probed the opening with my finger, felt the strands of rubber and above that, nearer the tread, the steel belts. I determined that the tire still had enough structural integrity to support the car. I was no more than five miles or so from home. All I wanted to do was get out there and get to a safe place. By this time, the sun had gone down completely and it was full on dark, so I put on my emergency flashers. I made it onto the expressway and, doing no more than twenty to thirty miles an hour in the right lane made for home. I was grateful that a utility company truck pulled in behind me. The driver must have noticed the tire and my distress. He put on the flashing lights on the roof of his vehicle as well as his hazard lights. All of that flashing drew a lot of attention and reminded me a bit of the scene in the desert, but I was glad for the escort to my exit.

As I drove along, the reality of what had happened started to set in. I'd fired weapons often enough to know that it's hard to hit a target from a standstill. To be able to fire from a moving car into another moving car and hit a tire was either a stroke of luck—good luck for me that they hadn't hit the gas tank or me, and bad luck for them if either of those two options had been their intent. And bad luck for me if these guys had been aiming for that target and had succeeded in hitting it. Those guys would have had to be pros. Maybe all they wanted to do was warn me. If so, the message was received. I got another bit of a fright when I pulled off the expressway. The utility truck followed me off the ramp and all the way home. I briefly considered not going home and trying to lose them. I told myself to just calm down, and proceeded to James Lovell. When I signaled to enter my driveway, the truck flashed its bright lights in recognition and went on its way. I breathed another sigh of relief and gratitude.

I slumped at the wheel for a few seconds, my forehead resting on it. When I looked up, I noticed that there was another car, one unfamiliar to me, parked alongside me. Startled, I wondered if it was best to just remain inside or try to run for it. A few seconds later, a woman in a dark suit tight-walked over to me, waving an envelope in her hand. What was this? A processes server notifying me that I was being sued for divorce? A female assassin? Publisher's Clearinghouse coming to inform me that I was a lucky winner?

It turned out to be none of those. As she got closer, I recognized her as Joyce, another real estate appraiser. I was supposed to have stopped by her office after Gene's. In all the commotion I'd completely forgotten. I got out of the car and stepped into a cloud of perfume.

"Bob, I hope you don't mind. I live nearby and I was on my way to dinner. A date, actually, nice place off the strip, first time going there, so excited."

"It's okay Joyce. I owe you an apology."

"It's fine, Bob. Just fine. Just that I'll need these tomorrow. You know how it is, Bob."

I'd begun to wonder if she was going to Bob me to death.

"Sure, I do." I went to lean against the car, casually but feeling a bit lightheaded after the rush. I staggered a little given the car's leaning lopsidedness.

She looked at me, then down at her feet and then at me again, appraising. "Bob, your car looks a little funny. Or is your driveway sagging? I'd get that taken care of. Professional opinion, of course, Bob."

"No. Driveway's fine. Just that someone shot my tire out a little earlier."

"Oh, well. Only in Vegas," she handed the envelope with the film roll in it. "I really, really need these first thing in the morning, Bob."

I nodded. "Enjoy your evening, Joyce," I said, realizing that I was seriously outgunned in the name game.

As she drove away, I marveled at how she had acted, as if someone getting the tire of a car shot out while driving was just a normal part of the daily commute. But it wasn't. I was completely unnerved.

After the brief mini-absurdity with Joyce, I returned to the other absurdity in my life. Despite Joyce's assessment that these kinds of things happen in Vegas, they didn't. I could have encountered some random act of violence, but in my gut I sensed something more sinister lurking beneath the surface of a coincidence gone bad. Dennis had said they would be in touch. Was this the form that re-establishing of communication was going to take?

Given the circumstances, I abandoned my no-phone-calls rule. I immediately got in touch with Gene and John. John was emphatic. I had to get over to his place immediately. It was the safest location for me, and thought that together we could formulate a plan of action. I decided it was best to do as John instructed. Gene had been in favor all along of me getting away from the house, and potentially out of Las Vegas. In the weeks since I'd met with Dennis, the observers had been a constant presence, though they had stopped tailing me. By the time I'd arrived, it seemed as if John had already formulated the start of a plan.

"You protected yourself by telling us about what you were doing, right?"

"Yes."

"And that worked didn't it?"

"Well, John, I don't know about *worked*."

"You're still above ground aren't you? Still breathing?"

"That's true."

"So what I think is, is stick with the formula. Except you tweak it a little bit. Go wider with the information."

"How do I do that? Stand on a street corner like some loon with a days until doomsday sign?"

"That would work, but you're sign painting skills probably aren't up to snuff. I've seen your handwriting."

"I think you're right about going wider. I was thinking of that myself."

"Mind if I call George Knapp? Get him over here? I trust him."

"Fine."

George Knapp was a reporter for the local ABC affiliate in Las Vegas. He covered the usual stuff, but also did some investigative journalism into political corruption and the like. He also was one of the first journalists in the city to really poke around and into Area 51. He'd interviewed John before, and if John liked him and trusted him, even considering John's lack of a filter, that was good enough for me.

John got me a drink and escorted me into his office. I was joined later by George and his boss, the station's manager Bob Stovall. It seemed a bit strange to be face-to-face with a man who was in my living room a lot of nights. George had a straight-forward demeanor and a candor that put me at ease. John, on the other hand, didn't. He was looking out for my best interests, but he immediately took the lead in speaking with the two TV guys. He was acting as if he was my attorney, not as if he was asking for a favor, essentially. He started out by making demands, telling them I'd go on camera, wouldn't reveal my name, and I would only answer yes or no questions. That last bit got to me.

"It's not like I'm on trial, John. You don't have to worry about me saying too much and possibly incriminating myself." I can't believe, now, how naïve I was to believe that I wasn't going to be put on trial, not in the legal sense but in the court of public opinion. I shared more of my story with George and Bob, and eventually we came to an agreement. We'd do a piece that they would broadcast the following day. I wouldn't use my own name, they would only show me in silhouette to protect my identity, and I would give them responses that went beyond simple yes or no replies. I was concerned about the guys watching my house, so they left it up to me to give them the location, and they would send a truck over. I'd think about it overnight and see who would be willing to have this all go on. We chatted for a while longer then shook hands. John and I sat up long into the night speculating about how my coming forward on television would be received by Dennis and the powers that be. We knew I would be stirring up a shit storm in the desert, but kicking the sand up in someone's face was better than having my tires shot out from underneath me. Maybe it was the product of such a long day, but it took me a while to realize that a shit storm and a sand storm weren't quite the same thing, but John must have understood, except for one thing: he'd dozed off.

I made my way to the guest bedroom and figured that a shower would help ease some of the tension from my body. I stood under the spray with my eyes shut, the sound of the water on my skull a soothing and sibilant tone. I wondered if I was doing the right thing, half-marveled and half-bemoaned the fact that it had all come down to this. I wasn't prone to the "if only's" but as the room filled with steam and I could finally feel some weariness overtaking me, I wondered what it would have been like if I had been content to remain Bob the Photo Guy. And later, as I drifted off to sleep, I wondered after tomorrow who I would be known as. One thing was for certain; I wasn't going to be Bob the _____.

CHAPTER NINE

THERE ARE TIMES WHEN YOU JUST HAVE TO TRUST your gut instinct. There are times when you have to trust other people. When the two of those are working in concert, then there's a pretty good chance that you're heading in the right direction. Though I didn't know George Knapp at all before I met him, I trusted that he believed in my story and he believed that it needed to be made available for wider consumption. I'd gone back and forth on the issue of the public's right to know about the presence of these craft and our government and military's involvement with them. I wasn't, strictly speaking, anti-authoritarian. I didn't rebel against people in positions of power just because they were in power. What generally got me a bit riled up was when authorities injudiciously wielded power or didn't account for the fact that not everyone was incapable of behaving rationally. Again, that's going back to the idea of a widespread panic or serious social disruption resulting from revelations that there were other life forms in the universe and we'd had contact with them. I'm not sure I bought into that idea completely. I also understood that whoever was able to decrypt how this astounding technology worked needed, as the phrase goes, to use their powers for good and not evil. And if no one but a select few knew we had that power, then they would have greater rein to do with it what they wanted and without impunity. That was not good.

But as I woke up in John's house and thought more and more about what I'd committed to—the on-camera interview—my decision was based more on self-preservation than altruism. The blend wasn't so highly concentrated with self-interest than it had been, but it was still predominant. I wasn't trying to be a hero; I was simply trying to still *be*, to continue to exist.

In the passage of time, I can recount these events in a much more reasonable, much more logical fashion. At the time, I was in state of panic, wondering if my decision to go wide was going to fan the smoldering embers into a fire or throw enough water on them to douse them completely. I knew that I had the full support of George, Gene, and John. That was helpful, but in the end, only one person should be responsible for decisions like these—the one whose life is most immediately going to be affected.

To say that I discussed my decision with Tracy would be a gross overstatement. I *informed* her of what I was going to do, but I didn't ask for permission nor did I consult with her about how I should conduct myself. To her credit, she and her sister both decided to accompany me on the "set" of the interview. I realize that using the word "set," as in a movie "set" implies some kind of fictional aspect of what was going on, but we did stage the interview at my request. I mention this only because of how the day proceeded. I had agreed to appear in silhouette and so the camera operator who accompanied George, the sound guy, and George himself all had to figure out how to achieve the visual effects they wanted to make it look interesting and to abide by my request that my identity be protected.

Because, as Gene kept reminding me, John Lear's house was a kind of secure compound, we chose to do the interview at his place. I don't care who you are or where you live or what kind of relationship you have with your neighbors, if you drive along in a remote coverage TV news van, with its call letters and insignia emblazoned all over it into someone's driveway, you're going to draw a crowd of onlookers, even if the van disappears from view. Most of the time when I'd gone to visit John, I'd parked in his driveway. It wasn't possible to be seen from the street, so it was unlikely that anyone would be able to identify me when I came out of the house and got positioned to speak with George.

For that reason, on May 14, 1989, my supporters and I stood around waiting while the technicians did their thing. George had briefed me very generally about the kinds of questions he was going to ask, and I had to appear on camera for a couple of test shoots to make sure that I was framed properly and the lighting worked out. I couldn't resist one more poke at those in charge of the program, and I instructed George to refer to me by my chosen pseudonym—Dennis. I'm not sure if that qualifies as gallows humor, but I did have a small sense of what it might be like to be facing a moment so heavily freighted with consequences. I chose to look at the interview as an example of one of the most widely known principles of physics: Newton's first and second laws of motion. I could still act on what I'd set into motion, and even up until the last moment, when I was standing on my mark and George stood off camera and signaled me that it was a go, I still wasn't certain if I'd let any words come out of my mouth.

I did, and I found that the release of one was easily followed by another and then more and more. We talked for perhaps five minutes. I gave George the basics. The interview was more about the existence of the craft, my role at S4 generally, and involved no critique of the program, no mention of why it was, or how it was, that I had become embroiled in a dispute with my now former employees. When it was aired, we all gathered to watch it at John's house. The interview was paired down to just a couple of minutes at most. It hit all the high points, and I got a big kick out of seeing the name "Dennis" appear on screen beneath my shadowy image. I was glad that George had kept in the part about me stating that I believed the technology needed to be kept classified. The fact that we obtained the technology from an extraterrestrial civilization should be public knowledge. I'd finally settle on that as my stance, believed it at that moment, and continued to believe it ever since.

I had no idea how the piece was going to be received. I felt some relief that, again to revert to the Newton metaphor, I'd set an object in motion. That would at least govern, to some degree, what they did next and what I did next. Like I said, I've always been a lover of big reactions and explosions and that sort of thing. Being inert and observing things that are inert is, for the most part, tedious. I wanted to see how all of this was going to be resolved and I hoped that it would be so sooner rather than later. Either let me go back to my "old" life, or at least let me deal with the new one.

After the broadcast was over, I returned to my home on James Lovell. I didn't feel like I was in exile from my shattered marriage or my possibly ruined career as a scientist. As I saw it, I'd done nothing wrong, and shouldn't walk around feeling like a leper or a pariah. That feeling didn't last very long.

The phone rang, and I picked up. "Bob, do you have any idea what we're going to do to you now?"

"No."

It was Dennis. He didn't say anything else, and in a second or two, the connection terminated.

So much for any kind of certainty. So much for the laws of physics somehow applying here. So much for feeling safe and secure in my home.

Without a word to Tracy, I slipped out the door and returned to John's house where I spent the night. I lay in the dark and thought about all the ways that this should matter but concluded as the first gray light of morning came through the windows that living or dying was seldom a question of choice. It was often governed by forces inside and outside of ourselves, but I was too tired and too distraught by all that had gone on to really care which outcome was produced. As before, I also gave some thought to fleeing, but knowing who I was dealing with and powers they likely possessed, running away seemed fruitless. They'd be able to track me down if they wanted to. At least in Las Vegas, I had my core group of friends who had my best interest at heart.

Though I woke up the next morning and initially felt like I wanted to just sleep the rest of the day and many others away, duty called. Despite being freaked out the previous night by Dennis's call, I had remembered to grab a packet of photos that Gene needed. I set out early for his office. I stopped at a café for a quick bite to eat, and sat there nervously looking around the room for any signs of suspicious activity. I stopped doing that pretty quickly when I realized that I didn't have a firm grasp of what might constitute suspicious activity. I finished up and left, scanning the newspapers in the box outside the café. Nothing on the front page indicated that my story had been important enough to be above the fold and maybe not even anywhere in the paper. I chided myself for thinking I was such a big deal and headed over to Gene's office.

I walked in and another man was with him.

"Bob," Gene said. "This is Gene."

"How you doing?" the man said. "I'm an appraiser too. Weird isn't it? Gene. Gene. Like you stepped into another universe."

In fact the man looked nothing at all like Gene, his olive complexion and thinning black hair splayed across his pate in dark isosceles clumps.

"I know, how about this? Gene, Gene the dancing machine. Remember that show with Chuck Barris? The Gong Show?" The man said.

"Sorry. I don't."

Gene #2 narrowed his gaze and cocked his head. "Hey, this guy," he said, directing his remarks to Gene, "Sounds like the guy I saw on the news last night."

"No. I don't know. I saw it, but I don't hear the similarity. He's just the Photo Guy."

I shrugged and said, "Just the Photo Guy."

When we were alone, I said to Gene, "That's a bit scary. I thought about them using a voice disguiser but didn't think it was necessary. Now this?"

Gene tried to reassure me that it was a fluke, and he was eventually proved right. Over the course of the next few days and into June, no one else claimed that I sounded like the Alien Guy on TV. In the years to come, I'd learn that the brief bit of footage that aired that night eventually made its way to various locations around the globe. I thought I had some sense of the power of the media and mass communications back then, but didn't realize what a greater force it would become and how it would directly impact my life.

My friends took on a more low-tech campaign on my behalf. Gene and a few others, suspecting that his phone line was now being tapped as well, started to talk more openly over the phone about what I'd been doing. He called friends and family, enlarging the circle of those in the "know." The idea was simple. If Dennis and others decided that I was such a security risk that I needed to be killed, and more people knew what it was that I was doing, having those individuals with that knowledge could either act as a deterrent or increase the number of people would know the truth of why I had been killed. The fact that I was eliminated, we believed, would confirm the truth of what I had revealed. In their conversations with others, my inner circle also stated that they had written letters to members of various agencies with the government, etc. Those communications—the letters—would be far more difficult to track and interfere with. We felt, that simply giving the appearance of having communicated more widely was enough to keep those who might want to do me harm at bay. We were going to keep them guessing as to who knew what, and created our own kind of disinformation campaign.

I was still being followed and monitored, and Gene reported to me that he believed he had been tailed as well. I can't confirm this, and neither could Gene, but it certainly was within the realm of possibility. The kinds of gamesmanship that was going on was never fun or thrilling. It was always frightening and aggravating. I don't know if Dennis felt the same way, but eventually, on the third Saturday in June, the 17th, after having been contacted by Dennis that a "personal level" meeting was needed, we set a time and place for a face-to-face meeting. At my suggestion, we were to meet at eight in the evening at the Union Plaza casino in downtown Las Vegas. The more public the location, the better, and a Las Vegas casino on a Saturday night was sure to be crowded. As it happened, a friend and former colleague from Los Alamos, Joe Vaninetti, was in town. He knew what was going on generally with me. Gene and I brought him up to speed on the most recent events, and Joe helped us with our planning for the meeting with Dennis.

I wasn't about to go there alone. I also wasn't about to expose my friends to any more jeopardy than necessary. I didn't think that Dennis or whoever might have been with him or at whose urging the meeting was set up, intended to do me harm at the casino. But, as a scientist, you know that you should never jump to any conclusions. We devised a plan so that Joe and Gene could keep me in sight at all times, or at least be able to track my location. The day of the meeting I spent quietly at home reading. I'd resumed my activities at the library. Prior to going to work out at S4 and for most of my adult life, I hadn't earned enough money to afford subscriptions to all the scientific journals and periodicals that I wanted to read in order to keep current on latest advances. As I recall, even back then, the wonderful, though not highly technical magazine, *Nature*, cost nearly twenty-dollars an issue. Some of the less popular ones, like the *Journal of Technical Physics*, were somewhat obscure but its peer-reviewed articles were fascinating to me. I struggled with my focus that day, but being in a library had a calming effect on me. But you know you've got a lot on your mind when an article on microsecond plasma pulses in MW range can't hold your attention.

I sat there watching the odd assortment of characters who frequent the library. I suppose that people who don't live in Las Vegas and only visit the strip don't really think about the daily lives of its residents let alone libraries. In the days before the Internet exploded with information that we could access so easily, libraries were still an interesting gathering place for those who pursued esoteric as well as commonplace knowledge.

I thought a bit about the site of our meeting. The Union Plaza Hotel sat on the site of the former Union Pacific Railroad station. The hotel still had a train station as part of its structure. I sought out photos of the older version, built in 1940, and admired its Art Moderne, or Streamline Moderne, façade, a later variation of the Art Deco movement. In the photographs I saw, the station's single story front that faced Main Street looked like the diner that Edward Hopper depicted in his great painting, "Nighthawks." Something always captivated me about that painting: the couple sitting at the counter, the lone man with his back to the viewer, the counterman going about his business. It sometimes struck me as desperately lonely and at other times hopeful. Those people had someplace that they could go, even if they didn't speak, some place where they could find some fellowship. We don't build great cathedrals much anymore and our artists don't tend to depict them in their works the way painters centuries ago once did.

I was in a reflective mood, obviously, while also trying to make sense of my present. I can't say that I ever uttered the words, "Why me?" when it came to my dealings with Naval Intelligence, EG&G, and Dennis as the representative of some larger body of people. I simply accepted what had happened. And in those moments as I flipped through that book depicting some early scenes of Las Vegas and its development, I allowed myself to feel some sense of loss at what had transpired. I'd been privileged to work at S4 and had seen a technological marvel that defied my complete understanding.

In a way, I was like some pilgrim who traveled to Chartres in France in the thirteenth century and had seen the marvels of the flying buttress and wondered how it was done, and maybe worried that it might all come crashing down on my head, but believed that no, God, in his goodness was holding that high ceiling aloft. Or maybe I was like someone who lived long enough to come to Las Vegas and arrive at the Union Pacific Railroad station aboard a locomotive, powered by a means that was a marvel of its time: internal combustion and electromechanical. The streamliners of that era were capable of an astounding one hundred miles an hour. We'd made progress to be sure, and who was to say what a race from a distant galaxy might have been able to accomplish if they'd been around longer or hadn't been around as long, but possessed an intellect superior to our own?

As much as I said that I wanted all of *this* to be over and wondered how my meeting with Dennis might lead me toward or away from that goal, I was saddened by the end of the opportunity I'd been given. I was sad about the end of my marriage. I don't know if I had any real innocence left to be able to lose it, but I wished that in all that had gone on, there was more time to just simply marvel at it all, to take in the beauty and elegant simplicity of the propulsion system, to appreciate it for what it was and not think of it so much as a problem to be solved but a gift that I could take pleasure in.

I knew that the time for my appointment with Dennis was nearing. The library emitted a signal indicating it was last call. I slid the periodicals back on their shelves—in defiance of the request to return them to the front desk. I went home and walked through each of the rooms, thinking a bit about Tracy and our time there together. Sometimes things change, but they aren't always a sign of progress. Before I knew it, it was time to leave. Joe, Gene, and I all climbed into Joe's vehicle, and headed into downtown Vegas. By the time the glittering lights of the hotel were visible, my stomach was in knots. As much as I was concerned for the personal safety of us all, I knew that there were a lot of forms that the "Do you know what we can do to you?" question could take. I could easily imagine having all my financial records messed with, my education and employment history disappearing, all of which wouldn't be mere inconveniences but could really damage me and ruin my life. I felt like I'd already lost so much and was going to have to start over in so many ways, that it felt overwhelming.

Stepping into the noisy cacophony of a Vegas casino was a bit of sensory overload, given my exile at John's house and my time at the library. I truly did feel like I was under assault, that the loud noises, the flashing lights, the loud thrum of conversations pierced by sharp laughter produced a physical pain in my head and throughout my body that had me gritting my teeth. As per our plan, I'd entered alone and began to scout for Dennis. At three minute intervals, Gene and then Joe would enter the casino, chose strategically advantageous seats at the slot machines, and keep an eye on me. I'd described Dennis to them both, but I didn't want them approaching him. I kept up my reconnoitering the slot room and other parts of the casino, dodging waitresses and patrons in the narrow openings between the games and tables.

After waiting fifteen minutes, and wondering if the rules applied about a professor being late to a class applied, I found a staff member to assist me with a request. I asked that someone page Dennis Mariana and ask him to meet his party at the entrance to the blackjack room. Several minutes later, I saw Dennis working his way from the entrance toward where I'd been standing along a sidewall. He was part of a large clump of people trying to make their way deeper into the casino. He looked exactly as he always did. Unsmiling, eyes focused straight ahead, his neat moustache, hair, and erect posture hardly allowing him to blend in with the more casual and relaxed patrons. I walked toward him, made myself very clearly present in front of him, my heart rate climbing and my throat tightening.

Dennis walked right past me, not acknowledging me in the slightest. My eyes darted from side to side and I noticed another man, his presence very much like Dennis's as he stood stiffly along the wall not far from where I'd previously been standing. I looked at him again, stared at him, and realized that he was one of the security people I'd seen while at S4. It made sense to me that Dennis wasn't going to be alone. I saw that Joe and Gene were both looking in my direction. I tried to point out Dennis to them both, but knew that given the cluster of people moving through the area it was going to hard for them to positively identify which man he was. I walked over to Joe and then to Gene, letting them know what I wanted to do next.

"Gene, I'm going to go in there, into the blackjack room, and I want you to follow me. Stay out of sight, but keep an eye on Dennis."

"I saw him, Bob. I spotted a guy who I think matched the description."

"Good. But I want you to be more positive, undeniably positive about the identification. I'm going to speak with him, and I need you to witness that, to have zero doubt in your mind that the man I was talking to was Dennis."

"Got it."

Gene trailed behind me. To my right, I saw Dennis sitting at a table. I inclined my head toward him to let Gene know I'd spotted the guy. I also held up my hand briefly to remind Gene that I didn't want him to approach Dennis at all. It was nearly comical to see Dennis in that environment. He was sandwiched between two curvaceous women who were both enjoying their drinks and the festive atmosphere. I watched Dennis for a minute or so, and he never took his eyes off his cards, a remarkable feat of self-restraint. I circled the blackjack pit and came up behind Dennis. For some reason, by this time whatever anxiety I was feeling seemed to disappear.

The dealer eyed me suspiciously. He's seen me circling and looking at the table. I could have easily been eying everyone's cards. A stupid and blatantly obvious cheat, but not one I was working.

"Well, Dennis, you said you wanted to meet and here I am. What's the deal?" I spoke loud enough to insure that the dealer heard me. He was some six to eight feet away. I saw his expression soften at my words. Dennis didn't react at all. I waited a minute before I said to Dennis, "Dennis, what the hell is going on? What is this shit?"

I stared at the back of Dennis's head, noticed an ingrown hair that had begun to boil the skin in a small but angry red welt. Aware that creating some kind of scene was not wise, I walked away. Gene joined me on the edge of the pit, edging out from behind a slot machine. Dennis remained in our sightline.

"What the hell was that?" Gene asked.

"I don't know. More bullshit from them, from him."

"What do you want to do?" Gene asked.

"I don't know what we can do. He doesn't want to talk. He hasn't told me where we could meet, nothing."

"Do you want to tail him? Follow him to his car? At least get his license number?'

I was so frustrated and let down at that point, so sick of Dennis's arrogance, that I agreed to that plan, knowing it really wouldn't do us one bit of good. But at least it was something.

Dennis stood up from the table and merged into the crowd. We all moved to follow him, but in just a few seconds the crowd swallowed him up.

Gene and I made our way toward the front entrance, near Joe's position.

"Did you see anybody who looked like Dennis leaving?"

Joe shook his head and shrugged, "Not really."

"Not really or no?" I sounded more agitated than I really was.

"Sorry, Bob. Nobody that looked like your guy came past."

"Split up and find him?" Gene asked.

"Why not?" I said.

We each took off in solo pursuit. I checked all the tables, the bathrooms, even stuck my head inside a bar and a restaurant nearby. Not a single sign of him. After twenty to thirty minutes, I wandered back to the entrance. Gene and Joe eventually showed up.

"I could use a drink," Gene said.

"Not here," I told him.

We drove to a small tavern nearer my house than the joints along the strip or downtown.

"Why would he do that?" Gene asked. "He asked for the meeting. He had backup with the other guy."

"I guess we shouldn't assume that the other security guy was with Dennis. Easy enough to think he was, but not necessarily true."

Joe's face lit up in recognition of what I was suggesting. "So, Dennis says he wants to talk to you personally. He shows up, sees the same guy you do, now he's thinking that he's being watched. He talks to you, he's in the shit with somebody else above him."

"Could be," I said. I swirled the ice in my glass and held it up to the light, "Things are about as opaque as this."

"I'm not a gambling man when it comes to people's actions, but I'm pretty sure that Dennis will be back in touch," Gene said.

"I wonder if he's as confused by all of this as you are, Bob," Joe added.

"That may be. I've got no evidence to say one way or the other. Until he tells me it's so, it's all conjecture."

I was definitely disappointed at how the evening turned out. I knew that there were a lot more unpleasant alternative endings to it, but as with so much of what had gone on, this felt far too incomplete, far too fragmented to really feel satisfying, far too much like how life generally makes us all feel.

CHAPTER TEN

I F I WAS HOPING FOR CLOSURE ON THE EXPERIENCES I had with the people who hired me to reverse engineer that propulsion system, including Dennis, I wasn't going to get it. Not then. Not ever.

Unfortunately for me, EG&G, S4, and all the rest of that drama was just one part of the ongoing unravelling of my life. Even though the failed meeting with Dennis resulted in the end of the surveillance teams watching me and my house, I still took some precautions in the weeks and months to come. Among those was taking different routes when I went to the gym, went to Gene's office, or just continued to do my job processing film and the pick-ups and drop-offs that entailed. Even though there were no more incidents of my car being broken into or someone running me off the road and firing shots at my tires, those incidents had already exacted a toll. I felt like I constantly had to have my head on a swivel, and the paranoia those experiences induced contributed to my stress.

In the aftermath of that first broadcast, Tracy moved out of the house. That made me terribly sad, but I wasn't going to stop her. She'd betrayed me by having an affair, but that didn't mean that I couldn't work toward forgiving her and ultimately get there. Though she moved out, we continued to talk. We decided that we should speak with a counselor and made arrangements to do that. We sat down in his office and he asked us to describe what had been going on and how it was that we ended up in his office. All went well for a while, but I could see a cloud of disapproval or disbelief move across his face when Tracy talked about her affair and how my working with alien spacecraft and having government agents watching us all the time contributed to her infidelity. We each got a chance to talk about what we hoped to accomplish and what we'd been through. At the end of the session the therapist sat for a few seconds. He bridged his hands in front of himself and leaned back in his chair. Then he turned away for a moment to make a few more notes. The only sound in the room was the scratching of his pen across the page. The light through the window cast a saucer-shaped shadow on the carpet.

He looked back at the two of us after that and said, "There's nothing I can do for the two of you. I don't want to see you again."

Tracy and I looked at one another completely stunned. We got up and slinked out of the room. Once we got inside the car, we sat there for a second before we both burst out laughing.

"We got fired," Tracy said. "That's not good when your therapist fires you."

We didn't try to find another counselor, but it eventually became clear that as I got involved in doing more interviews that whatever damage I'd done to the relationship and what she'd done wasn't something we both were truly invested in fixing. We decided it was best to just go our own ways and not make anything more difficult for the other person.

I'd lost one wife to disease and now a second one to work and infidelity. No matter how strong you think you are mentally, or how amicable the parting, a divorce preys on your sense of self and self-worth. A few of the "why me" moments—why did I have to accept that job, why did she have to have an affair and mess up my chances, and a host of others—brought me pretty low. I understood why some people chose to end their lives. I didn't feel any of suicidal impulses, but on an intellectual level, I could understand the desire and the despair that might lead to it.

As much as was possible I tried to resume my normal activities as soon as I could after the initial broadcast George had done. I knew immediately, even in those pre-Internet days, the power of the media. Two days after that first interview aired, George was on the phone with me.

"Listen, this story is getting picked up around the world. A Japanese station called me. They want to fly you over there and conduct their own interviews with you."

"Japan," I said dully. For an instant the thought appealed to me. I could get away from here and escape everything.

"Yeah. Pretty amazing. What do you think?" George asked, his voice twinged with a mixture of curiosity and trepidation.

"No. Not interested."

"Mind if I ask why?"

"You're the investigative journalist. You figure it out."

George laughed. "That's good. That's funny."

At the time, I wouldn't have been able to tell you if I was joking or not. I was feeling deadened. It was like I was an unplugged electric guitar. Someone could pluck at a string but it produced a pale imitation of the sounds it was capable of producing.

"Well, Bob, I'm kind of glad to hear that. I mean the exposure is great. That's one of the reasons why I called."

I knew he expected me to respond to that some way. Ask what he wanted or something to that effect. I just let the silence linger.

"So, the response here has been great. We'd like to do more interviews with you."

Again, I didn't respond. At least not directly.

"I've thought about the Japan trip. I think I'd like to do it. Maybe it would be good to get away."

"That's great. I'll see about the arrangements. What about my offer? More interviews?"

George waited and then filled in the gaps. To his credit, he didn't sound exasperated or panicked.

"As you can imagine, some people are a skeptical. Anyone could have made up what you said. We didn't get into much detail. To be honest, I need to be convinced. I need better evidence. I need to verify some things."

"Yeah."

"As I'm sure you know, this story has implications beyond you. Beyond me, for that matter. This is big stuff. Government cover-ups. This is the kind of thing I do. If you're going to trust anyone about this, I'm your guy."

We both let the pause go on.

I understood what he was saying. At first, coming forward was an act of self-preservation. I had only briefly considered what this all meant beyond me being able to get back to my life. I did understand that some people believed in a greater degree of transparency within the government than I did. I had a long history of working with and inside government agencies. I was comfortable with the "need to know" mentality that drove much of the control of information. I'd seen that in action in the private sector as well.

But with this, I was leaning in a slightly different direction. I didn't really see the harm that would come from revealing this information. It sounded to me as if the Russians were already in the know. I thought that people would be receptive to knowing about alien life and technologies. After all, we had these craft in our possession and it was like they were shot down in some large scale invasion. I still didn't know how we got them, and in truth, how didn't matter. We had them, so let's just move on and see how this technology could assist us. Maybe if I put a bit more pressure on the powers that be, they might be willing, or might be forced to, make this information more well known.

I thought about *Sputnik*. When we learned the Russians had a space program and then that they'd gotten a man to orbit the Earth before we did, the result proved to be a net positive—a huge net positive—down the line. Obviously this situation was different, but I thought that the principle still applied. An open exchange of information could be a good thing. I'd seen firsthand at S4 and in other work what happened when compartmentalization of information took precedence over other considerations: progress slowed. I had a pretty good idea of the weapons potential of this antigravity system. If we had it and others knew we had it, maybe our having such an enormous advantage would be a real game changer. It had been ten years since the Strategic Arms Limitation Talks II treaty had been signed. Talks to end the Cold War were ongoing, various Soviet Republics had been rising in opposition to Kremlin rule. Momentum seemed to be in our favor, but we could have had a heavy weight to tip the balance in our favor.

I told George that I would think about it and get back to him.

"Don't think about it too long. I'd rather do this with your cooperation. News business is tricky. People forget."

I didn't tell him that people forget, but that groups and agencies like I had been dealing with most likely didn't.

After George and I spoke, I called John Lear to let him know that I was going to take him up on his offer. I told him that I would meet him for dinner not later than six o'clock that evening. John understood that agreeing to dinner at six o'clock meant I would be staying at his house that night. A minor precaution against eavesdropping, but a necessary one at the time.

I spoke briefly with John about George's offer to do more interviews. He thought it would be wise to do them. I consulted with a few other friends, but to be honest, by the time I did, my mind was already made up. I was going to go forward with the plan to talk about what I'd seen. I also agreed with something that George told me.

For this story to have any credibility, I needed to step up and allow myself to be interviewed on camera and have my likeness and name be clearly on display. The first interview I'd been silhouetted and not identified. That wasn't going to pass the test this time. I told him that I would allow that.

George also said that it was his professional responsibility as a journalist to do more inquiring into my past.

I agreed to that.

George said that he would like to travel with me to Los Alamos to meet some of the people I worked with and to see where I had worked.

I agreed to that.

George told me that if at any time I was uncomfortable with any of this, up to the moment it was sent out over the air, he'd agree to pull the plug on it.

I heartily agreed with that. In fact, I told him that if he hadn't specified that part of the agreement, I wouldn't have agreed to any of the other stipulations.

I also spoke with Gene about the Japan trip and asked him if he would like to join me. In the intervening day since the offer was first presented, I'd had second thoughts. But Gene seemed to think it was a good idea. At that point, I was easily swayed and was likely just looking for reasons from other people to do or not do many things.

I got another offer that I couldn't refuse. George had contacted the people in Japan to let them know that Gene and I agreed to accept their offer. A few days later, we bought tickets. That same night, I was at home and the phone rang. I picked up and a voice I couldn't identify said, "If you take this trip to Japan you will never return. Understand?"

He then hung up.

I didn't know if this meant I was still being followed or if my phone line was still being tapped. Not that either of the two mattered. Gene and I talked it over and since neither of us was all that excited about going in the first place, we turned in the tickets and didn't go. The Japanese later threatened to sue for the entire cost of their TV special, which I just ignored. I never heard from anyone about it again. I guess it's a reflection of what my state of mind was at the time that the threatening phone call unnerved me only to the point that I canceled the trip. I didn't pursue any other course of action—no going to the authorities, no going on the record about it and telling George.

The reason George wanted me to accompany him to Los Alamos was to speed up the process. For him to get the clearances himself, to make all the necessary arrangements would have taken a lot of time. He wanted to strike, as the saying goes, while the iron was hot. I also met with his boss, the station manager, and both of them struck me as consummate professionals. They'd agreed to the conditions I'd mandated for the first interview and everything had gone off as planned.

As much as I was drifting at that point in my life, as much as I was feeling unmoored from the life I'd once led just a few short months before, it was great to have the two of them and my friends in my corner. No matter what happened as a result of the additional exposure, I was comfortable with accepting whatever consequences came my way. I was never going to blame anyone else for being complicit in what happened. I made my choice, and I was going to stick with it. It felt good to be able to trust someone and to trust myself.

The trip to Los Alamos was relatively brief. Because I'd worked there and had contracted with them, access to the facility and to the people who could verify my presence there was a relatively easy matter. I took George around and introduced him, showed him places I'd worked and various aspects of the accelerator. George was clearly pleased to see that I hadn't fabricated a bit of my back story. We returned to Las Vegas and George said that he would be in touch to schedule the next taping session. We scheduled it for the last week of May, just prior to Memorial Day.

We filmed in an office at the studio. I decided that since I wasn't going to have my image disguised, I should make myself as presentable as possible. I wore a gray shirt and a matching thin gray tie. I didn't want to dress too businesslike, so I wore a pair of jeans and white running shoes. Gene didn't prepare me ahead of time with a list of questions and that was fine with me. We talked at length, but I knew that he was going to have to edit it all down to about ninety seconds for broadcast. I wasn't nervous. Blocking out the camera operator and the sound person wasn't easy. We were in a fairly tight space, but I kept reminding myself that I was having a conversation with George and no one else. He was somebody I trusted and all I had to do was answer honestly and talk to George.

I stuck around the studio while George went into another part of the facility to do his editing. I'm not sure why, but I wanted to be there in person when that evening's news, with my interview included, went out over the airwaves. As time went on, I thought more and more about what I was doing. I was kind of like a groom sitting in a church office while his bride and everybody else bustled around making the last preparations before a wedding. All of them were occupied with doing something, but all I could do was sit there and think. I started off picking up a pencil and drumming it on the table top while considering everything I had done and what this new revelation might mean for my future. I had no idea really of how far-reaching the consequences might be.

I didn't believe I'd be in any real physical harm; that was the initial thought I'd had in coming forward, that I should do this and protect myself. I wondered a bit about my reputation in Las Vegas and dismissed that as a concern pretty quickly. Las Vegas was a place where many people led a transient existence. They came here with some kind of dream or ideal or last-resort mentality. It was kind of like the gambling that drew a lot of people to the place. You started out with high hopes, went bust, and limped home.

I was concerned about making a living beyond photo processing. I'd signed up with EG&G hoping to get back into the scientific community. I wasn't sure how I'd be perceived if I came forward. I'd damaged my chances at getting the security clearance I'd needed to work at S4. But what about at other places, like Los Alamos? What about in the private sector? That was a real unknown. That concerned me. I knew that I'd take a kind of fuck-it-all attitude in the wake of Tracy and the rest of it, but I had to be practical. I was gutted, that was for sure, but there was some instinct inside of me, some sense of survivorship that kept nagging at me. No, I wasn't going to kill myself, but was I doing something that would cripple me in a way? Was I committing some act of self-sabotage? Was I looking for an easy way out, a way to later explain whatever state of failure I'd entered into?

I had been pacing for a good amount of time as I waited for the countdown to air time. There was something inside me that hated the idea of having Dennis and the rest of them, whoever they were, win. I thought that maybe by coming forward in the manner I had, I'd done just that—beaten them. But in reality, that wasn't likely to be the case. I was pretty sure that as I sat there in that room, there was someone else out at the facility filling out the same paperwork I had, submitting to the same physical, getting introduced to Barry or some other version of Barry. The golf ball would still be hitting the ceiling and raining down bits of acoustical tile. The high performance tests would still be going on every Wednesday night.

I wouldn't be a part of any of that. What would I be a part of? I felt a gnawing in my stomach and envisioned a gaping black hole that was my future.

Someone knocked at the door.

"We're about to go with it," someone I hadn't met before told me. "If you'll follow me."

We walked through the aisles of the newsroom along rows of desks and half-walls. I spotted George a few yards ahead of me. He carried a large format videocassette in his hand. I took off after him.

"George! No! You can't!" I yelled.

"We have to go with this," George said. His eyes darted from side to side and he'd edged around a file cabinet and a potted plant wobbled unsteadily.

"You said up to the last second just say the word."

George was still on the move. I accelerated and caught him around the waist and chest and tackled him.

"I'm saying the word, George. NO!"

We rolled around on the floor, George holding the tape just outside my grasp. I looked up and saw a few faces staring down at us, though no one stepped in to intervene.

"We have a deal," I said, sounding very much like a pre-pubescent boy shouting at his older brother.

George scrambled to his feet.

"I'm doing the right thing, Bob. You know. Cold feet is all. This is going to be okay."

With that, he went into the studio and I sat on the floor with my head cradled in my hands and my knees drawn up, wondering not for the first time or the last, "What have I done?"

In the last twenty-seven years, I've asked myself that question many, many times. Unlike a lot of people who've lived through a really rough period and come out on the other side who say, "If I had to do it over again, I wouldn't change a thing," I would.

I wouldn't have come forward. I should have probably just waited things out after learning about Tracy's affair. I would have been more patient and lived with the hope that once that matter was all settled I could have gone to work at S4 full time. I had a once-in-a-lifetime opportunity to work at the forefront of science, and I pissed it away. Sure, there are the ethical implications of what was going on out there, but at heart, I'm a scientist. I seek knowledge and understanding. All that's been left in the wake of my time at

S4 is other people's doubts and uncertainties about me, and about the program there. As a scientist, on one level you have to accept that there's going to be uncertainty, that your theories and findings could always eventually be proven to be less than one hundred percent certain. Simultaneously, you also have to believe that you've cracked the code, solved the riddle, advanced human understanding.

At S4, I didn't do any of that.

I have to live with that regret and a bunch of others as well.

This is no fairy tale. We all didn't live happily ever after.

A lot of innocent people got hurt as a result of me stepping forward. People lost security clearances, jobs, possible futures because of their association with me and my revealing to them what I did. That's hard to deal with, but is in no way as hard as the consequences they had to face. Over the years, I've tried to do what I can to make amends, but words fall short and gestures fail.

As for me, I picked up the pieces eventually. Over the long haul, things did get better. I've spoken a few times at conferences and done some interviews. I've had Hollywood film and TV producers contact me. In the scripts they've written, they've tried to show me as some kind of action hero, leaping onto the hoods of cars escaping the bad guys. I'm no action hero. I wasn't then and I'm certainly not one now. I'm no kind of hero.

I've settled into a quiet life running a scientific supply company. I can't do science anymore now that my reputation has turned to crap. I've tried sending out resumes but they produce no results, not even a polite decline, just silence, the deep silence of space.

I chose to come forward again and hoped to set the record straight by writing this. I didn't seek someone out; he sought me. I knew that this was going to be a part of a larger overarching project, and I liked that idea. I was just one small cog in a larger machine at S4 and S4 is part of a larger story as well. I can't, and won't comment on that because it is not something I witnessed or experienced myself.

I'm a modest guy with modest ambitions. All I've ever done in interviews and appearances is to comment on and to relate what I know and I what I did.

I see no reason to change that now.

I imagine that anyone caught up in events that feel larger than themselves feels the impulse to create a story that is in ways as large as the forces that are acting on them. I guess that's a part of physics. But I know that our understanding of the nature of the universe is, to put it mildly, incomplete.

Maybe there's some comfort in knowing that this story is over. Maybe some will take comfort in feeling that there are some questions yet to be answered.

That's life.

In the end, though, I was rewarded for my patience and my perseverance. I'm married now and have a lovely wife and step-grandkids. None of them know me as the Area 51 Guy. And that's just fine with me.

WHEN KINGS RISE

VI CARTER

Copyright © 2024 by Author Vi Carter
All rights reserved.
No portion of this book may be reproduced in any form without written permission from the publisher or author, except as permitted by U.S. copyright law.

WARNING

This book is a dark romance. This book contains scenes that may be triggering to some readers and should be read by those only 18 or older.

NEWSLETTER

Join my newsletter and never miss a new release or giveaway:

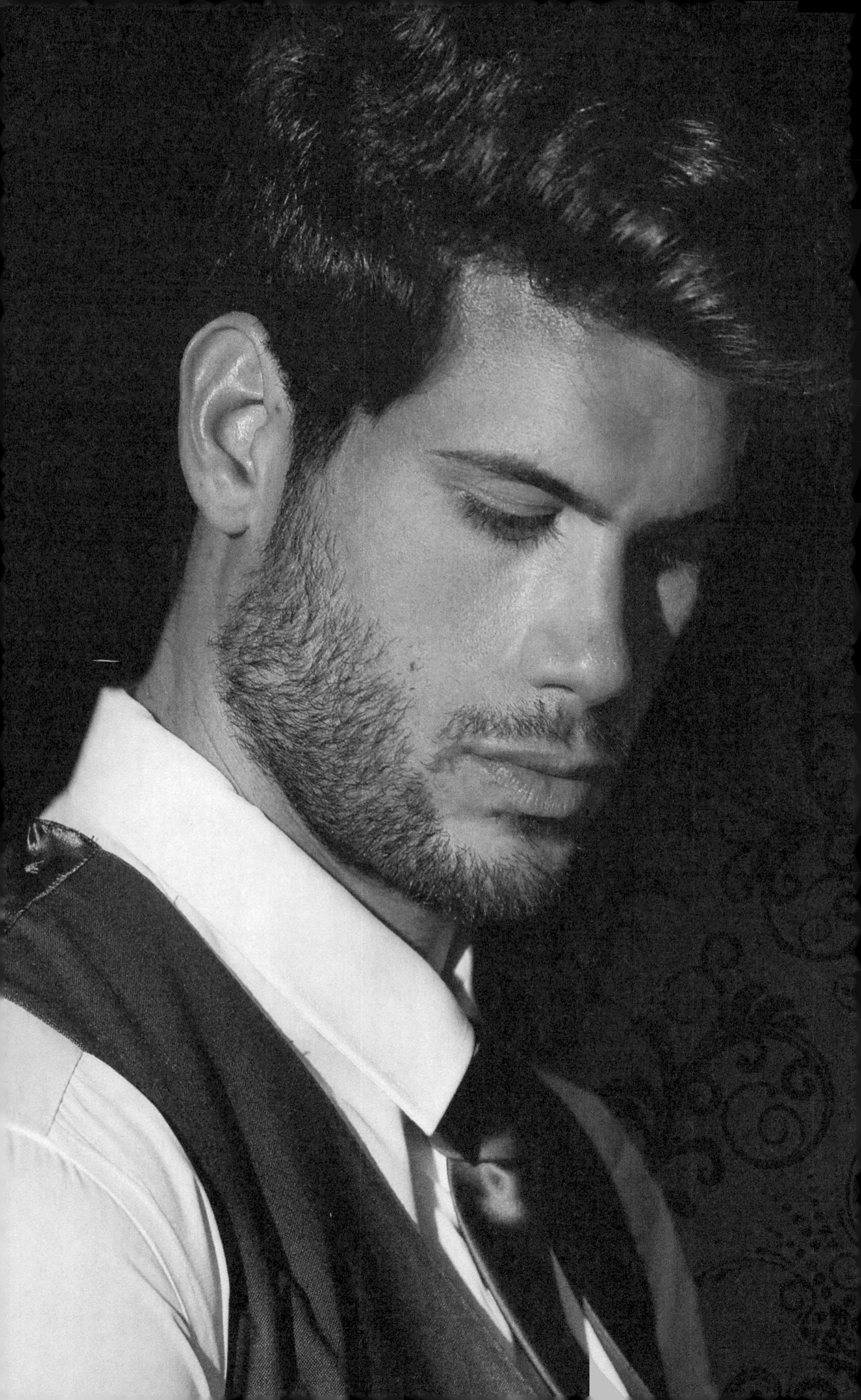

CHAPTER ONE

Diarmuid

HANDS OF THE KINGS EDICT ONE

The Hand of Kings is not a political movement, rebellion, or cult. It is a natural order of life. Just as the moon and sun command the heavens, the Kings command the Earth.

THERE IS ALWAYS a sense of peace in chaos.

Quiet chaos, that's what I walk into in the grand ballroom on the top floor of the mansion. A part of me wished this could be done in my home, but that would be unheard of. The showing of the brides was always at the Hand of the Kings' mansion.

The long red velvet curtains have been drawn. The gold weights that keep the curtains in place, still shifting along the oak flooring inside their lining, tell me they have only recently been pulled to plunge the room into a romantic darkness.

Nonsense really.

The candles along the walls have been lit—hundreds of them—more nonsense, but this is what the arriving brides are accustomed to— or so I've heard. A room shimmering in romance, but their shaking figures scream anything but that.

That is their way, their duty. I run my thumb along my lip as I think about our traditions. Every King is given three candidates who must show obedience at all times. How many kings are there? That I'm not sure of.

But right now, my three brides are obedient.

All their gazes are downcast, which is what is expected. They will only look at me when I request it. I take my time glancing at the portraits of all the past leaders that hang along the walls. Their eyes follow my every move. They don't intimidate me; they are the past, and I am the future.

The final three pictures, however, do give me pause. The first is Andrew O'Sullivan, who was the head of the family until his recent disappearance. A twinge of a smile dances along my lips, but I suppress it as I stop in front of the final two paintings. One is of Richard O'Sullivan, my father, whom everyone assumed would one day take over.

I chuckle. *"You know what they say about assuming things."* Beside him is my mother, Elise O'Sullivan.

I stare at her face, the steel gray eyes that I inherited from her. All else I received from her was hate. Hate for how she allowed men to take me, shape me, and damage me. She never protected me. No one did. But I would have expected some form of protection from her as my mother.

I place my hands behind my back and walk past the row of servants. Seven, to be exact. Once I reach the final one, the first turns, and the rest fall into line, leaving the room. Leaving me alone with my prizes. They are not just servants; each is chosen carefully and skilled in a variety of ways to take a life. Working in the mansion of the "Hands of the Kings" requires knowledge of how to kill—wolves in sheep's clothing.

I continue my walk to the waiting brides.

One of them I will have to marry, but until that moment, I get to play, and like my brothers would admit, I don't play nicely. I keep walking the distance until I'm in front of the three naked ladies. The one in the center has her hands folded across her private area. Like she has any right to shield herself from me. These women have been bred for this, so she should know better.

"Place your hands at your sides." Her response is instant, and her dusky Mediterranean skin flows along her graceful arms that hang loosely, fingertips grazing her thighs. "It's not a good sign when you have to be corrected already." I let out a bored sigh, and she flicks a glance up at me before focusing on the floor at her bare feet. She may be the troublesome one.

The troublemaker.

I hide a grin.

"Troublemaker, what is your name?" She glances up.

"Selene." Her voice is soft. Her eyes aren't the only part of her that is hostile. The shape of her shoulders and how they slouch forward like she can shield herself from my gaze isn't lost on me.

"I think I prefer Troublemaker," I say.

She holds my gaze for a beat more before diverting her attention to her toes. I follow her line of vision, and her toes tense along the hard oak flooring. This room is accustomed to polished shoes and dancing heels, not bare feet. She shivers, and I wonder if it is the cold or fear. Long dark brown hair is neatly arranged on top of her head. Pinned back almost severely. Nothing can shield her face from me.

Fires have been lit in the room, all three send out a soft wave of heat. One reaches the side of my face and I almost want to bat it away.

I prefer no fires in my own private rooms. But this isn't my home, so I don't have a say.

I move on to the next girl. She reminds me of a statue with how she holds herself so still. Her fingers seem to move involuntarily along her side. Her nerves are getting the best of her. I pass her and stop at the final bride.

She looks at me directly. "What is your full name?" I ask.

"Amira Reardon." She has soft brown eyes and an oval, innocent-looking face. She won't be very innocent when I'm finished with her. Her complexion is pale, yet under the glow of the light, it appears slightly tanned. Once she says her name, she averts her gaze, but not before I catch something dark and intelligent hiding behind her eyes.

My darkness recognizes something inside her. Damage.

I go back two steps and stand in front of the woman who is first. She hasn't moved a muscle. She reminds me of a beautiful statue. All angles and posture. She has an athletic structure. Her long blonde hair hangs loosely around her shoulders. "Look at me," I say.

She keeps her gaze downcast, but she squeezes her thighs together. I reach in and touch her chin, tilting her face toward me until I'm looking into brown eyes.

"You must be Niamh Connolly. Ms. Connolly, my name is Diarmuid O'Sullivan, and it would be wise to listen to me when I give you a command."

I release her chin and walk in front of the three of them, taking in their beautiful bodies and faces. These are the brides chosen for me by Victor Madigan and Wolf O'Sullivan. I will get to know all of them, and in the end, I will only choose one. They were selected very carefully. I don't particularly like the idea of Victor or Wolf selecting anything for me. I can't stand either of them. But it's our hierarchy, and Victor is the head of the Hand of Kings, so he isn't someone I can ever question. Even if I was dissatisfied with his choice.

I'm not. All three are stunning, perfect...hopefully, they are obedient, too.

I return to the troublemaker, Selene. "Look at me," I command. She does, but there is no longer a fire in her gaze. She has tucked that away. She learns quickly. Now, she wears a blank look on her face, like she is facing her execution. I take in her beauty, allowing myself as much time as I wish to study her body before bringing my attention back to her hair.

"Let your hair down."

There is a flash of fear, but Selene raises her hands and removes each pin. One pings along the oak flooring. The noise has Amira glancing in our direction. When I catch her eye, she quickly looks away.

Selene holds all the pins in her hands, and I watch her beauty transform and grow as silky, wavy brown hair cascades down her back.

"I bet that feels better?" I ask.

She gives a quick nod of her head, but her pulse flickers wildly along her neck.

I'm ready to move on but the fire in her gaze grows again. I don't like fire. I want to extinguish it. Fire is uncontrollable, and to survive in our world you need control. My bride will need control. This lesson will anger Selene, I'm sure, but in time will teach her a lesson.

"Touch yourself," I command.

She uses the hand that is fisted with pins and slides her knuckles along her private area. "Use your other hand," I order.

She swallows and closes her eyes.

"Keep looking at me while you touch yourself, Selene."

Her eyelids flutter open, and her free hand runs along her mound. My cock grows instantly. I close the distance between us and grip her hand. With my other one, I grab her thigh, and she spreads her legs, giving both our hands access. I hold only one of her fingers and push it inside her opening. She inhales sharply. I let the tip of my finger follow hers, and the moisture that greets me makes me want to take more. To explore the warmth that tightens around my finger. I pull out and sink back in while looking into her eyes. Her cheeks are tinged with pink, but her gaze is steady. Control. That's exactly what I want to see. I withdraw. She doesn't. She's smart.

"Keep going, Selene," I order, and she takes her finger out before pushing it back in.

"Amira," I say. Her childlike features remind me of a porcelain doll. She doesn't belong here, maybe on a shelf in a child's room. A part of me can't wait to corrupt her, yet that darkness that I saw earlier is there and makes me curious about her. "Watch Selene." Amira does, and I stop in front of Niamh.

"Do you want to join Selene?" I ask.

I like watching her brown eyes swirl with fear. "Do... I have to?" She has a stammer; it's a flaw, but I like flaws.

I run my thumb along my lip. "You answered a question with a question," I bite back.

She moves instantly, and I take in her toned oval ass. She must be athletic. I wonder what sports she favors. She moves quickly to Selene but pauses in front of her. Selene is making great work of stimulating herself. If I were to guess, I'd say she was enjoying it.

Niamh peeks at me before she reaches out with a shaky hand and touches Selene's breast. She squeezes it, showing she has never touched a breast before.

I walk back to them, and they both pause. My gaze darkens, and they continue the show for me.

I grip Selene's free breasts at the base and drag my hand upwards, tightening until I reach the nipple and squeeze. Selene hisses with pleasure.

"Just like that, Niamh."

Niamh copies me, and I can already tell she is going to be a very good student. I reach down and cover Selene's hand. Pulling her finger out of her opening, they are soaking, and I push her wet finger against her swollen bud. I keep my hand over hers as I circle her sensitive area while Niamh repeats the action on Selene's breasts.

I look at Amira, who's watching. Her mouth forms a small *oh*. She catches me looking at her, and it's clear in her gaze that she wants to be part of this, but for that reason, I don't invite her. I don't stop until Selene shudders on her hand and her orgasm ceases.

I step back and give a short clap.

"Very well done, Selene; maybe you aren't a troublemaker after all." My cock bulges against my trousers.

Maybe Amira could have her first lesson in pleasing me. As I take a step toward her, my phone rings. I scoop it out of my pocket and turn my back on my three brides.

It's Wolf. I assume he's ringing to see what I think of his selection.

"I can't fucking believe it." That's Wolf's opening line.

I step toward the red velvet curtains.

"I don't know who would do this." He sounds upset. Almost distraught.

I don't care. I like hearing him this distraught.

"They found my father, Diarmuid. He's dead."

I knew this moment would arrive. "Are you sure?" I ask.

"What?"

"Are you sure he's dead? Did you see his body?" I ask.

"Yes, he was pulled out of a shallow grave. Yes, he's very fucking dead," Wolf snaps.

I turn to my three brides, who are back in formation. Selene looks flushed. Perfect in the light.

I hadn't time to dig any deeper, but a shallow grave is all that Andrew O'Sullivan deserved.

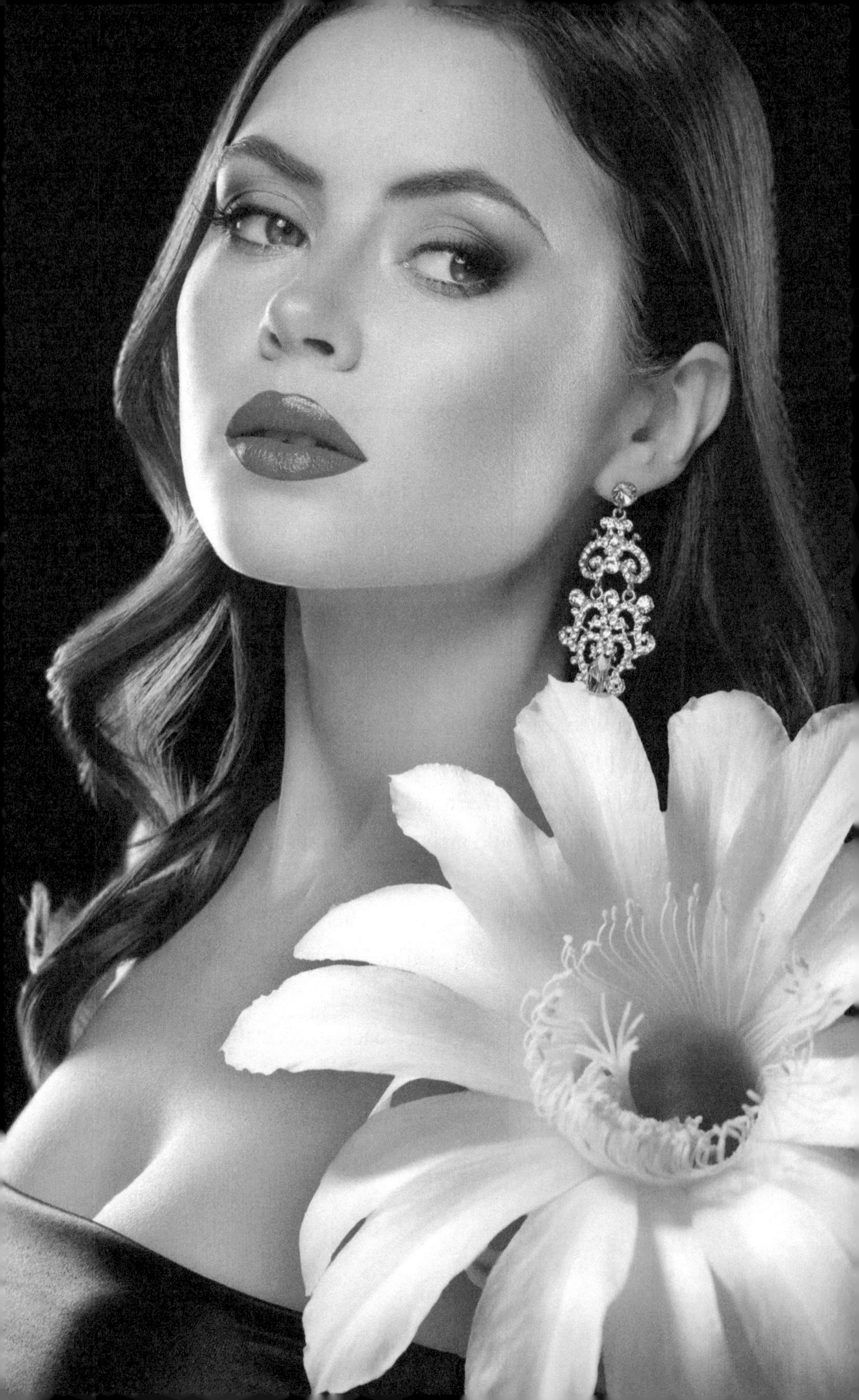

CHAPTER TWO

Amira

A WAVE RISES hard and fast, and I clench my fists so I don't strike my own face like I want to. I can't react, not with one of my guards' thighs pressed firmly against mine.

My vision wavers, and I sneer at how weak I am. I failed today.

"Don't you always." A voice whispers in my mind.

I straighten my spine, my mind ruminating on the moment that Diarmuid had given all his attention to the other two girls and not me.

I want a mirror to check my face. I'm pretty—I know that—so what didn't he like? A lot of people see me as angelic and innocent, and I play that part so well. Maybe not well enough this time. Diarmuid had looked at me with a tilt of his head like he had seen through the porcelain skin and bright brown eyes to the real me.

I divert my gaze to my lap and flex my fingers. Next time, I need to do better.

After seeing Diarmuid O'Sullivan, I know with every fiber of my being that I want to be his. I will win my place as his bride.

The darkness around the vehicle sinks deeper inside, casting too many shadows. I'm tempted to reach up and flick on the small overhead light, but I remain still. I know if I make a move, I will lose the last shred of my control.

My heart thumps as we travel down the winding road that opens a bit wider. Trees bend toward the vehicle, their long, bare branches like claws reaching out to me.

The vehicle slows, taking a left turn, past wrought iron gates that once shined with a polished black varnish, but now, the peeling paint makes a line across the driveway that gets crushed under the wheels of the heavy vehicle as we pass.

I don't want to go home. I smash my eyelids tighter and conjure the image of Diarmuid O'Sullivan. My heart rate slows, and a sense of peace flows across my chest. The cogs of my heart loosen, and a smile plays on my lips.

He's so handsome, like a prince from a fairytale who has come to take me away.

Thump, thump thump. My heart threatens to start racing again, and I will it to settle. I only have seconds before we reach my home, and I want them to be quiet seconds.

Diarmuid's gray eyes are like the marble tops of my kitchen counters: soft, pretty, and soothing. My core tightens as I remember his large frame entering the room. He would be powerful on top of me. His shoulders wide and his arms strong. The vehicle comes to a stop, shattering my moment, and I open my eyes.

I glance up at my home. The estate house shrouded in darkness is beautiful. A light shines from the second floor. It's my welcome home.

The light of my sanctuary, my bedroom, that I left on. It's the only welcome I will receive.

The pressure against my thigh is relieved as my guard gets out. The driver opens my door, and I step out into a soft but sharp breeze. I roll my shoulders and imagine armor materializing on my shoulders and spreading across my chest.

The front door is opened for me, and I turn to see the guard sink into the darkness as he takes his station beside the front door. The driver doesn't enter but returns to take the car to the garage.

I close the door, and the sound is loud. I turn to the darkened hall. I must win Diarmuid's heart; I must be his bride. The words repeat in my mind with each step I take into our estate home. It once was beautifully maintained, but it crumbled when my father fell apart. He left our family in ruins, and I intend to turn this all around for us.

I nod to myself; I will fix what he destroyed.

My foot touches the first step of the staircase that leads to my room, but I pause and listen to the low hum of the house. Pipes gurgle somewhere deep in the mansion. A breeze touches my back, and I turn to the main drawing room. Inside, it's dark, but I find the source of the breeze. One of the windows has been left ajar.

I push the heavy gold drapes aside, and a scattering of dust flutters down on top of me, making me cough. I ignore the assault on my lungs, grab the handle of the window, and yank it toward me. It doesn't budge but creaks in resistance. A chair is in my way, and I push it aside to get closer. Using both hands, I pull, and the window slams shut.

I push the handle down and lock the window. Another cough erupts as I step back from the window that overlooks the garden. Weeds merge with once-blossoming flower beds, and the hedge line coexists with trees and hangs down onto the overgrown lawn. I want to pull the drapes so I don't have to look at the offending state of the grounds, but I'm sure the dust would suffocate me. I turn to the rows of bookshelves. They, too, have their very own coating of dust. The books that line the shelves are not for reading pleasure but for decoration. My father bought them for their visual appeal, not the words that are printed between the hardbacks.

I crave the privacy of my room, away from the dust and musty smell of our home.

I pause in the hallway as my stomach rumbles. I haven't eaten. The nerves earlier today didn't allow me to even have a sip of water.

The kitchen has a lamp on the counter that spreads a small amount of light. The overhead chandelier is in darkness, its bulbs long blown, and no one to replace them.

I open the fridge and find some fresh ham I had bought two days ago. If I hadn't ventured into our local village, we would starve. I close the fridge door and pause. There is something different. It's not something I can see but smell first. Vodka has such a distinctive smell to me, and I scan the darkened corners and stop when my mother appears from the other side of the kitchen.

My mother stares at me, and my fingers tighten around the ham.

She snorts before she speaks. "Eating again, I see?"

The glass hangs loosely from her fingers as she walks toward me with a snarl plastered on her face. "A fat bride isn't really a good look."

I'm tempted to look down at my frame. I always make sure I stay close to 1000 calories a day to remain slim.

"I haven't eaten today." My voice is so frail, and I hate it.

"You look like you eat plenty to me. Put the ham back."

I don't want to fight, and I do as she says. I'm ready to walk away as she laughs. It's cruel, slurred, and intended to hurt.

It finds its mark, and my stomach clenches.

"I never thought my own daughter would be a whore."

I think of how Diarmuid favored the other girls, how he never touched me. I wish he had. I wish he had made me feel something beyond what I always feel—Insignificant.

"Isn't that what you raised me to be, Mother?" My bold words give me a moment of satisfaction. I didn't choose to be a bride; my parents handed me over on a platter. Just like they had my brothers. The thought of my brothers sends another wave of grief through me. Grief that I have never been allowed to deal with. My father is mafia, my brothers were recruited into the Hands of the Kings, and being part of the organization took my brothers' lives. It's our world. Now, it might take mine.

She marches to me, and my bravery dies quickly, it becomes a pool around my ankles on the grimy kitchen floor.

"I gave you everything. The best parts of me." She runs a hand through her thinning hair. She once was a woman men stopped to look at for her beauty; now they look for a different reason. They look down on us, the family who fell from grace by the hand of a king.

"When you have a daughter, they say she takes your beauty away, and didn't you do that to me?" She grips my face and tightens her fingers painfully along my jaw. "You don't deserve my beauty."

I pull my face away from her. "Goodnight, Mother." I've heard this since I was a child. How I stole her beauty and grace. My brothers, of course, didn't—only me. I took everything from her as I grew in her womb. When my brothers died, her grief manifested into a deeper hate for me, along with her indulgence of alcohol.

"All of a sudden, you think you can treat me like this." Her anger grows. "You think because you are a whore to the O'Sullivans, that you can just walk away from me."

I stand my ground as she rants, and when her glass flies from her fingers and smashes against the counter, I flinch.

"Disgusting." She's irate, and I want to leave but try to make myself small so I don't anger her further. My shoulders hunch closer to my chest, but I'm never going to be small enough.

My silence has her anger swelling; her hand strikes my face once, twice before she falls into a heap at my feet, sobbing. My face burns, and I step around her.

"Don't walk away from me," she calls, but the violence has left her words, and sobs rack her frame. I don't stop walking. I make it to the stairs when I hear cabinet doors banging. She's looking for more vodka. She hides bottles throughout the house but forgets where she left them.

I've found them in most corners of the house and enjoy pouring each bottle down the toilet. I often hear her frantically searching the house and smashing things in her search for her poison.

My bedroom is well—lit, and the bed neatly made. I close the door behind me and turn the lock—not that she would ever come into my room, but I won't ever allow her to spoil this space for me.

I walk to the window and draw the curtains before entering my small ensuite. Taking a fresh face cloth from the top stack, I turn on the hot water, but only cold water pours out. I soak the cloth before rinsing out as much moisture as possible.

Taking a seat at my vanity that's tidy and polished, I curse her as the red welt on my face burns. My stomach rumbles but I won't venture back downstairs. I'll have to wait until morning.

I meet my gaze in the mirror. "Let's fix you up." I smile at my reflection and dab my cloth against my face. The burn intensifies, and I remove it.

A child's nursery tune comes to mind, and I hum as I clean my face. When I can't do anymore, I open my jar of night cream and rub it carefully on the welt as I continue to hum. When the cream dries, I paint on my makeup; using a soft brush, I apply blusher across the wound before blending in the makeup. Placing a small dab of gray eyeshadow on my lids, I stare at my reflection as I apply a coat of red lipstick. I don't look so innocent now. I look fierce. I smile at myself. This is the color I will wear the next time I meet Diarmuid. He needs to see me, and these bold colors will make him take notice.

I lean across the vanity and kiss my reflection. I giggle at my lipstick mark on the glass. Using the cloth, I wipe the lipstick away.

He will notice me next time.

"Yes, he will," I say to my reflection.

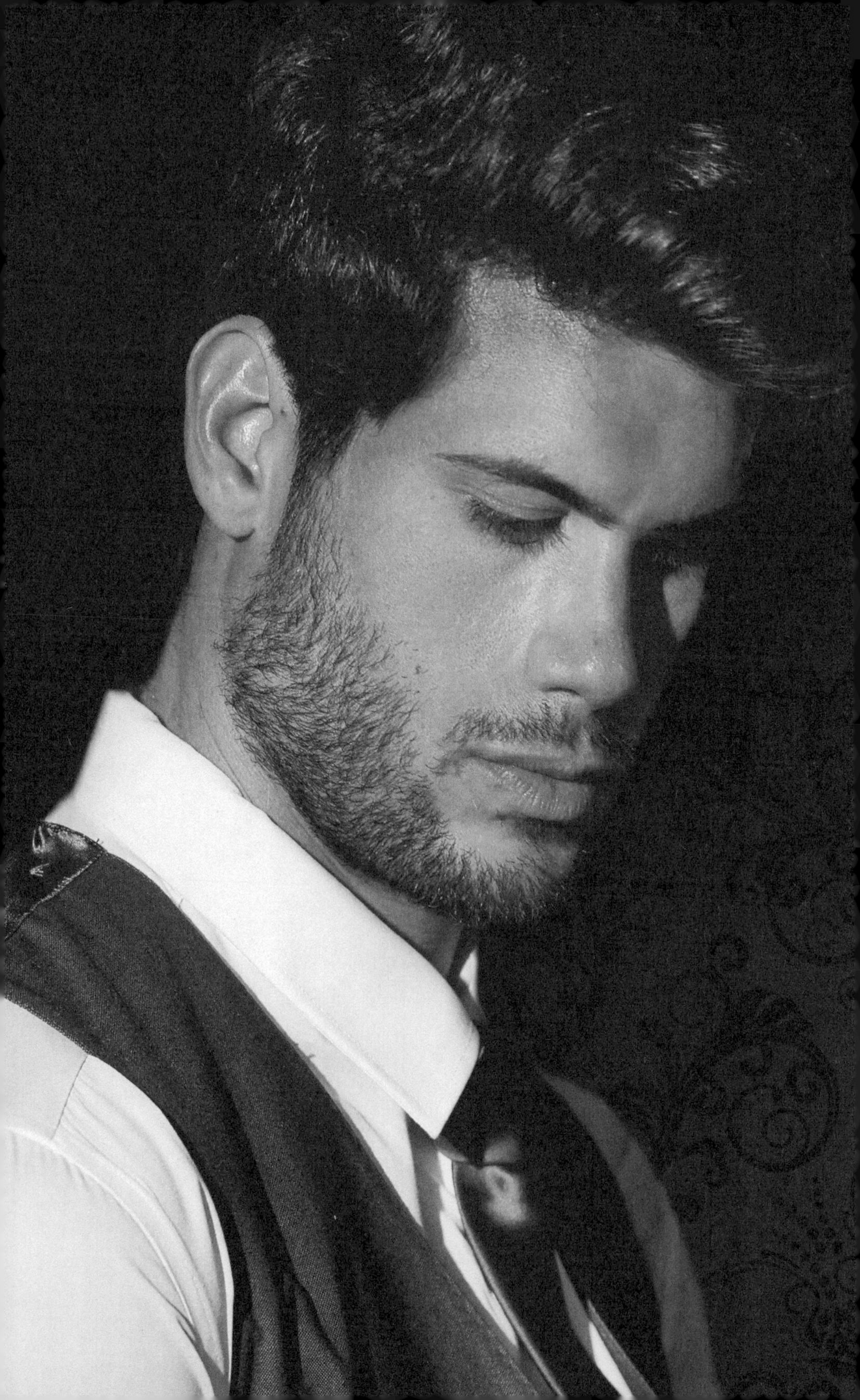

CHAPTER THREE

Diarmuid

HANDS OF THE KING EDICT TWO

Any sin can be forgiven except for the sin of abandonment. The wrongdoer will feel the abandonment of the order for three generations.

I TURN OFF the engine and climb out of my car. It's low to the ground and when I rise to my full height, I can easily see all around me. The gardens here sprawl further than my eye can see. They are manicured yet hold a wildness to them that I know is intentional. It's a sham. My tolerance for fakeness is low.

I laugh at the thought, considering my life is one huge charade rolled in a thick carpet of fakeness. I glare at the valet, who waits for the keys to my car. I place the keys in his hand, and he folds his fingers around them. No fear shines in his gaze. It really shouldn't. No person here is what they seem. I know the valets can kill with their bare hands. Just like the other servants, they're trained killers. Everyone who works for the Kings is, but respect shines in his gaze. He knows who I am; he knows my capabilities. As the Hand of the King's assassin, I outrank him.

The Tudor—style structure before me was built to display wealth, and it didn't disappoint. I like luxury, but I don't like being at the Hand of the Kings' headquarters; this is my second time this week.

The purr of my engine has me looking back at my car as it disappears out of sight. The large driveaway has a wide arc that sweeps into the trees. This property is also designed to hide its guests from prying eyes, not that anyone would get onto this property without approval. It's more protected than the Vatican.

I would imagine the pope has a say here, too. I know that Victor controls the Kings below him, but I don't think he acts alone. That information was never given to any of the O'Sullivan's, or Kings, for that matter. We are on a need-to-know basis as far as who the Hands of Kings really are and how far their reach truly goes. I don't think it's just Ireland that they control; I always get a sense that it's a wider web that's cast across the world.

Edward, the doorman, offers to take my coat, and I hand it over with a raised eyebrow. He grins and tilts his head. He's a poison expert who has assassinated many important guests by hiding "treats" in their coats. Sometimes, the poison is on the coat and sinks into the skin; other times, it's a surprise in their pockets. I know I haven't anything to fear, so we jest.

I watch as the maid takes my coat from Edward and places it in a closet that's large enough to be a dining hall, but is lined with mostly empty rods for guests' coats, bags, and other belongings. The maid, who is also Edward's apprentice, disappears from view. She is skilled at turning her wrist and allowing something deadly to fall from her sleeve into your drink. A handy trick.

I walk through the foyer of the house, knowing there are four sets of eyes watching me at all times from the watchtower where the cameras are located. I'd been there as a child and had to map out this entire property with its hidden hallways and secret tunnels. It's how I get around easily if my skill is required, but mostly, I'm sent out into the world to take down anyone the House of the Kings deems an enemy. That's how I know their hold on the world is wider than just the Irish landscape. I've been sent almost everywhere in the world to take down their enemies.

I pause and glance up at the winding staircase that leads to the room where I had my first glimpse of my brides. Excitement curls in my stomach. They were all delicious, and they did as commanded so well. They were well-picked and clearly informed that their obedience was expected, no matter what I asked. The first meeting was a test, and one they all passed.

Usually, meetings are held in the grand dining room, where drinks and a meal are served. But not today. We are being ushered to the private study. As I step across the threshold, I brace myself for what is to come from this meeting.

It is late September. If the old man had ordered a fire lit, it would be stifling here. If a disagreement were to take place, someone in that room could end up with his face smashed against the grate.

I glance at the fireplace and am relieved to see the fire is out. The study is large considering there are a few couches, but not enough for the people attending the meeting.

A chair is vacant, and Edward's apprentice offers it to me in deference to my position. A part of me wants to sit to show my power, but I decline with a wave of my hand.

Standing is safer in a room filled with such deadly people. I won't make myself vulnerable. It's one of the lessons that is ingrained in me.

The crowd consists of the most powerful members of the Hands of Kings, excluding my brothers, who aren't here. I do, however, see my cousin Wolf across the room. Wolf catches my eye and nods toward me. I nod back and take a drink offered by another maid. I raise the glass in a toast to my cousin and bring it to my lips, but I don't drink from it. Wolf notices, as he has been watching me carefully, and gives me a wicked grin that I ignore.

The murmuring stops when a far door, a door not used by guests, opens. I brace myself, expecting to see the one person who can strike fear in me, but someone else steps out.

It's Michael Reardon, a page for the Hands of Kings.

"Isn't this a bit above your pay grade?" Wolf sounds offended as he turns to Michael. No one else speaks, and Michael's face reddens.

This gives Wolf more glee as he continues. "You have a lot of balls to use the crisis room when you are barely a member."

Michael can't respond. It's forbidden. A page is the lowest rank in the cult. Wolf is a Duke, son of the recently deceased King. It is an insult for a page to speak to a Duke.

Michael dips his head in a plea for forgiveness before he speaks. "I do apologize, Duke. The Hand himself would have come, but the situation is uncertain in the group."

Michael speaks out of turn, in my opinion. He should never make it sound like the Hand of the King is hiding. But everyone is as curious as I am as to what has made our leader hide.

"Uncertain?" Wolf barks before waving his hand across the waiting crowd. "Do enlighten us, Page," he sneers.

Michael looks ready to bolt from the room and glances back at the door he came from. It seems as if it's a reminder of the task he was given, so he addresses the room. "As you all know, almost a year ago, Andrew O'Sullivan disappeared," Michael speaks to the rest of the room and doesn't direct his attention to Wolf, who is fit to kill the page for just mentioning his father's name.

When Andrew disappeared, everyone thought he had either been killed or had gone into hiding as he controlled an illegal gambling ring, but nothing was ever confirmed.

"Three weeks ago, we got confirmation of his death," Michael says, and there is an audible intake of breath in the room.

I keep my hand relaxed around my drink, not reacting to the news. Three weeks ago? How had I not heard about this sooner?

"It's Andrew for certain?" One of the members asks in a small voice. They're trying to pretend like they care, but already, the scramble for power has begun. Who will align with whom, and what positions will become available with this turn of events?

"Yes, the gardai struggled to place his identity at first." For the first time, Michael looks at Wolf.

"Get on with it," Wolf barks, like we aren't discussing his father's brutal but well-deserved death.

"His head, fingertips, and toes were all removed. But luckily, one of our contacts in the gardai department reached out to the Kings. Richard O'Sullivan, Andrew's brother, provided a DNA sample. The body is that of Andrew Sullivan."

My father helped solve the crime and never informed me. I get many glances from the crowd, but when I continue to focus on Michael, they look away.

I was foolish. I should have buried the body deeper or done what I do best—make the body disappear. But anger and the need for revenge had made me sloppy.

"You can be assured that the Kings will be using all their resources to find the killer and find who set up the killer." Michael appears pleased with himself.

Set up the killer? I tense and try to relax my frame again.

Wolf places his drink down on the study desk, liquid sloshing across the rim of the glass.

"Wait, someone knows who the killer is?"

"A woman, currently unidentified, was found on top of Andrew's grave. She was placed perfectly above it. It's either a coincidence, or someone within the organization is telling the killer that they know who they are."

Once again, I control my body's reaction. This is all news to me. A guilty person would fail to make eye contact with the people around them. So, I look around the room at everyone. I'm assessing them, for a gleam in their eyes that might tell me they know my secret, but their greedy eyes shine as their own futures spin away in their minds. Some who have known Andrew a long time show rage, and some show fear. If someone in this room knows my secret, they are just as good at playing the game as I am.

"Thank you all for your time," Michael says, and the crowd starts to talk amongst one another. Michael meets my gaze and makes his way to me, dipping his head in respect at everyone he passes. I place my full glass on a nearby table.

"I have a message from Victor. He requests you to be at church on Sunday. And he wants you to bring one of your brides." His voice is low, the message for me and me alone, but the few around us are listening intently.

I nod in agreement, and Michael quickly departs. I don't linger in the room but leave. Edward has my coat ready when I arrive at the main door.

"I hope you had a good meeting?" He helps me slip into the coat.

I watch him, wondering if he knows. Someone placed a woman's body on top of the grave I dug. That is no coincidence. I'm getting a message; someone knows what I have done, and that's why they placed the body there, so it would be found along with Andrew's. I need to be careful and treat everyone as suspects. Once I find the person responsible, I will silence them. Making them completely disappear is my only option.

"It was eventful," I answer. "Tell the valet to bring my car."

"It's already waiting." Edward slips his hands behind his back. "I know how you like to leave as soon as possible."

He does, but that doesn't douse the suspicion that has risen in me. I nod and exit the mansion. Sure enough, my car is waiting. Once I'm inside, I take my phone out and look at the names of my three brides. I have to bring one of them to mass on Sunday.

I'm thinking of Selene, but remember she's a troublemaker, and I don't need to draw attention to myself. Amira has a darkness that I will explore, but not now. So I choose Niamh.

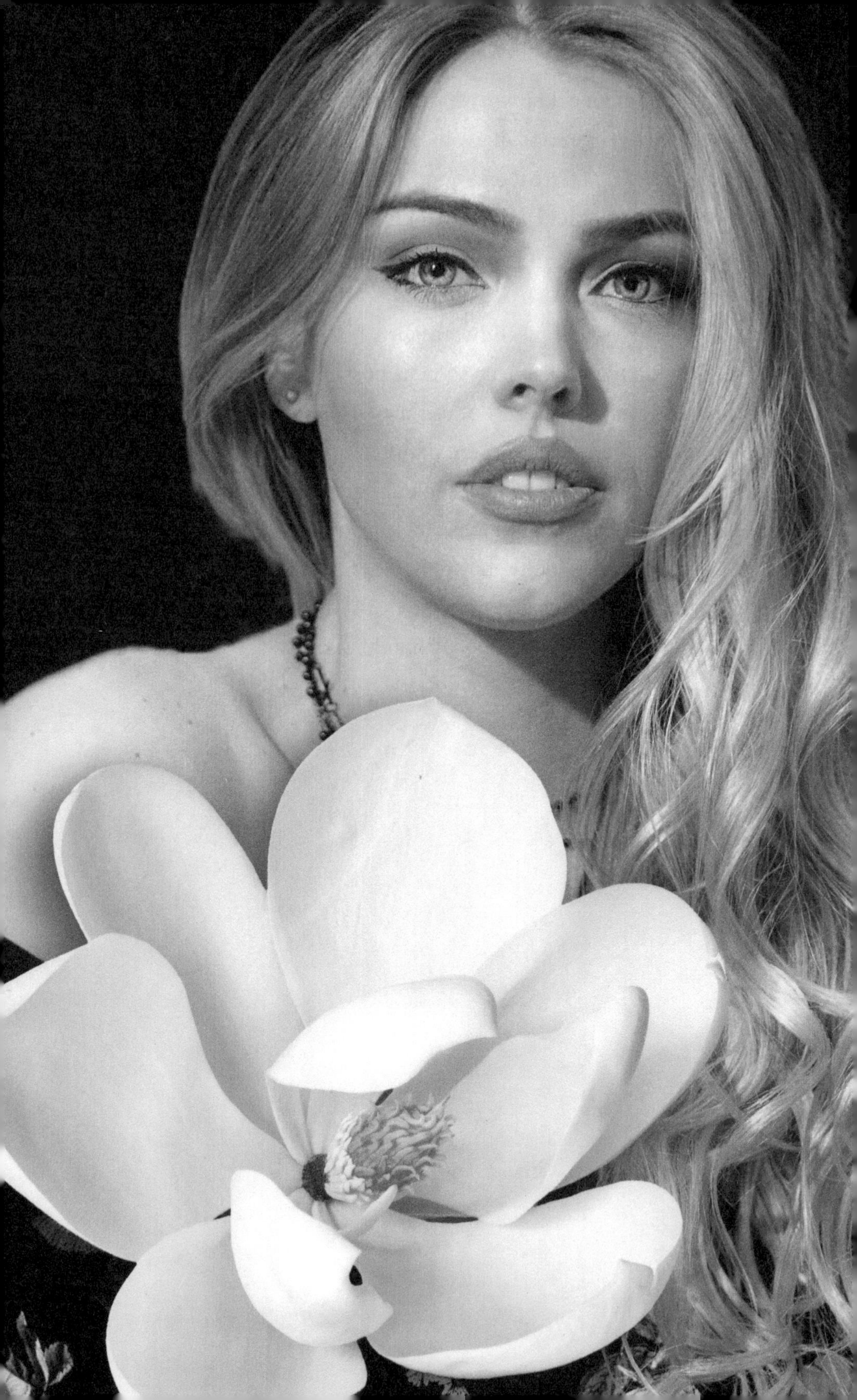

CHAPTER FOUR

Hianh

BLUE! The color I adore, as it reminds me of the bottom of the swimming pool, a place where I can get lost in. There are no demands from the water; sounds are muffled, and at times, it silences my mind. My fingers tighten around the material, but the dress is too short for Sunday service. I fling it over my shoulder, and it joins the piles of dresses behind me.

The next item of clothing is a cream blouse, one my mother always hated as she declared it showed off far too much skin with its transparent material. This one I linger on longer, imagining her features pinching in complete disapproval.

I shake off the rebellious thought and throw the blouse onto the ground. A screech sounds, and I turn to see Scamps race out of the walk-in closet with the blouse covering his furry body.

I'm tempted to chase after the cat, but the sound of soft laughter has me staying put. "Come here, Scamps; what has she done to you?" My sister's soothing voice reaches my ears.

"Is she okay?" I call when I spot a silky blue scarf on the ground. I will take a piece of me to Sunday Service. I scoop it up off the floor and exit the walk-in closet with the only dress that falls below my knees. It will have to do. I nearly tumble across the piles of clothes that I have discarded in my search for appropriate clothing.

"She's fine," Ella speaks from my bed. She's lying on her stomach with her phone in hand. She doesn't look up at me as she grins and continues to scroll.

"You better not be posting that on social media," I warn.

Ella has turned Scamps into a cat star, or so she likes to think.

Ella still doesn't look up at me. "I won't," she lies.

I bet she snapped a picture of Scamps wearing my blouse.

I slip the dark blue dress over my head; it falls perfectly below my knee. I've already put on pantyhose, and I take the light silk scarf and tighten it around my neck. I have to return to the closet for my white gloves and hat. Once everything is on, I return to my bedroom.

"What do you think?" I ask.

For the first time, Ella looks up at me. Her eyes widen, and I feel like I've nailed it until she bursts out laughing. "What on God's green earth are you wearing?"

I glance down at my dress and brush imaginary wrinkles away. Dolores, our housemaid, would be appalled to think everything wasn't perfect; it always is, but my nerves are getting the best of me.

Ella rolls onto her back, her chuckles coming to a stop when she sits up, but she still wears a goofy smile. Her soft brown eyes and sandy blonde hair are identical to mine.

"Church clothes!" I hold my arms out as if to say, isn't it obvious?

"It's not the Christ child's birthday, is it? No one goes to church like that on a September Sunday."

I place my hands on my hips. "How would I know? We are not exactly church people. Besides, how would you know?"

Ella raises her phone, and I quickly try to grab the contraption before she takes a picture of me. She pulls it out of my reach, but I'm satisfied when she places it on my bed. "Every time I sleep over Riley's house, her mother insists on us attending Mass in order to save my poor soul. Just put on a nice sweater and brush your teeth, and everyone will be cool with you. Trust me."

I allow my hands to run across my dress one more time. Maybe this is too much. "Even if I am going with someone like Diarmuid O'Sullivan?" I hate how I stammer over his name. The memory of what I did with Selene turns my face red, and I dip my chin, the hat hopefully hiding my burning cheeks from my sister.

"I mean…maybe keep the dress? Definitely get rid of the gloves and hat. You look like Nan." Ella's voice has softened.

I pull off the hat and the gloves and sit down beside Ella on the bed. There is a comfortable silence between us. I imagine she's thinking about how one day she may be handed over to a strange man just like me. It's not the future I want for my sister. I wrap my arm around her shoulder and lean my head against hers. "What would I do without you?" She is the reason I am going through with this. I will not allow her to be handed over to a strange man. If I do this for our family's position, then she won't have to.

"Probably end up with a husband with no teeth, especially if you dress like that."

I release Ella as she chuckles again. The dress isn't that bad. "Oh my God, let it go," I warn her as I return to the mirror.

"Sorry, sis. I have a visual memory, and that was quite a visual."

I meet Ella's gaze in the mirror. She's no longer smiling; she has picked up her phone but hasn't turned it on. I spin, leaving my own worries to the side, and focus on Ella.

"How many hours today?"

Ella shrugs her shoulders before she speaks. "Four. I got lucky. My recital went well, and Mother is pleased. Sunday is supposed to be my one day off from ballet, but she still wants four hours in the basement."

Guilt churns heavily in my stomach. It should be me, is all I can think. "That sounds rough." I finally say. But I remind myself I can't save her from everything. So, the tradeoff isn't so bad. She must do dance instead of being handed over to the hands of the kings.

Ella sighs and looks at me sadly. "It could be worse."

She's sixteen; she should be having fun, playing on her phone, hanging out with friends like an ordinary teenager, but we aren't ordinary. I don't think we ever will be. Our mother demands perfection in the form of extracurricular activities that she craved during her youth. She lives her life through us. Ballet was her love, and I had to endure years of training, but now that I'm being married off, I was allowed to step away, but only at the expense of Ella picking up the exhausting training.

"When will he be here?" Ella changes the subject, reminding me how wise she is beyond her years.

I glance at the dainty gold watch that wraps my wrist, a gift from my father on my twenty-first birthday.

I exhale. "Soon. I need to get downstairs." I pick up a pair of small black kitten heels and slip my feet into them.

"Niamh?"

Ella's soft voice has me picking up my clutch and pausing before I leave the room. "Yeah?"

"I hope he is kind to you."

My throat tightens at her words. "I hope so too, kid."

I leave the room before I start to cry. I won't break. I go down to the main floor and walk to the back of the house, where I can see through plate-glass windows. In the distance, the Irish Sea laps gently against the shores of Dublin. I close my eyes and think about the taste of the salt on my lips. The weight of the water against my body, the freedom the ocean offers me. Freedom that I can't find in this world.

The doorbell rings, startling me out of my meditation. I leave my favorite room and make my way gracefully to the front door. Years of being a ballerina and advanced swimmer have given my footing grace and poise that makes me look sure and calm. I am neither on the inside.

I take one final glance at my reflection in the hall mirror as the doorbell rings again. I'm not one to wear makeup, so I've kept it light, with a single coat of gloss across my lips and a thin application of mascara. With my hair swept up in a knot at the nape of my neck, I look composed and respectful. I open the door, expecting to see Diarmuid O'Sullivan's driver, but I'm taken aback to find the man himself on my doorstep. The night I met him, a driver had collected me from my home, so I expected the same today.

His gray eyes take me in from the tip of my toes all the way to the crown of my head. I hold still, remembering the level of respect we must show him. The obedience we must give. That part was drilled into my head by my parents. I hate it, but to keep Ella safe, I will do what is necessary. His Armani gray suit is almost the same color as his eyes. I don't want this marriage, but that doesn't stop me from admiring how handsome he is. I smile like I'm at the start of a show and compose all my nerves.

"Good evening, Mr. O'Sullivan." I don't stutter, and for that, I'm grateful.

"Miss Connelly." His voice is deep and sends shivers across my flesh. I grip my clutch and step out onto the porch; he turns his back on me as I close the door.

Diarmuid walks to his car and opens the passenger door for me; he's driving us himself. This will make us very close. I get into the passenger seat and thank him. His large frame walks around the front of the car, and when he gets in, his cologne sends butterflies erupting in my stomach. I wave away the unwanted attraction. I don't want this. I don't want to be going to church or anywhere with him.

I had hoped that I would not be picked as his bride, but that hope was squashed when I got the message that he would like me to attend Sunday church with him. I had been quiet when we first met; I even stuttered. I had thought that would make him not want me. It was small things that I hoped he didn't pick me for, so when my parents found out, they would know I was obedient, but I just wasn't his taste.

Amira was stunning, and when he hadn't gotten her involved in our first meeting, I had thought maybe he favored her more than myself and Selene, like boys are always mean to girls they fancy. I glance at Diarmuid. He is a far cry from a boy. He's a man and one who clearly knows his power.

I'm wondering if he thought I was easy prey. That maybe, I would be so alone with him. My stomach churns again at the thought.

"How has your week been?" Diarmuid's voice pulls me out of my musing.

"Very well, thank you, and yours?" I ask.

"Interesting," he states as we leave my family estate. His voice holds disinterest. I'm not one for making small talk, but I know for my parents' sake, I need to make some kind of an effort. Everything will be reported back to them.

"It was very kind of you to pick me up," I say politely.

He takes a quick look at me like he's seeing me for the first time. A smile plays on his lips but doesn't form.

"My pleasure."

"It's a beautiful day," I say and focus out the window. We can't get to the church quickly enough. He shifts gears, and the car moves faster like he wishes this ride to end as much as I do.

"I'm sure it will rain at some point today."

I continue with the small talk until Diarmuid starts to shift, like he can't take another second of this. Maybe this isn't a bad thing. He slows down and takes a left-hand turn into the grounds of the churchyard. It's lined with high-powered cars. This mass will only be open to people with an invite.

He pulls into a reserved parking spot and turns off the engine.

He doesn't speak as he gets out, and I stay where I am until he opens my door. I thank him and climb out. I only have a moment to breathe when the junior priest walks with a quick gait over to Diarmuid and takes his hand, shaking it several times. "Fantastic to see you at service, Mr. O'Sullivan."

Diarmuid removes his hand from the priest. "I'm looking forward to the service." Diarmuid makes it sound as though he's dead bored of everything. He reaches over and touches the small of my back, sending my spine into a tense straightness. If he notices the tension, he doesn't show it as he walks me to the door, where two more junior priests wait and shake Diarmuid's hand before we are led to our seats. Only then does Diarmuid take his hand off my back. We are fashionably late, and the church is full. Only a moment after we are seated, the mass starts.

Victor, who I know of, walks out onto the altar. Diarmuid tenses beside me, his jaw growing tight, but it's like a flash of silver when a fish surfaces in the ocean. And like that fish, it buries itself swiftly back into the darkness of the water. The tension is gone, and I'm left wondering if it even existed in the first place.

Victor. He's in his late fifties or early sixties. His hair is gray, black, and white, the gray mostly brushing the sides which are turning gray with age. He has heavy-lidded eyes that look both sincere and stern. He is a perfect man of the cloth—exactly what I would expect an old-world priest to look like.

Why would a man of the cloth make him uncomfortable? Recently, I've been educated about the world I might be marrying into. My parents wanted me to be Diarmuid O'Sullivan's bride so they could gain access to this world. Being part of this world means being controlled by priests like Victor.

Everyone kneels, and I move to do so, a step behind everyone; my mind is reeling. Diarmuid stares straight ahead, his disapproval unclear.

The O'Sullivans, too, are dangerous. In the late 1800s, they were announced as a mafia family. I shiver at that, just like I had done when my father educated me late into the night about the family I may be joining. Right now, the O'Sullivans claim that era is over for them. I take another peek at Diarmuid as we rise to join in the prayer, that I don't know the words to. Maybe they only declared they are no longer mafia as they are currently running for a position in the Dail Eireann, a political party who makes decisions with the people of Ireland as their interest. Most of their decisions are based on greed and climbing higher up the social ladder. My father said that one of them may be destined to be the president of Ireland.

We kneel again, and I'm glad when the service ends. Everyone files out and bends their knee to the altar before leaving. Once again, I'm grateful for my poise and manage to genuflect easily.

While I walk down the middle aisle, Diarmuid's hand finds the small of my back again. For the first time, I'm aware of so many people watching us. We stop at the exit as Victor himself shakes hands with the departing patrons. Diarmuid's fingers stiffen on my back, but his other hand encases Victor's, and they shake.

"Thank you for coming," Victor says, releasing Diarmuid's hand.

"It was my pleasure." Diarmuid's words are polite, a complete contradiction to the pressing fingers into my back. Victor's attention swivels to the next family as we leave the church. I'm very aware of how I am somehow invisible to these men. I don't mind, and to be fair, I'd prefer to go unnoticed. Diarmuid doesn't take his hand off my back until he opens the car door for me.

"That was a beautiful service." My voice is chirpy, not because I thought the service was good but because I survived my first outing, and the idea of getting home and maybe taking a swim makes me smile.

Diarmuid doesn't smile. "There is a map in the glove box." He juts his chin forward, his eyes focused on the glove box.

Okay. I open the glove box and take out the small map that has many grid lines crisscrossing it, and making it impossible to read.

I hand it to Diarmuid. He opens the map, and a small, cream-colored piece of heavy paper falls out. I glance at the paper and see what seems to be coordinates. Diarmuid runs his finger along the map, glancing at the scrap of paper before he taps the map twice.

"You can put that back." He tucks the paper into the map and hands it back to me. I've placed it back in its home when he reverses out of the churchyard. He doesn't say where we are going, but it isn't my home. I slouch in the seat but then remember that while he might have his eyes on the road, he is surely aware of my every move.

I open my clutch and take out some hand sanitizer, rubbing it on my hands. I'm tempted to offer some to Diarmuid but think twice and place it in my bag, which I leave sitting on my lap. I exhale at the thought of not going home.

"Are you bored?" Diarmuid asks.

Shit.

I glance at him, and a smirk plays at his kissable lips.

"No."

"If you are having trouble occupying yourself, you are more than welcome to pleasure me."

My heart thumps in my chest. I'm wondering if I heard him wrong, but I know I didn't. "I… I…" I have no idea how to respond. I don't want to pleasure him, but I also know I must do as he commands.

When a low laugh bubbles from his chest, my cheeks heat.

"You see, Niamh Connolly, I could make you do it." His laughter is gone. "I won't, but I could. Remember that."

I nod, just glad that I don't have to pleasure him. I know I won't always be as lucky. But today, I'll count my blessings.

We both remain silent. Diarmuid starts to slow the car down at an abandoned house in the Stepaside area. He comes to a complete halt. He leans closer to me, but his attention is out the window, as he points to a rusty old mailbox. "Will you retrieve the parcel from that mailbox?" He's parked along the sidewalk on my side.

I unclip my belt, happy to get out and take in some fresh air. The mailbox doesn't look like it's been used, so I don't expect to find anything in it. But a manila envelope held tightly shut by packing twine sits in the center. I turn to find Diarmuid watching me, and I raise the envelope to show him I got it and climb back into the car. When I hand it to him, I'm expecting him to open it, but he places it in the console between us.

We are silent again, but he's driving in the direction of my home, thank God.

"How do you feel, Niamh, about becoming an accessory?"

His words startle me, and I look from the passing scenery to Diarmuid. "Accessory to what?"

Diarmuid glances down at the envelope before his gray eyes land on me for a moment, sending a shiver racing down my spine. He refocuses on the road as he speaks. "Victor Madigan is the unholy priest of Dublin, my bride. In the envelope is the name and location of a person I have been commanded to kill, and that, my dear, is why we had to attend church." He pauses. "You're just here to look pretty."

When Kings Rise

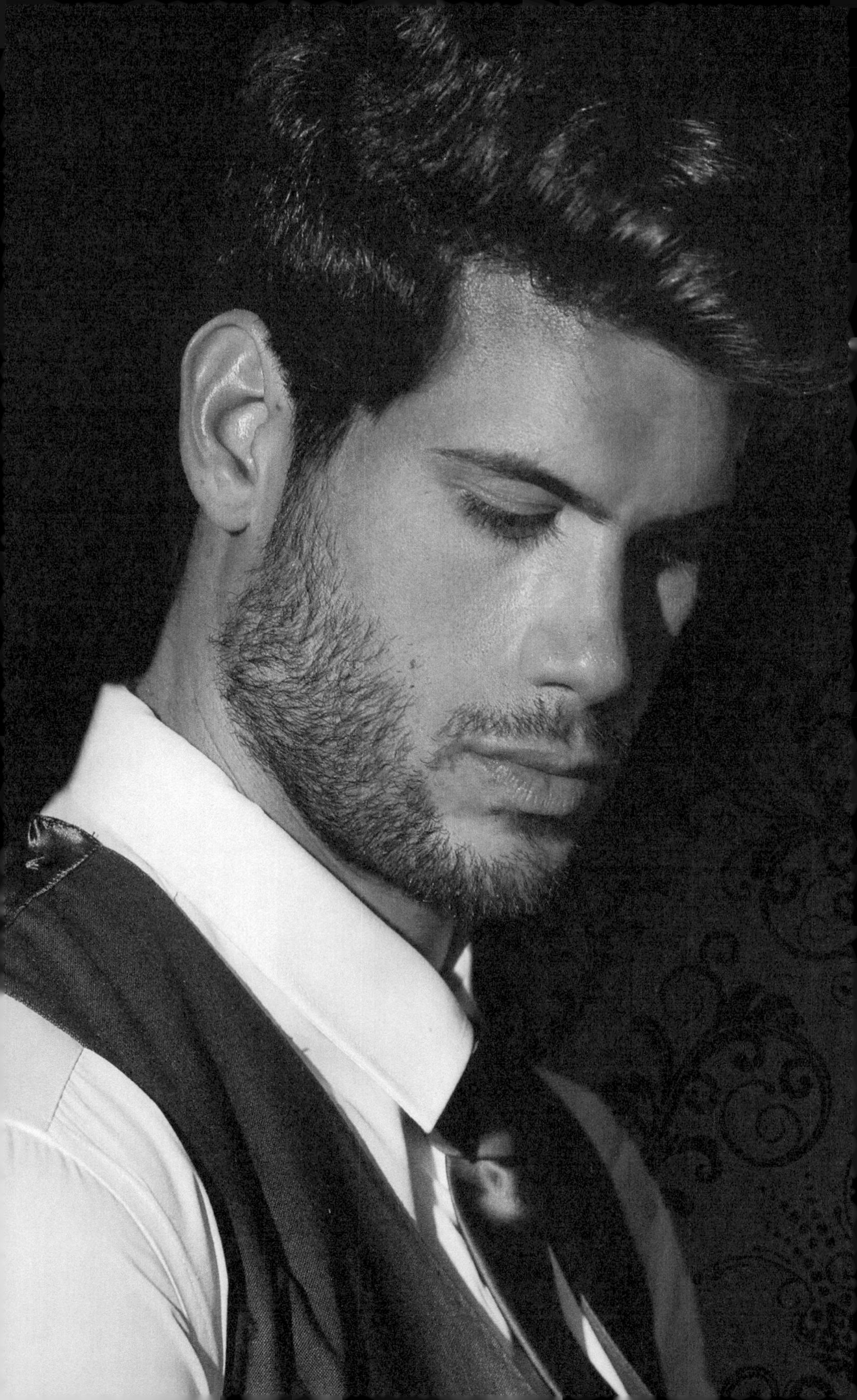

CHAPTER FIVE

Diarmuid

HANDS OF THE KING EDICT THREE

The order will perform its duties to humanity, regardless of the laws of nations and average men. Kings are above all other men.

I CAN'T SETTLE myself. It's an odd feeling for me. So, it sends a thrill through me that I must be on the lookout for a threat. And there is a threat. I've never been the mouse in a cat chase, always the cat instead, so the role intrigues me.

As I pull into the public garage, I see Lorcan's and Ronan's high-powered cars. They're parked six cars apart, and I pull into a distant spot, ensuring we've all hidden our vehicles among the older models. We'd be far more noticeable if we parked side by side.

Stepping out of the fluorescent-lit garage, I look left and right before I start to walk.

A woman across the road walks her dog; she glances at me but quickly looks away while dipping her head. I don't sense danger from her, but she appears to feel it from me. That is wise.

It's a great instinct to have. We all have it; just most people fob it off as paranoia. I never ignore any gut feeling. I'll kill on instinct, and it never fails me.

I've been bred to kill; it wasn't exactly my first choice. But Victor saw a killer when I was a kid and made me into one. The training I received was some of the most intense and definitely not any kind of UN-sanctioned training in existence. What Victor made me do from a young age had turned me into the killing machine he needed, and I did it without question.

But I wanted more. I wanted to rule. I could be a King. But I couldn't say no to Victor—no one could.

I enter the "Church." A fitting name for the bar where I know my brothers await my arrival.

It's not the only property in the building. It acts as a multipurpose structure. A gold plaque beside the elevator lists the businesses here and which floor they are on. A doctor's office and hair salon are on the second floor. The first floor is occupied by a pet store, and the third floor is a tax agent. I'm going to none of them.

The key that I scoop out of my pocket presses into my palm as I hit the silver button on the elevator in the entry hall. Stepping in, I wait until the door closes before inserting my key into the elevator panel. The elevator shows the main floor, a few upper floors, and a basement level. The elevator goes two stories underground, one story farther than it is supposed to go. I turn the key fully, and the elevator starts to move.

We had modified the panel so that only our families' keys will take us to the final level of the building.

The elevator doors open to reveal a white brick wall with a single door in the center. I pull my key out of the elevator panel and place it securely in my pocket before I produce another key that slots into the door in front of me. The door opens to a semi-lit underground bar. There is a main room and several more private areas branching off. This is the throbbing heart of the O'Sullivans' enterprises. We were all sent into different ventures for the sake of the family, but everything comes back to the Church. Business deals, buying one-night companions, meeting political rivals—everything happens at the most exclusive bar in the country: The Church.

"Hello, Brother." Lorcan greets me with a wide smile that I don't return, but this doesn't faze him. Lorcan is the face of our political empire, and smiling and appearing friendly is part of his job that he never shakes off. He's almost animated in his greeting. It's always good for a mafia enterprise to have people entrenched in whatever political party we need to control, and Lorcan has been molded to perfection for the role.

Behind the smile is a man as equally dangerous as I am. Lorcan leads the way to a back table that has been partitioned off for privacy.

Ronan is already seated at the table. I knew he would be here, but the sight of my younger brother sends my fingers curling into fists; I grin at him, remembering how it felt to slam them into his face. Ronan picks up his drink and raises it at me as I take my seat. The fight we had was deeper than the excuse we gave everyone.

We said it was over a woman, but we never squabble about women; we never have to. They are always there at our disposal.

No. It was over who would rule.

Whispers that Ronan would be the leader of the O'Sullivan family didn't sit right with me as he is the youngest. The right passes first to Lorcan, who doesn't mind as he may lead Ireland one day, then to Wolf, who, in my opinion, doesn't deserve to lead a pack of wild dogs, nevermind people, and then to myself, who has been placed in a box by the Hand of the King, one that I want to get out of, but it's not looking great for me.

Lorcan sits across from me, and the waiter arrives to take our orders. Lorcan orders vodka straight, and I opt for a coffee. Both brothers look at me curiously, but I don't explain why I'm not drinking alcohol. I have a job after this, one that will require me to be clear-headed.

"Everything is looking great for us." Ronan kicks off the meeting. He is responsible for gaining legal sources of income for the family, with part of it going to the "Hand of the Kings," of course.

"My political party is a favorite right now, so I will make it into Dail Eireann and then on up." Our drinks arrive, and Lorcan takes a sip with a smile. "To ruling," he says, Ronan joins in with our brother's positivity.

I don't.

Both of my brothers look at me. Both with the same gray eyes, and dark hair, and dark suits. We are built similarly, but our rearing was all different.

Lorcan wasn't around Ronan and me much growing up, as he spent most of his time in a prestigious boarding school being educated on how to rule Ireland one day. That day seems to be growing closer and closer.

"I've made several deals on the black market for weapons, both in Ireland and the mainland of Britain," I say. Our enterprise is untouchable.

Just like us.

Ronan nods, and he leans closer on the table. "What have you found out about Andrew?"

I sit back in my seat and grin. "If you had showed up to the meeting, you would know."

Ronan glances at Lorcan before he speaks. "We arrived as soon as we could, but you, brother, were already gone. We will have a chance to speak at the annual Diners of Influence party."

Another outing. Great.

"I know about the event. I've been commanded to go with my brides."

Ronan smiles. "I'd like to get a good look at these brides of yours."

I release my cup of coffee before the porcelain cracks under my tightening fingers. "I will take your eyes out if you do," I say.

Ronan laughs, but there is no humor there.

"Now, no need for that. Ronan, you are going to get your chance soon." Lorcan says, always the politician.

Ronan shrugs. "You may be next. Doesn't it look better for your constituents if you are married? Aren't you into your image? Bleached teeth. Kissing babies and shit." Ronan doesn't know when to stop. His mouth has a way of running away with itself.

"You mock me, but having someone in the family in the upper tiers of government will open up avenues for our family that we couldn't even dream of ten years ago." Lorcan fires back in his defense. He's been brainwashed into thinking this is beneficial for just our family.

"You won't be benefiting our family. You are a stooge for Victor," I say.

Lorcan glares at me. The beast rises inside him. We all have it, but he controls it the best. He has to. "Our alliance with the Hands of Kings has given our family power like it hasn't had since Interpol started getting on our asses." Lorcan defends them.

I know I should stop, but I can't. Having Ronan across from me giving me a smug smile that I want to wipe off his face has me continuing.

"I'm not seeing it. Everything we do is for *them*." Bitterness I had thought was deeply buried, raises its head. It's my next job that is bothering me more than anything I have been asked to do before. I've never questioned a kill, but this time is different.

"Careful, little brother. It isn't wise to question my loyalty to me and mine." Lorcan states.

"You seem to be paranoid, Diarmuid. Are you experiencing some disloyalty with your new swines?"

I rise, having listened to enough of Ronan's shit. Lorcan places a hand on my arm. "Sit, Brother." He glances around like someone might be watching.

I don't sit because of that; I sit because I know hitting Ronan once wouldn't satisfy me, and I'm not sure I'd stop.

Ronan grins like he has won.

"So, what was the news about Andrew?" Lorcan swings back to his earlier question.

"He was found dead three weeks ago, a woman's body laid upon his shallow grave."

I pick up my coffee and take a long sip. Shutting my brain down, not thinking about how I killed him, how nobody but the person who laid the body there knows. No one can ever know.

"Who was the woman?" Ronan asks.

I place the cup carefully back on the saucer. "They didn't say. A page delivered the news."

"Why?" Lorcan asks.

"Security reasons," I answer and watch both my brothers.

Ronan grins like this is a game. "They thought the killer was in the room." Bingo. He's right.

I shrug. "They didn't say."

He tuts. "It's clear as mud, Diarmuid. It's served me well to have Andrew out of the way, so I won't pine over him."

"Careful, Brother, you sound like you might have had a hand in his death." It's my turn to smirk.

He grits his teeth and then relaxes. "I'm looking forward to seeing your brides."

I don't know my brides well, but they are my brides, and Ronan better not push me.

"Ronan, show some respect," Lorcan states.

Any man who touches what is mine won't live long enough to brag about it. I make that promise through narrowed eyes at Ronan. It's a good thing that I am supposed to be meeting with my brides tonight. I need to let out some tension. I hope Niamh wears her church dress; it would be my pleasure to destroy it. After a bit more speculation about Andrew, I depart. I must do my job. Duty always calls.

I sit in my car, adjacent to the front door of the school. The graffiti on the wall next to me reads "Stey in Skool." I touch the envelope in my lap, the envelope that contains the identity of the person I am supposed to kill.

I open it again and glance at the picture. It's not a shady-looking teacher or the stern headmaster. I'm looking into the blue eyes of a young boy, Brien Cahill. He is eight. The son of Kane Cahill. Victor wants to hurt Kane, so Brien must die. That's my order. I've never disobeyed an order, but Brien isn't the guilty party here.

Why should sons pay for their father's sins?

I stare at the picture, brain spinning. When I was Brien's age, I made my first kill; Victor made me finish off one of his hits. I hesitated. Victor and Andrew punished me for this later when they found out I shivered and cried. I never hesitated again.

I place the image of the boy back in the envelope when movement along the sidewalk catches my eye. Brien meets two people, who I assume are his parents, and the three walk down the road together. I wait until the family is almost out of sight before I follow them all the way to the church. They enter, and I step onto holy ground only a moment later.

The church is deserted at this hour, and I take a seat at the back. I watch as the parents kneel to pray. I assume the man is Kane. His belly rolls out over his trousers; he's obviously well fed. The strain on his shirt buttons is unappealing to the eye, but his wife still looks up at him with love in her eyes. Love can be blind. In this case, you would be better off having no sight.

The boy runs a fire engine toy along the top of the pew. I could follow them home and wait until the boy's mother starts dinner. While Brien is playing in the yard, I could snatch him, and make him disappear. Kane wouldn't be fit enough to catch me with all that extra cushion.

That's what I could do. Instead, I watch as a priest walks to the family with a bucket and a mop. Brien tucks his toy in his backpack and takes the bucket without question. The priest smiles fondly at the boy and rubs his hair before speaking to Brien's parents.

Of course. Nothing looks better on a college application than a lifetime of volunteer work. This boy has a planned future.

One that I am supposed to take away.

I rise from my seat and walk up the center aisle. The priest glances my way and smiles. When I reach him, he holds out his hand. "Welcome. You are new to our church? I'm Isaac Waryn."

I don't take the priest's outstretched hand. The boy is far enough away not to hear as he mops the mosaic floor of the church.

"We need to talk." I glance at Brien's parents.

I look at Kane, who has paled. "You are Diarmuid O'Sullivan."

I nod. "Yes, I am, Kane."

I don't turn to the priest as I speak to him but keep my gaze fixed on Kane in case he decides to run.

"Father, close the main doors." The priest doesn't act, but Brien's father nods while he swallows. "Do as he says."

The priest doesn't look happy but closes the main church doors. I take a seat behind the couple, and they turn to look at me.

"What is this about?" The wife asks.

"A hit has been placed on Brien's head," I say.

The mother's wild eyes seek her son out as the priest returns.

But I'm focused on Kane. "Why would that be?" I ask him. I know why, but I want to hear it from his miserable lips.

"A gambling debt, I can assume," he says.

The wife starts to rise, appalled at her husband.

"Sit down," I warn her. She slowly does, but not before checking to make sure her son is still there.

The priest stands and watches me, not with fear but disgust.

"I'm the hitman," I say, so they understand the gravity of the situation.

"You will not harm that boy in the house of God." Isaac, the priest, grips his rosary beads like they would save the boy. If I wanted Brien dead, he would be dead.

"I won't, Father, that's why I'm sitting here telling you this."

I glance at Kane. "I don't think a son should pay for his father's incompetence," I say through gritted teeth. Every father figure I had never showed a shred of goodness.

Maybe this is why I got this job. To save someone from a future like mine.

"Brien needs to be sent to relatives in the United States," I say, and the mother covers her mouth, tears pouring onto her hand. Her husband reaches out to comfort her, but she moves away and glares at him.

Good.

"You will have a closed-casket funeral," I inform the priest who sits across from us like the weight of my words are resting on his frail shoulders.

I will try, and hide it the best I can from the authorities. I know this is a huge risk, but I won't take the life of a child. I can't say no to Victor, so this is the only way. I have my own people with the police, but Victor doesn't let anyone know all of the members of the Hand of Kings. One of the authorities could be one of Victor's men, another risk I'll have to take.

"Your sister lives in Texas," The father says.

The wife glares at him again but looks at me. "Can't you renegotiate? We have money," she says and then pauses. "You spent it all, didn't you?" The accusation causes her husband's face to darken.

"This is no place to air our dirty laundry."

Her hand connects with her husband's face. "You brought this down on your son's head."

"This is the only way it can be," I say.

The priest nods. "I can organize a closed casket. He glances at Brien, who is oblivious to how his life is about to change.

I never have sympathy in these matters, but the mother's devastation is bringing out my softer side. I want to offer her comfort, but I don't have much to give.

"You must wait at least a year before following your son to the United States." The mother's sobs grow, but she nods.

The priest looks upon them with shock and sympathy, but he is also scared, and I need to lean into their fear so they do exactly what I say.

"If you don't do as I say, someone else will come to kill Brien, and trust me, they will do it."

I rise from my seat and stare down at Kane. "Walk me to the door," I tell him.

He's afraid, his tail between his legs as he numbly walks toward the main doors. Once we reach the small porch, I stop and look at the man who has done this to his son.

My fist connects with his face. His nose cracks under the weight of the punch, and he cries out, but I cover his mouth with my hand.

"Swallow that fucking pain, you piece of shit. If I hear of you gambling one more cent, I'll come and kill you myself." Blood from his nose soaks my hand. "I won't make it quick. Trust me." His eyes widen, and I release him, leaving the church.

I am so tired of parents who don't think about their kids.

When Kings Rise

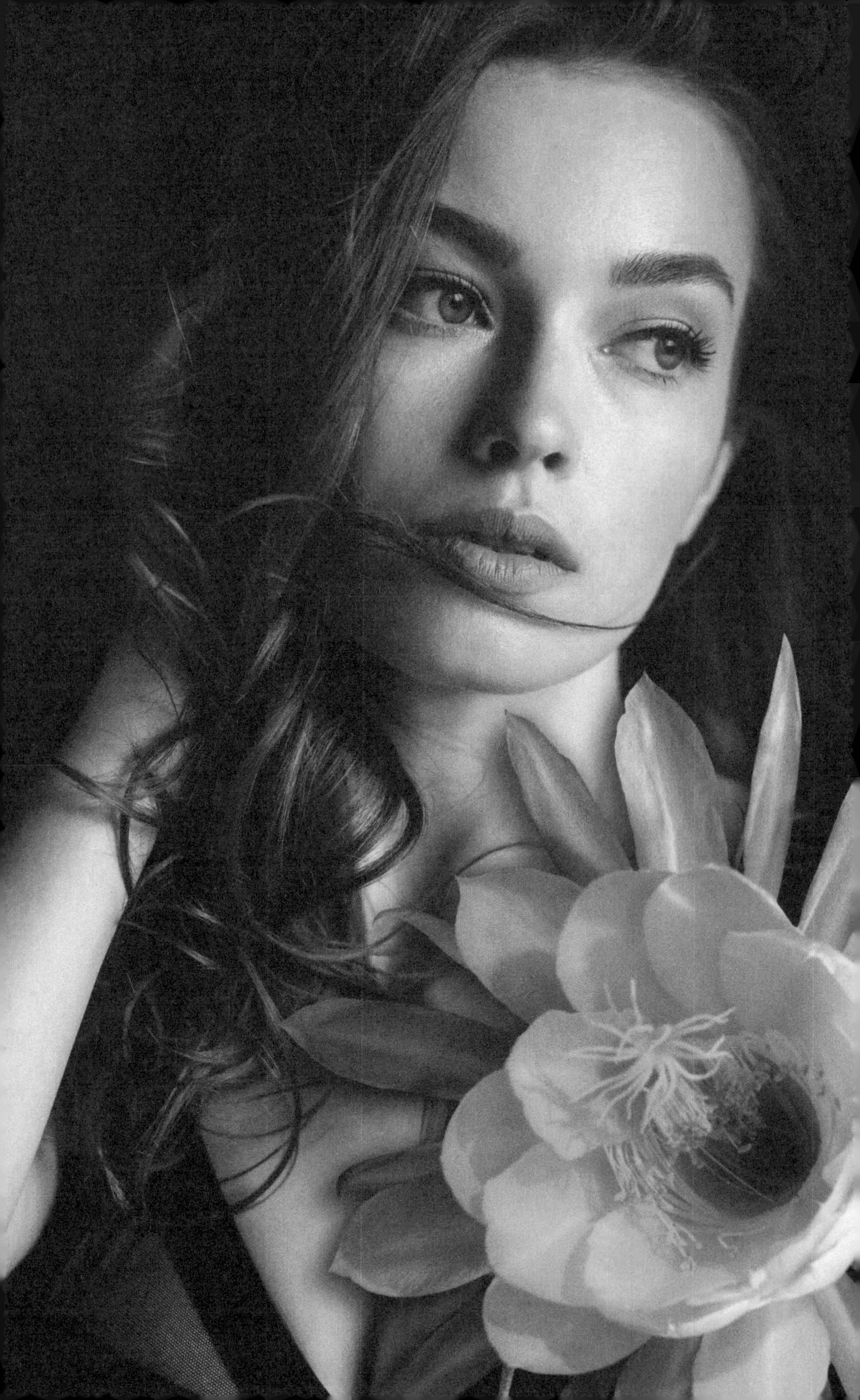

CHAPTER SIX

Selene

P**LEASURE.** SUCH A small word but one I can't find in the one-hundred-word crossword puzzle. A cup of tea that had been piping hot when I started has cooled in the mug. I swallow the liquid and continue my search. I have only two more words to find, and it will be complete.

A groan from my grandfather has me glancing at him.

"You know the rules. I intended on finishing off that puzzle." He slips a small pair of glasses from the breast pocket of his navy shirt and places them on his nose. He stands over me to see how much I have completed.

I continue my search as I speak. "I know the rules. You have twenty-four hours to complete it before I get in." And his twenty-four hours are up.

"Well, the new newspaper hasn't been delivered yet, so you are doing the puzzle early."

With my eyes still glued to the puzzle, I point to the newspaper at the end of the dining table with my pencil. "Your new newspaper is there." I hadn't broken our golden rule. He steps away from me and removes his glasses. I hide a smile as I watch him leave the room.

A large arch joins the dining room with the kitchen. The kettle buzzes away as he clicks it on.

Found it. I circle the word and move on to the final one. I can smell pudding and hear the microwave come to life.

"You know Grandmother doesn't like her pudding microwaved." I smile softly as my grandfather tuts again.

"It's a good thing it's only for me," he says as he re-enters the dining room with his small timer. He pauses at my shoulder and reads the word I'm looking for. "Hmmm," he says but continues to his side of the table and sets up his newspaper, opening it up to the puzzle page.

He only gets the papers for the puzzle. I told him so many times that he could buy a puzzle book as he doesn't read anything else in the paper, but he likes the newspaper and not the small booklet of puzzles.

The microwave still hums in the background as he takes a seat at the head of the old mahogany table. It has some secret drawers that I loved playing with as a kid. I had found the panel when I was coloring under the table on a stormy night. Whenever we had thunder and lightning, my grandmother would make me hide under the table with a flashlight and tell me to color. One particular night, the storm raged for longer than normal, and I got bored of coloring, so I started to touch the table and discovered the secret drawers. Old twine, a pair of scissors, and some clippings of newspapers were all I found. To me, they were treasures.

"I ran into your parents yesterday," he says, and I look up from the crossword puzzle, leaving my memories behind.

"Let me guess…they didn't ask about me." I know the answer already; it should stop hurting after all this time, but I still feel the twinge in my gut at how they don't care.

"No, they didn't," he says. He is never one to sugarcoat things, and I love that about him.

I'm tempted to return to my puzzle but can see in my grandfather's deep brown eyes that he isn't finished talking. I take another sip of my cold tea and regret the decision straight away.

"I know that your mother is my child, but I promise I won't play favorites."

"If you did that, I would be the winner." I smile at him.

"Yes, very true. I just meant that you could tell me about this tiff that has separated you from your parents."

I don't want to talk about it, and it peeves me that I always have to be the one to explain what happened with them.

I don't, of course. How can I tell my grandparents that I only exist to be married off? Discovering that at the age of sixteen was life altering, to say the least. I'd had everything as an only child and could do anything I wanted. A picture-perfect childhood until the reality of what my future would be destroyed it all for me.

I won't be the one to destroy my grandparents' world. They would object, but doing so would only cause a rift within our family, and I'd still have to marry. They loved their daughter, and I didn't want to put my mother in a bad light with them.

"You will need to ask them about that," I finally say.

"I did, and they wouldn't answer. That's why I'm asking you." Of course, he did. He hates to see me separated from my parents even though I've lived with my grandparents for years. Simply saying that my parents traveled a lot, and I wanted company or making the excuse that my mother and I didn't get along has worked for the past few years, but I know the older I get, and my parent's cold response to me, makes my grandparents question this arrangement all the more.

"Móraí, you know I'm not an idiot. If I am not speaking to my parents, there is a good reason."

His eyes soften, "I guess I can trust that reasoning."

"You can," I answer just as the doorbell rings.

I rise at the same time as my grandfather, but my grandmother calls out to us. "I'll get it." We both sit back down, and I focus on finding the final word. I can hear the low, distant hum of conversation at the door. Grandfather gets up as the microwave beeps to let him know his pudding is ready for consumption. When the talking ceases, my grandmother enters the dining room with an envelope in her hand.

She looks so much like my mother; only there is kindness and laugh lines on her face, whereas, my mother's is perfect due to all the Botox she has had over the years.

"It's for you, dear." She hands over the envelope, and my stomach tightens. I recognize the seal instantly. I force a smile and get up, the puzzle forgotten.

"I'd better get ready for the day." I press a quick kiss to her rosy and freshly washed cheek and clutch the envelope to my chest.

"Will you be back for dinner?" she asks, her gaze darting to the envelope that she is curious about.

"I'll try." I press a second kiss to her cheek and inhale the scent of her moisturizer.

"Bye, Grandfather," I call.

"See you later, love. I'll have this new puzzle done in no time," he shouts back from the kitchen.

My grandmother purses her lips and shakes her head. I laugh. "Twenty-four hours, Grandfather," I remind him and leave.

I walk outside and across the small space to the apartment that my grandparents had converted for me above a detached garage. I didn't mind staying in their home, but they believed a lady my age needed her own space. Their kindness never ceases to amaze me.

I climb the stairs and enter my apartment, which is always unlocked.

I told my grandparents they could come in whenever they wanted, but they always knocked and never entered unless I told them they could.

I smile at how lucky I am to have them.

The apartment is cozy and always warm. It's a bit of a sun trap. I appreciate the Velux windows that line the roof, allowing all the sunshine to pour in.

The small two-seater couch is scattered with cushions that my grandmother and I knitted over time. I love each one of them.

Stopping at my small kitchen table, I sit down and turn the thick envelope over, breaking the seal. I allow the contents to fall out, already knowing what it is.

Birth control.

Three months ago, these deliveries began. Part of my agreement with the Hand of Kings is to make certain that no unexpected princes are created. Tonight, I have to go to the house. To perform my duties, duties that I was born to perform. I've had years to let this knowledge sink in, but it never did until now. This is the destiny my parents decided for me before I was even born. It was the only reason I was born.

I was sixteen when they sat me down. I'll never forget my mother's cold exterior. They gave me everything but never affection. I never craved it, as my grandparents filled that void. I never understood why there was no "I love you" at bedtime. No morning hugs. I just assumed that's how parents were with their kids. They were not our friends, but our parents. But it ran so much deeper than that. They literally had me so I could be raised to fill a role that terrified me.

"You are old enough now for us to tell you about your future." My father had started. There was a nervous energy in the room. That energy was mine, but at the time, I couldn't pinpoint it.

"Okay," I responded. Was this the part where I got serious about my future? Would they push me into politics or law? But I nodded, ready to hear what path had been laid out before me.

"The O'Sullivan family have just given us great news." My father glanced at my mother, and for the first time, she smiled with real delight.

I'd never heard of the family and at the time wondered if they owned a firm or some business.

"You will be a bride to Diarmuid O'Sullivan." My father had said while looking back at me, but the smile he shared with my mother was gone.

"A bride?" Confusion at their words had me squirming in my seat.

My father shuffled further in his seat. "Yes, he will have three to select from, and we know he will pick you."

"What if I don't want to be a bride?" What sixteen-year-old thought about marriage? I know I didn't. My dreams were further from that path than they could ever imagine.

My mother spoke this time. "It's not about what you want, dear; it is what is going to happen. Don't get awkward. It's already been agreed upon." She seemed agitated.

"No." I shook my head.

My father rises and walks to the fireplace. "Like your mother said, it's already been agreed upon, and you don't say no to these people."

His words were deadly. These people? "I don't understand."

My mother rose and joined my father at the fireplace. "You will be his bride when the time comes. You will obey him, and you will earn your place at his side. Make no mistake, Selene, there is no getting out of this."

They had left me alone, trembling, and I had run. Run to my grandparents, never revealing the real reason that I couldn't stay under my parents' roof.

The driver opens the car door for me, and I try not to gasp at the sprawling mansion before me. I've been brought up with wealth and have a healthy trust fund, but the sheer size of this place leaves my mind scrambling. I've already seen the Hand of the Kings' mansion, but it still takes my breath away for more reasons than one.

Its sheer size is intimidating, but I know what waits for me on the other side of the door. My first meeting with Diarmuid comes to my mind, when he made me touch myself in front of the other girls, and how he made Niamh touch my breasts, but the part that sent waves of pleasure coursing through me was when he helped me.

I've never had a man's hands on me. It was forbidden. Before I was sixteen, I had some stolen kisses, but that was it. I've never been touched. I hate how I liked Diarmuid's touch so much.

It's cold, and the breeze has me tightening my beige coat around me. I dressed for the autumn weather in warm pants, a cream polo neck jumper, and heavy black boots. My dark hair flows down my back, and I dip my head as I'm led into the hallways of the mansion. Our footsteps echo loudly as a wordless maid guides me through the house, and we don't stop until we arrive at double doors that open into a bedroom.

Amira and Niamh are already inside, and both of them look at me when I enter. Amira can't hide her disgust with me, but Niamh offers a warm, shy smile. I like her and can see kindness in her brown eyes. I wonder if she was introduced into this world the same way I was. Or if she always knew what she was.

Amira, on the other hand, has waves of hostility pouring off her. I try to ignore her as the maid closes the door behind me.

In the room is a large queen-size bed, and my stomach tightens. All of a sudden, I'm feeling hot and shrug out of my beige jacket.

Diarmuid is very attractive, more so than I could have hoped for, and I spent years trying to picture the man my parents handed me over to. His brother Lorcan was easy to find on the internet, as he is into politics, and his face is plastered across so many articles. He's extremely handsome, too, and after meeting Diarmuid, it is easy to see they are brothers.

No matter how attractive Diarmuid is, though, this entire arrangement is so off-putting to me. I feel like a farm animal being led to auction, and I don't know if I or that farm animal would be the better for winning.

I look around and place my coat on a chair where another one sits. I have no idea which girl it belongs to.

Awkwardness fills the room now. Amira folds her arms across her chest. A bold red paints her lips, and she raises a brow at me.

I don't like her at all.

The door opens, and we all shift our stances but relax when a maid enters. She doesn't speak but lays out three parcels all wrapped in deep green paper. Once she leaves, I walk to the bed and open the first one. What in God's name are these? They can't be called clothing. They would barely cover me, and the more intimate parts of the black underwear have Velcro that can be ripped open. The noise of pulling it apart is loud in the room. Amira's elbow rubs against me not so gently as she opens the second parcel. It's identical to the one I opened. She holds it up, and with a cruel smile in her gaze, she turns to me, stretching out the material.

"Hmmm. This one is kind of large. Might be for you." I don't take the garment, and she tosses it in my direction. It lands right beside my hand. I'm not heavy by any means, but I do have far more curves than Amira.

Amira opens the third one and holds it up. "This seems like it's my size. I can't picture either of you fitting in it." She grins and begins stripping out of her clothes, unfazed that she's naked in front of us.

I glance at Niamh, who looks uncomfortable.

Amira slips into the bit of material and does a spin for us. "What do you think?" She smirks.

"Save that for Diarmuid. We're not interested," I snap, hating how she grins at us.

"How stupid of you. This is a competition." Amira walks to a full-length mirror and assesses herself.

When Kings Rise

"If you say so," I remark, and my fingers toy with the undergarment. I am not shy, but I won't parade myself before it's necessary.

"When you fail, I want you to remember what you are seeing right now. This is everything that you won't be." Amira spins from the mirror.

"If I remember correctly, Diarmuid left you to the side the last time," I bite back.

Amira marches to me with fire in her eyes. "Boys are always mean to the girls they want the most." She continues to the bed and picks up a set of silver bracelets that have a small hook on them. She slips them on. "I don't think he is into fat girls."

I bite my tongue, not wanting to argue, but she grates on my last nerve.

Amira looks at Niamh. "Dear God, you look petrified." She laughs. "He won't want something like you."

I spin toward Amira. "That's enough." I defend Niamh, as Niamh doesn't fight back.

I hate bullies.

"Or what?" Amira asks.

I don't answer her; I'm not the fighting type, but I also won't stand for her abuse. We are all here for the same reason and being cruel to one another isn't helping.

Niamh walks to the bed and picks up her own set of identical garments. Amira goes back to looking at herself in the mirror.

"Thank you," Niamh says gently.

I smile at her, but sadness pours into me. "You are welcome." I get a sense no one has ever defended Niamh. No matter the outcome of this, I make a promise to watch out for her.

I strip with my back to the girls and get into the underwear. Just like Amira, I slip on the bracelets. Robes I hadn't noticed are laid out at the top of the bed, and I pick up one, happy to cover up my skin. The silky material is cool against my flesh that has started to feel like it's burning.

57

My bare feet sink into the cream carpet under my feet. Niamh quickly gets dressed and does the same as me, donning the robe.

When we are ready, Amira marches past us and puts hers on, too. She doesn't tie it but leaves it open.

"May the best girl win. I'm certain that's me." She grins.

I want to snap back at her, but the door opens for a third time, and the same maid that had led us here and brought in the garments looks at us all over. The stern look she has worn since our arrival doesn't leave her face. Her nose is pinched, her lips downturned. She wouldn't be getting employee of the month. There is no warmth in her green eyes.

"Follow me." Her lips barely move as she speaks. We file out of the room as she leads us all barefoot up another flight of stairs. We don't meet anyone, and I'm too focused on trying to settle my pounding heart to take in my surroundings.

She opens the third door on the left, and Amira pushes past and enters first. When we are all in the room, I pause. A metal pole is erected in the middle of the room. Off to the right, a wash station has been set up.

Niamh is frozen like a deer in headlights, and the maid roughly shoves Niamh forward and guides her to the pole. Amira is standing at the pole. The maid gives me a stern look, and I join the girls. There are chains dangling above our heads.

"Put your hands up." The maid barks like we are a bunch of disobedient toddlers.

Niamh raises her hands, and I watch as the maid loops chains through the small hole on our bracelets. Amira is next, and for the first time, fear shines in her gaze.

"I'm not putting my hands up there," she says.

I don't blame her; I'm not exactly excited to be chained like Niamh.

What I didn't expect was for the maid to slap Amira across the face. The sound of flesh hitting flesh makes me flinch.

I think Amira will attack the maid, but she surprises me by raising her hands. It is shocking to see how Amira acts like a kicked dog over this.

I raise my hands as I'm next, and once the maid has us all tied up, she goes back to Niamh and takes out black silk blindfolds from a fold in the front of her apron. She places a blindfold over Niamh's eyes. Next is Amira, who has a mark on her face but doesn't object this time. When the maid stops at me, I dip my head, but my mind is reeling. What the hell is this for?

The last thing that I see is the stern look on the maid's face before my world turns black, and I'm left dangling with the other brides as we wait for what is next.

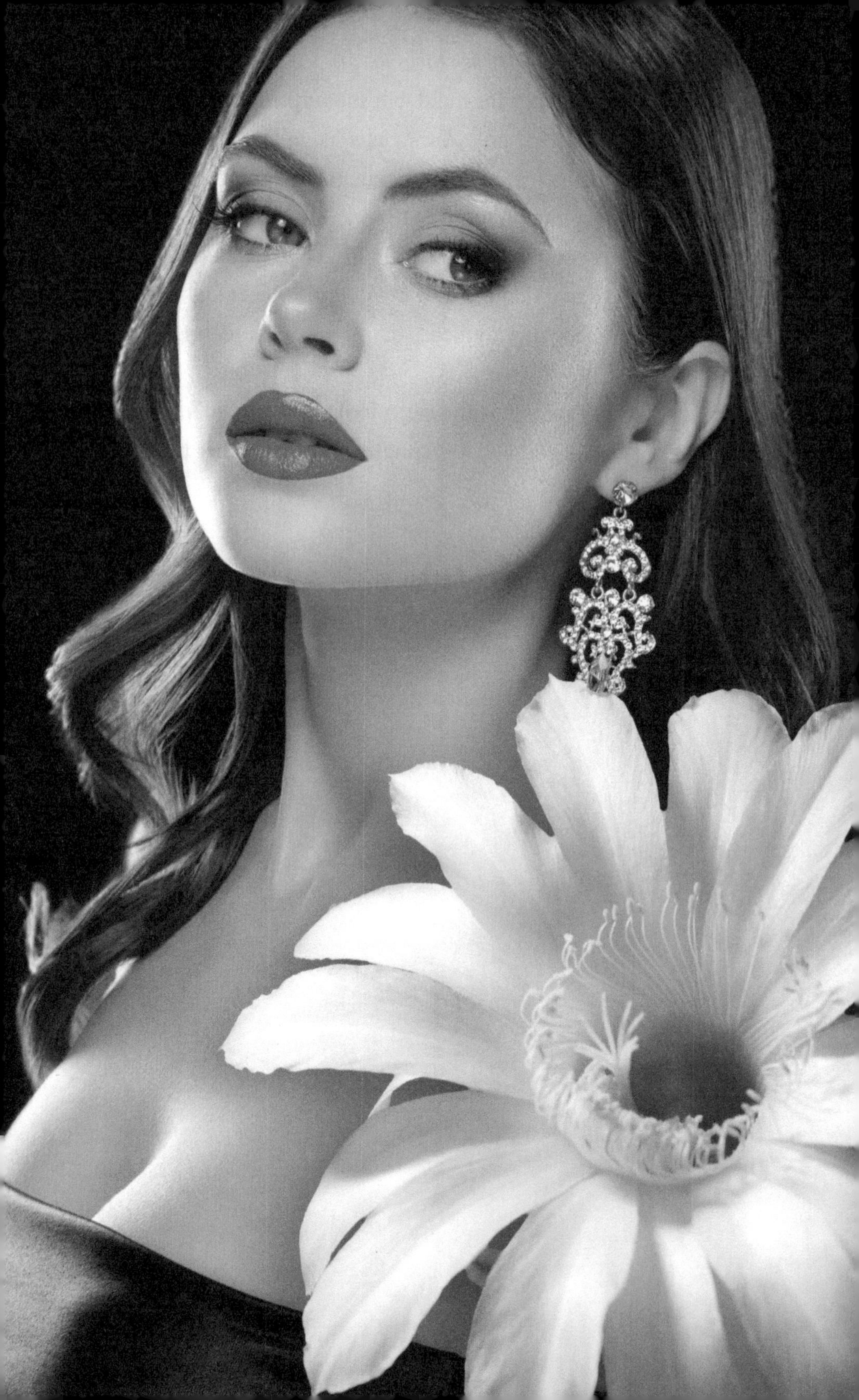

CHAPTER SEVEN

Amira

MY FACE BURNS along with my skin. I want to lash out; I want to hurt someone—the maid, maybe, for putting her hands on me. The moment her hand had struck my face, I was back home with my mother, knowing answering back or retaliating would lead to something far worse. The maid had hit me right on the spot where my mother's hand had struck, and I ached to touch my face.

I pull against the chains, and they rattle. I hear a large intake of breath, most likely from Niamh, who hangs beside me. Is she also thinking of how the maid had hurt me?

Humiliation makes me hot. Shame hammers through my system. I should have hit the maid back; she had struck me in front of the other two, who already thought they were better than me. I had seen the look on Selene's face; pity had filled her gaze. I didn't want her pity; I wanted her gone from here with her smart mouth.

I vow, as I yank against my chains, that the next time the maid touches me, I will claw into her and won't stop. I release more anger on the chains, but they are tight, and I don't get any release.

My mother hit me so much, but I could control myself. She never struck me in front of people and definitely not two girls who were out to get me. They are already teaming up against me. But they don't know how strong I am. I will myself to stay still and try out a smile.

It's all right. The next time I come across that maid, she will wish she had never laid a finger on me.

I stop my musing when I hear the door open. It closes with a soft click. I can't see, and I am tempted to rub my blindfold against my shoulder to see who has entered. My other senses kick in; I can smell him. His cologne brings me back to that first meeting with him. Seeing him walk into the room, watching him instruct the other girls to touch each other. He had left me out once, not this time. I lick my red-painted lips and relax my mouth. I want to smile but tell myself I'll appear too eager, so I keep my lips slightly parted in what I can picture as a sensual manner.

His footsteps are slow, and my body comes alive, wondering what he will do. Will he touch me? I grow moist between my legs at the mere thought.

He walks again, and the air seems to shift as he moves. A breeze brushes my navel, I left my robe open intentionally to entice him. It must be working. A sense of disappointment rushes as his footsteps echo past me, but not before I inhale him deeply.

Silence grows, and then I hear a whimper. It's Niamh; I'd recognize her voice anywhere. She whimpers again, and the shuffle of clothes has me tightening my thighs together, making my chains rattle. A slurping sound has me tingling, and then Niamh's whimpers turn to moans.

I want to shift again; I want it to be my turn. I imagine him with her legs spread, his handsome face buried between them, tasting her. He

When Kings Rise

will taste us all like wine; only I will be the finest. I've shaved my pussy and used some rose-scented shower gel just for him.

The waiting is sending me forward, and my chains rattle again as Niamh continues to moan, and the slurping sound doesn't stop; it only grows faster and louder. Is he using only his mouth or his fingers, too? I want to see what is happening so badly.

I want him to do to me whatever it is that is making her cry out in ecstasy. I tighten my thighs again as she continues to cry out until her moans turn to shudders and the lapping sound ceases.

Diarmuid's footsteps grow close to me, and I spread my legs, letting him know I'm ready and want him, but once again, he walks past.

I hear water splashing and think of the wash station I had seen set up in the corner of the room.

It's agony waiting. *Please let me be next.* But once again, the scent of his cologne flutters past me in a teasing breeze. I get a whiff, and then he's gone.

Selene's breathing grows harsh, and I know he's moved onto her.

She isn't as loud as Niamh, and I don't hear any slurping, so I'm not sure what he is doing to her. But her rushed breaths turned to groans like she's trying to keep them in but can't any longer. Who could? I can assume that Diarmuid is a man who knows exactly what a woman wants. My breasts swell with need. I'm soaking, and I know the mere touch from him would have me releasing the agony between my legs.

It feels like an eternity until he pauses in front of me. I'm pushing against the restraints, yearning for his touch. And I get what I want. A large hand slides between my parted thighs and stops at my soaking, throbbing core. I inhale a sharp breath as he touches my clitoris. His breath brushes my cheek, and I throw my head back and let out a moan as he slips one finger inside me. I've never had any other fingers, only my own, inside me, and this is so much better than I could have dreamed of. A second finger enters, my head shoots up, and the pressure inside my walls grows around his fingers. I push myself down on his hand, wanting more. I moan into his face. His hand lands close to my chin before it trails up to my cheek, and he pauses. I want to scream at him. *Don't stop...* I'm so close to coming, but he removes his fingers and then proceeds to remove my mask. I blink several times at the light before his face comes into view. He's far more handsome up close, his jawline so strong, and his lips are red and swollen. His finger prods my face, and I recoil.

I glance at the other two girls, who still have their blindfolds on. Why has he gone off-script with me? I want to beg him to continue.

His finger circles the ache in my cheek.

"Who did this?"

The slap the maid gave me must have left a mark. I'm sure it's worse since my mother got there first with her own slap.

"The maid," I whisper, looking into gray eyes of steel. "She slapped me."

He nods once; his features appear carved from stone. I'm surprised when he reaches up and undoes my chains. Does that mean this ends? I won't get to come? I want to say something to him but let my hands hang at my side; as with professional efficiency, Diarmuid releases the other two girls, also.

I watch as he strides across the room like an angel who just escaped Hell. He's beautiful but deadly, dressed all in black. He disappears through the door, but I still yearn for his touch. I find myself stumbling after him.

"Do not follow him," Selene warns, but she doesn't understand. It was different with me; he removed my blindfold and not theirs. So, there is more to Diarmuid and me than them. I leave the room and race to the balcony that looks down on the second floor, where a commotion has broken out.

He approaches the maid with slow, controlled steps, but the rise and fall of his shoulders is a warning that he's angry. His guards stand along the wall, their gazes fixed ahead.

"You put your hands on one of my brides?" He questions, but before she can gather her courage to respond, he holds two fingers up, silencing her.

"I should have your hands for that." His statement is so off-hand, yet it sends a deadly thrill down my spine.

"Or maybe your life." He tilts his head like he's really thinking about doing just that.

The maid no longer has that stern look on her face; she appears shaken to the core. I don't know what has her looking up, but her gaze clashes with mine, and I grin.

"Take her out back." Diarmuid addresses one of his guards, who moves toward the maid. I lean over the rail to see what will happen next.

"It's your choice; he will take your hands or your life." The guard grips her by the arm, and she starts to plead, but there is no forgiveness shown for what she just did.

"NO ONE touches them. No one. This is your only warning. I don't care who you are. I will kill you." His anger swirls and grows, yet his voice is low. Deadly. I clutch my heart in joy that brings tears to my eyes. He's protecting me.

I've never felt more powerful as I walk back to the room where Selene and Niamh are waiting with their robes back in place.

I grin at them but don't tell them what just happened. They will have to ask nicely first.

This is exactly the kind of man I need in my life. I have to win his heart, and I will.

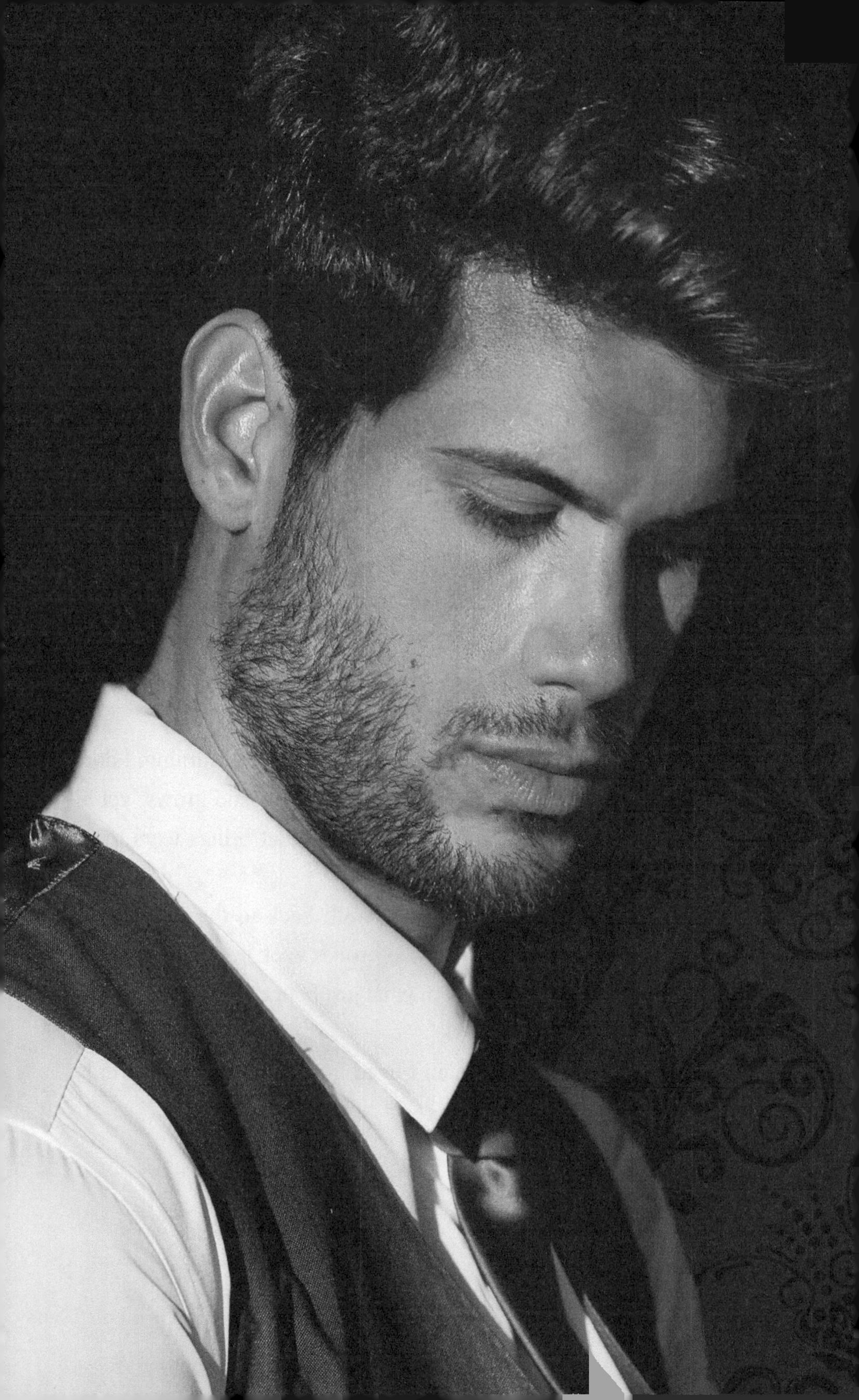

CHAPTER EIGHT

Diarmuid

HANDS OF THE KING EDICT FOUR

Kings are made to lead our world, and they must also lead their homes. Kings are required to take on a Consort. Three candidates (Brides) are chosen for the examination, exploration, and exploitation of the King. One will be chosen as his Consort.

THE TASTING DID little to satisfy me. On the contrary, I find myself wanting more. Needing more. Every shiver and reaction from them had sent a thrill to my cock. I want to take each of them one at a time or all together. Each of them was exquisite. Choosing just one is proving to be a problem. If only I could keep them all…but I know that is against the rules.

The churchyard that I drive into is deserted. But that doesn't fool me; I know from the moment I pass the large wrought iron gates and move down the long, winding driveway that I am being watched.

St. Gertrude's church is the perfect location for a man like Victor to hold his private meetings in. Who would suspect a man of such high standing in our society could be so calculated? Truly, a wolf in sheep's clothing.

Aren't we all? I muse as I park the car and get out. No service is happening, but several people mill about the sanctuary. If this truly was God's house, we would all combust into flames for our sins.

Even the people cleaning the church are Pages and Barons assigned to protect Victor with their lives. Even if someone successfully took out the priest, they wouldn't make it out of the building. Killing Victor meant killing yourself, which is the only reason the priest is still breathing. But a man can fantasize about ending Victor's life. His day will come, just not today, apparently.

I nod at the cleaners as I walk to the confessional boxes. There is a row of them at the back of the church. Each one has its red curtains closed. The one I select has a small, red light above the door, telling people it's occupied. It's my cue to enter this one. I step into the small box and draw the curtain behind me. I can hear a creak of wood from the other side and know Victor has been waiting for me.

I won't kneel as I make my confession. I never do.

"Bless me, Father, for I have sinned," I start.

"When did you sin?" Victor's voice is clear through the wood. I can't see him; I don't want to see his face.

He's asking me if the job was done. "Yesterday afternoon." I think of the boy whose life I was meant to take. He's on a plane heading for a new life in America. My stomach squirms when the mother's look of pure devastation enters my mind.

"With God as your witness?"

"My only witness," I respond. There is a long pause.

"You've done well, my son." Victor gives me his unwanted praise. More silence drags out.

I've always been sent to kill men, but a boy… I know it's wise for me to question it so as not to raise suspicion.

"I wonder if this kind of sin will happen again?" Basically, will I have to kill any more children?

"Not right now."

I glare at the wood that separates us. I can't see through it. It's not like in the movies when you can see the silhouette of the priest through the wooden partition with its small cutouts. This one is pretty much solid, with a few small holes to allow our voices to pass through.

But I don't need to see Victor in order to know what he is doing. He moves, and I hear the worn wood creak as the priest shifts his weight. The sound of the silky fabric of his robe against the rougher fabric of the seat is another noise I can hear. I can almost see Victor's nostrils flare as he lets out a deep breath. Is he annoyed at my line of questioning?

"This is your greatest sin."

"This sin surprised me, I admit with genuine emotion.

"You have done it all for the sake of a greater purpose, a greater world." Victor recites, and I detect no emotion in the old man's voice.

I roll my eyes at that. Brien Cahill's father had a gambling debt. With the wealth of the Hand of Kings, this could have been forgiven without affecting business at all. There was no greatness to this act. It was unnecessary. I bite my tongue, not saying what I truly want to say.

"Have you ever wondered why I didn't send you to boarding school like your brothers?" His voice is closer to the partition. It is something I had wondered. I am as intelligent as my brothers, yet Victor had sent me to run weapons and kill enemies, and now I've been lowered to killing children.

Once again, I seem incapable of answering.

"I plan to make all three O'Sullivan brothers Kings. Do you know why we need Kings, Diarmuid?" Victor doesn't seem put off by my silence.

I already know the answer to this because it has been drilled into my head repeatedly since I was a child. It was part of the plan: repeat the creed until the children believed nothing else. I didn't buy into everything they taught us; sometimes, I saw beyond the curtain that hides the greed, monsters, and madness.

"Yes." I finally answer.

"This world is filled with manmade wonders, ancient and modern. The greatest of these wonders exist only because the right person led an entire nation of people. These men didn't have to heed the whims of politicians. They didn't have to worry about maintaining face for an election. They only needed to obey themselves. They accomplished great feats. The human race was made in such a way that the majority of people are followers. They are made to bring to life the dreams of greater men. Whether through evolution or divine right, some men are just made to be Kings."

I roll my eyes again at his spiel. He loves the sound of his own voice as he spews his poison. Like we are still kids eager to please or terrified to fail.

"We put those Kings in the right places so that humanity can continue to achieve great things. Your brothers will be Kings like the other Kings I have made, but you...you are a once-in-a-generation type of King. You are my Warrior King."

He really loves to talk, and all he is doing is grating on my nerves. He's trying to praise me, so I don't think about the child he thinks I've just murdered. He continues to speak about my brothers and me as if he owns us. I can sense the weight of my gun in the band of my trousers. My fingers twitch. I can detect exactly where Victor is sitting. I could end it all.

"Richard the Lionheart, Alexander the Great, and Charlemagne.

These were Warrior Kings, Diarmuid. Warrior Kings are made to fight the battles that turn the stomachs of other Kings. They are...special."

He must really detect my displeasure at killing the kid. He's never tried to praise me so much in one sitting. Is it panic that I hear in his voice? I think how one single bullet could fulfill every dream of revenge that has ever woken me from sleep. There are guards in the sanctuary, but none of them have my training. I could move quickly. I might be able to get away

"When your father left us, I felt great sadness. He thought that your family could survive without the Kings. He didn't succeed. We didn't let him. When he came crawling back to us, I felt as if the universe had given me another chance to make a difference in this world. I hope that I didn't make a mistake."

All my thoughts cease. Something in Victor's tone has changed. Is it from my lack of response earlier?

I hear the crinkle of paper and wait to hear what he has to say.

"This is the autopsy of Andrew O'Sullivan. A curious document, if I may be honest. The head, hands, and feet have been removed. Without the use of DNA, the coroner may have never identified our dear Andrew. I will say that Andrew didn't have a peaceful exit from this world. One of his lungs was punctured. Ribs broken. Femur snapped. Burn marks on his chest. Obscenely brutal, his death. I imagine that whoever killed your uncle harbored a great deal of resentment toward him." I hide a smile at each one of the wounds I gave my uncle. Pride swells in my chest.

"This line of work does that to a man. You can throw a stone in any direction in Dublin and hit a man who wanted to kill my uncle," I reply.

"Yes, but their want would have never made them actually do the deed. No one is foolish enough to do this. Unless they felt they could get away with it," Victor responds.

I grit my teeth and then relax my jaw. "Obviously, they won't get away with it."

"I don't imagine they will. It's just strange. No one I know would have ever left a body like this in a grave; it would have been destroyed," Victor says simply.

"It sounds like they were stupid," I respond.

"Or that they wanted to have a grave to spit on."

He knows. He knows I did it; that's exactly why I gave my uncle a grave. So I could return to it and remember his brutal death. The way Victor is talking suggests he suspects me, but I don't think he is the one who set me up. He isn't the one who placed the female body on top of my uncle's grave.

"I know how you feel about me, my son. I know that one day, you will disregard your own life and take mine. I can see it in you. I just want you to realize that there are worse monsters than me."

My heart hammers at his confession. The creak of the wood and the flash of light tells me he is gone. I sit for a moment, listening, and when I step out of the confessional box, the movement of the cleaners catches my attention. They don't look at me, but I now know I was surrounded the entire time.

Victor is leaving no room for an attempt on his life.

As I walk out of the church, I realize I may have far more people watching me than I suspected.

When Kings Rise

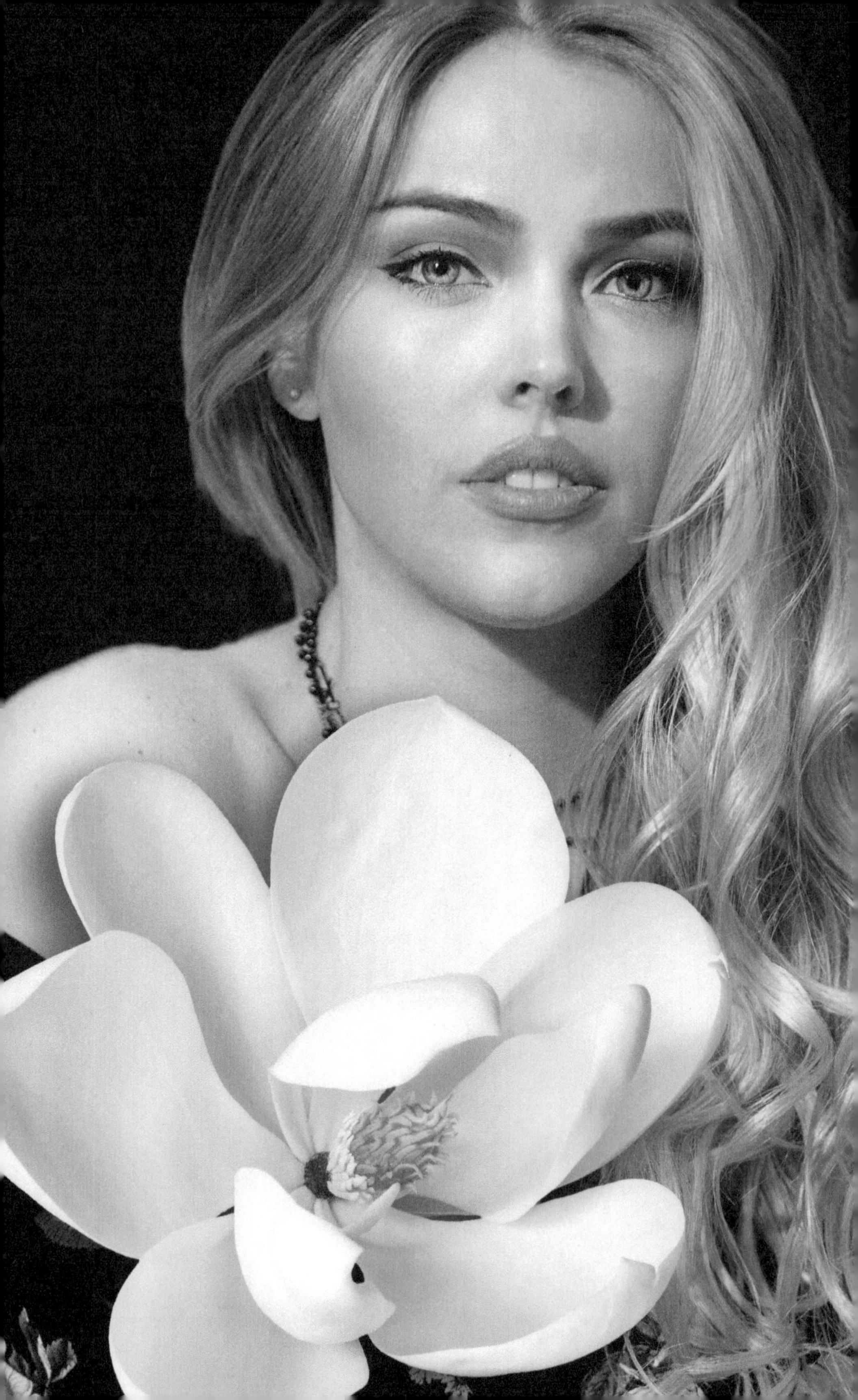

CHAPTER NINE

Niamh

THE SEA IS calm today. Waves crash along the shoreline, some trace across the heavy man-laid rocks and splash onto the concrete slab that warns people not to get any closer. Over the years, people have been dragged into the violent sea when storms erupt. They once were rare, but the weather here is growing more violent as the years pass. Global warming is what people blame.

To me, the sea is freedom; it's a mass of the unknown, so much not discovered. I fill my glass at the sink and continue to watch the relay teams that have crossed the stretch of sea between Ireland and Wales. The groups are large, as no one has ever done it alone. It's dangerous as the rough currents and low temperatures scare away anyone who thinks they are brave enough to try it. Also, the idea that the Irish Sea is home to thirty-five species of sharks makes groups feel far more secure swimming in teams than going it alone.

Today, I'm going to jog along the sea. I'm dressed in a light zip-up sweater, yoga pants, and running shoes. I don't bring music, as I love the sound the sea makes. I always wanted to be the first to row it by myself. Maybe one day.

My father's voice jolts me out of my musing. I empty the remaining water out of my glass into the sink. A kiss is pressed against my cheek before my father speaks cheerfully. "There is my breadwinner; I hope your night went well?"

I cringe internally. What an awkward way to ask me if I got intimate with someone so I could secure my father's business interests. I turn to my father. He's wearing a business suit that fits him perfectly. He's lean for his years and works out most days. "It was interesting," I say before zipping up my sweater fully to give myself something to do other than think about how I was chained or how Diarmuid made me climax.

My God, I didn't want to enjoy it, but it was pure ecstasy. I had tried to fight all the feelings and focus on how he had made me an accessory to murder. Was he serious when he told me that? Did Victor—a priest—give him a command to kill someone? I had so many questions.

"For my sake, spare me the rest of the details. I just hope you do well. We are counting on you." My father pours himself a coffee, and I look at his wide back. No, Ella is counting on me.

My mother wants me to be a ballerina, and my father wants a daughter to give to some sort of cult that promises favors if I'm chosen as a consort. If I fail, I bet that my father's desire for success will weigh more heavily than my mother's dream of a prima donna. A part of me wants to ask my father more about the O'Sullivans, but I know that will raise too many questions, so I keep my mouth shut on the topic.

"I'm going for a run," I say. My father takes a sip of his coffee before assessing what I am wearing.

"You should wear something for the rain; it is expected in an hour or so."

Rain sounds refreshing to me. "I won't be long," I say.

I leave the kitchen and exit the house through the front door that faces away from the sea. The yard has high concrete walls, making our residence private. The yard is small and kept free of plants or anything that would give it color. It's just for cars. A small door to my left brings me out onto the street. To the left will bring me to the sea, and to the right will take me into the small village that has a bus stop. I know what makes me jog right: curiosity and the fact that I might get some answers about the O'Sullivans my own way.

I jog to the small, sheltered bus stop. The local news has been covering the story of a body identified on the outskirts of Rathcoole, a suburb of Dublin. It's the body of Andrew O'Sullivan, uncle of Diarmuid. The obituary states that his funeral is today. So today would be a perfect opportunity to snoop around without fear of running into any of them. I've been curious about where the body of such a high-ranking member of the O'Sullivan family was found. I want to know more about Diarmuid, and this is the only lead I have. I don't want this life, but for my sister's sake, I need to satisfy my parents' hunger so they won't turn to Ella.

No one else is waiting for the bus, and it approaches in the distance. I walk to the edge of the sidewalk as the bus slows down. Using my card tucked in the pocket of my phone case, I purchase a return ticket to Rathcoole. A few people are on the bus, all consumed with their phones. I take a window seat as the bus pulls away, and I have a moment of excitement at going on an adventure. I'd never been brave enough to do this before, but since I was selected as a bride for Diarmuid O'Sullivan, my parents have loosened the leash they normally keep as a chokehold around my neck.

I watch out the window as we pass fields, and before long, the sea opens up and disappears as we enter a more built-up area. I have to switch buses here, and soon, we stop at the small area named Rathcoole. I've never been here before; it's a small village, even enchanting in a way. The buildings are old, and nothing has been updated, but whoever lives here takes great pride in its appearance. All the buildings appear to have a fresh coat of paint on them, each one a different vibrant color, from blues to greens, and I even spot a small pink shop.

The area where I live is suburban; my parents wanted us away from the noise of the city, but the quiet here is almost unsettling. Two people stand at the door of the only supermarket, chatting, and when I jog past, they wave with friendly smiles. It doesn't seem like a place where someone was recently murdered.

I jog to the outskirts of the village. I have no idea of the exact location where the body was found, just that it was here in this sleepy village. I pass a few lone houses, mostly cottages, and I try to see if I can spot garda tape through the sparse undergrowth that grows behind the houses. It grows thicker, and I leave the main road and start my way through the trees and underbrush. A light rain starts to trickle down, and I pause, looking up at the angry sky. The trees rustle around me, birds chirping a song that has me inhaling a lungful of fresh air.

I continue making my way through the tree line. I don't see any tape, and the longer I walk, the heavier the rain becomes. Before long, I'm soaked and thinking of how my father warned me about the change in weather. I should have brought my raincoat. I consider turning back, thinking how foolish this was. What would I find out anyway? Even if I came across the burial site, it wouldn't give me any information. The ground beneath my feet is laced with fallen leaves, and with the recent downpour, it grows slippery. I spin at the sound of a male voice and nearly lose my footing. A hand reaches out and grabs my waist, stopping me from face-planting into the ground.

"Let me go," I say through sheets of rain. The man steps back and raises his hands. "Sorry, I didn't mean to frighten you." He smiles and removes his glasses from his face. Pulling the hem of his sweater, he cleans his glasses before putting them back on. "I'm Rian Morrissey."

His voice is light and happy. Maybe he's a local.

Niamh," I offer up.

"What are you doing out here, Niamh?" he asks, looking around at the trees that surround us. The rain ceases its onslaught, stopping as quickly as it began.

"Looking for something," I say.

His smile widens. "Me, too."

His joyful voice and relaxed stature eased me a bit. He rummages in his pocket and extracts two frube yogurt tubes. He offers me one, but I decline with a shake of my head. I remember having them as a kid, but I haven't in years.

He shrugs and places one back in his pocket before he flips the other around. "Why did the yogurt go to therapy?" he asks, reading the joke off the back.

"I don't know," I answer.

"It had too many cultural issues." He grins and rips the top off before sucking the yogurt from the tube.

"Funny. So, what are you looking for?" I inquire.

He grins. "I could ask you the same thing. But I'm not one for secrets." He glances around us, his brows drawing together. "I'm here to see the burial site of Andrew O'Sullivan." He pushes the empty wrapper into his pocket.

That snippet of information surprises me. "Are you a detective?" I ask. He looks too young to be one.

He continues to smile, but once again, I get a sense that he isn't forcing it; he's just a naturally happy person.

"Kind of. I run a podcast on unsolved crimes."

"How interesting," I say.

He starts to walk, and I fall into step beside him. What kind of luck would it be if he knew something?

"It really is fascinating," he fixes his glasses. "I've loved unsolved crimes ever since James Reyos was proven innocent after spending forty years in prison for a murder he did not commit, thanks to a podcast. I've been trying to achieve the same kind of feat."

"Forty years, really?" I shiver at the thought. Imagine being wrongfully accused and suffering for that long.

Rian nods and smiles. "Yes, forty years of the man's life gone. Poof." He raises both of his hands up, presses his fingers together, then opens them wide. "Just like that. So, I like to keep an eye on unsolved murders. I have a police scanner and fire scanner in my home. It helps me keep up with the emergency calls in the area."

"That's neat," I say, not sure what else to add.

"I arrived at the scene of Andrew's body before the investigators got here. So yeah, it's really neat," he grins again.

"What do you know about Andrew's death?" I probe.

His grin widens. "So, you have an interest in unsolved crimes."

I nod. "Yes," I lie. My interest is only in this crime, in finding out more about Diarmuid, and now Victor is of interest to me.

"You didn't say where you were from." He's suspicious of me now.

"Neither did you," I fire back.

"Touche. Okay, so the scene was discovered by a pair of mushroom pickers. This case is high profile, as the family that Andrew O'Sullivan belongs to has a long-running history of being involved in organized crime."

He glances at me, and I raise my brows as if surprised.

"They say they have left that life, but they all say that. Whatever the O'Sullivans are into now is even bigger and more secretive."

"So, are you a conspiracy theorist also?"

"I don't close off any avenue of investigation. That's what makes me so good."

I want to ask him how many crimes he's solved but resist, as that doesn't really matter here. All I need is information. I don't even have to ask any questions, as he seems happy to offer up all he knows.

"On the day that Andrew was found, he was not the subject of the emergency call. A woman's body was found lying on Andrew's." Rian stops walking, and so do I. "I saw two bodies come out of these woods. I have been trying to see if there are any clues here that the investigators missed. Might have missed on purpose, if my theories are right."

"A woman's body? I never heard about that." Which I hadn't. I want to ask him if he's sure, but he's claiming to have seen it with his own eyes, and if he has a Gardaí scanner that he listened to when the call was made, there is no reason for him to lie to me. But I wonder why no one else knows about the woman's body. Maybe we shouldn't be here; this seems far more dangerous than I thought. Not just one body but two.

"You need to be careful," I say to Rian. He seems nice, and eager to solve this crime, but people like him might get hurt. If someone like Victor knew he was snooping around, would his name end up on some piece of paper in an abandoned post box? Would mine?

"Maybe stick to Gardaí scanners and the internet to do your research."

"What would be the fun in that? Besides, I've already been arrested twice, but I'm determined to find the truth."

The truth. Is it worth the price of someone finding out he has been poking his nose into the burial site? I'm questioning myself for even being here.

I shiver as my damp clothes and hair soaks into my bones. I had forgotten about the rain, too enthralled in what Rian had to say.

"I better go," I say, giving a little wave before turning away.

"You came here for the same reason as I did. You want to know what happened."

The truth is, I don't care what happened; I was seeking some kind of information on Diarmuid and his family, not to find out who murdered a woman.

I don't say anything, and Rian takes a step closer to me. He presses a business card into my hand. "You can call me if you ever want to find out the truth."

I want to hand the card back, but it's Rian who jogs off before I can do so. I glance down at the heavy paper.

When I glance back up, Rian is out of sight, swallowed up by all the trees and underbrush. I pray his curiosity doesn't lead to his ruin.

I stuff the card into my pocket and start back toward the village. I'll dump the card, as I don't want Diarmuid to find out that a man gave me his phone number, not after what happened with the maid. He's far more dangerous than anyone truly knows.

When Kings Rise

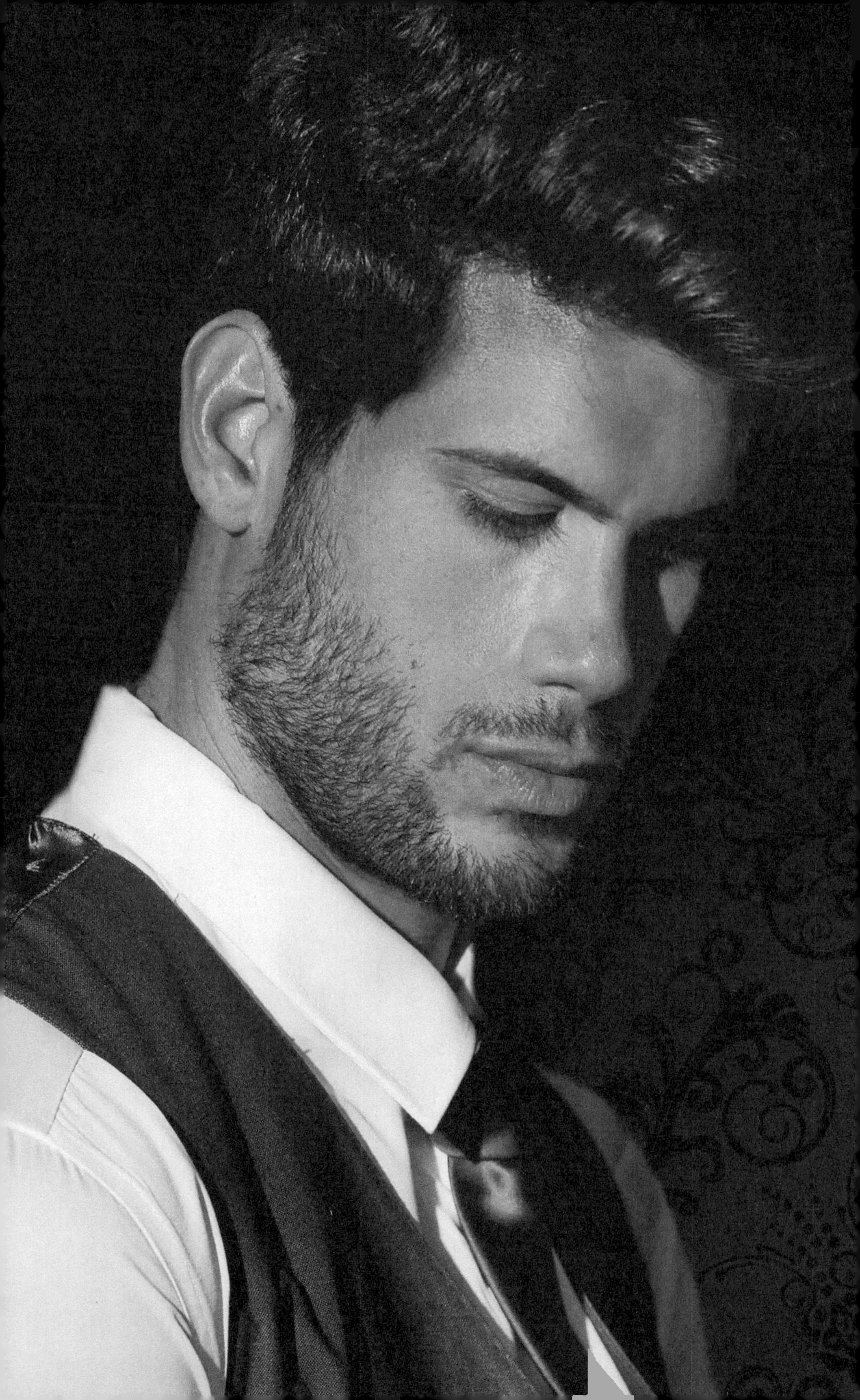

CHAPTER TEN

Diarmuid

HANDS OF THE KING EDICT FIVE

Any member of the order must be committed fully to the order. Personal goals must be cast aside. We live for the order. We live for the betterment of humanity.

THE RAIN FALLS in that persistent, unyielding way that only Irish weather can muster. Standing in the cemetery just outside of Dublin, my men cluster around me, their faces somber, their suits clinging to their bodies as the rain soaks everything. It's a scene straight out of a cliché film, yet here we are, living it in real-time. Only in Ireland would we stand in the rain like this and not think it odd.

The service at St. Gertrude's was a grand affair. Victor's voice, steady and somber, had filled the church, recounting tales of a man whose life had been as complex as the family he left behind. The pews were packed with the O'Sullivan clan and our captains. Among the sea of mourners, I caught glimpses of the Hand of Kings members, their presence a stark reminder of the dual life my family led.

This gathering of secret societies, one hidden within the other, was a testament to the complicated legacy my uncle had woven around us all. And there, amid it all, was Aunt Alicia, her tears flowing with a practiced ease that bordered on theatrical. I knew her grief was expected. Andrew had never shown her kindness, ye, here she was, mourning him as though they'd been close.

Now, as we wait for his final arrival, I can't help but think of my uncle's last demand: a tour of Dublin before his burial.

A final power play from beyond the grave, I think bitterly. He always had to have the last word, even in death, ensuring we'd all be here, drenched and waiting, bound by duty and respect for tradition.

I'm pulled from my thoughts by the approach of one of my men, a newcomer whose name I'm still committing to memory. He rushes over, his apology for lateness nearly lost to the sound of the rain hitting the canopy of umbrellas above us.

"Everything came in just fine, Diarmuid," he says, catching his breath. Rainwater drips from the brim of his hat.

"And who is receiving it tonight?" I ask, shifting my focus to the matters that never seem to pause, not even for death. It reminds me of a poem by W.B. Yeats: in death, he had demanded that all the clocks stop, but time stops for no man.

"O'Boyle's on it," he replies, oblivious to the immediate frown his answer draws from me.

"O'Boyle is on my shitlist," I snap. "Send him for collections. Have Hayes receive instead. We can't afford slip-ups, not now."

"Yes, sir," he responds quickly, turning to relay the instructions.

As he walks away, I turn my gaze back to the cemetery gates, anticipating the arrival of the hearse, my thoughts wandering back to my uncle's life and the intricate web of loyalty, betrayal, and power plays that define it. Even in death, he is still commanding us all. As we stand here, waiting to lay him to rest, I can't help but wonder about the future of the O'Sullivan clan and the secrets we're all bound to keep. Secrets I can't let anyone find out about. My mind reels, thinking that maybe someone in this churchyard knows exactly what I have done.

The rain, relentless in its pursuit, turns the world into a blur of grays and greens by the time Victor arrives. His entrance is, as always, marked by an air of command. He is surrounded by his personal guard, a small group who are loyal to him to a fault and could take out most of us here.

Victor seems to move with deliberate slowness, or maybe it's caution. The ground beneath our feet is slick with rain, glistening like a treacherous carpet under our footfalls. The slope that leads down to the O'Sullivan plot shouldn't be used today, but we have no choice. Andrew has to be laid to rest.

I hope he never rests; I hope the demons are chasing him relentlessly through Hell like he deserves. I watch Victor navigate the slippery descent, and a part of me, dark and unforgiving, wishes to see him falter, to witness him fall and snap his fragile neck, yet the thought is fleeting, chased away by the deeper, more insidious desire that when Victor falls, it will be by my hand and not Mother Nature's.

As Victor passes, the crowd parts, heads bowed in reverence or perhaps fear, whispering "Father" with a mixture of respect and obligation. The title, one he wears as both a mantle and a shield, grates on me. It's a reminder of the power he wields within our family, a power that has shaped our lives in ways both seen and unseen.

The moment is broken by the arrival of the hearse, a sleek, somber vehicle that seems to absorb the light around it. Following close behind is a limo, from which Aunt Alica and Wolf emerge. I step forward, Lorcan and Ronan at my side, to meet them. Aunt Alica's face, usually so composed, betrays a hint of the turmoil beneath, her eyes red-rimmed behind the veil of mourning. Wolf appears stoic, and he offers me a nod of acknowledgment.

As a further insult, Andrew named me as one of his pallbearers. I'm not sure if he knew all along I would take him down, and this was a final slap, or if he truly trusted me enough that he wanted me to carry the weight of his death.

Either way, my men step back as the coffin slides out with ease from the hearse. I'm ready to get this done and over with. I walk to my position, and when we are all ready, I heave the coffin onto my shoulder, and I begin the march toward the grave with the other pallbearers.

Rain drips into my eyes, and it drips off the coffin into the neck of my shirt.

As we make our way down the steep, grass-covered hill, the world seems to tilt beneath my feet. The weight of the casket on my shoulder is too much to bear with the slippery surface beneath. My footing completely slips.

Panic flares within me, hot and immediate. The casket lurches, threatening to escape our grasp and turn this procession into a farce. My legs strain against the sudden imbalance, muscles screaming in protest. In that heartbeat of chaos, a memory crashes through the dam of my consciousness.

I'm back there again, on the cold, unforgiving ground. My uncle's voice, a harsh, grating sound, bellows at me to rise. "GET UP! DAMN YOU, GET UP!" he screams, each word a lash against my already battered body. Pain is my world, a relentless sea in which I'm drowning. The threat of unconsciousness looms, only to be shattered by the cruel cold of a bucket of icy water.

In the shadow of that memory stands Oisin Cormick. He, the hitman whose quiet voice once suggested mercy might be mine. But his words were always lost on my uncle, drowned out by the roar of his own rage.

The beatings never ceased, each one a test of my resolve to remain on my feet, to not give in.

But here, on this hill, with the weight of my uncle's casket threatening to drag us all down, something shifts within me. My foot turns sideways. My leg, the one still loyal, pushes against the earth with all the strength borne of years of enduring and overcoming. Muscles I didn't know I could still summon bulge and flex, and miraculously, the casket steadies.

The moment passes in a blur of effort and adrenaline, allowing the other pallbearers to regain their footing. We continue our descent, a bit more wary, but intact. The irony of fighting so hard to prevent the man who taught me about pain from tumbling into disgrace isn't lost on me. As much as part of me would have relished the fall, I can't draw suspicion to myself. Everyone would wonder why I allowed it to happen.

And especially not today, with eyes watching. Always watching.

Someone among the gathered mourners knows the truth of what I did to my uncle.

As we finally reach the bottom of the hill, the cemetery gates loom before us, a threshold between the past and the present. For now, I focus on the steps ahead.

I don't reset until we no longer bear the weight of the coffin and it's lowered into the ground.

As we gather our family and Andrew's friends around the grave, the world seems to hold its breath.

"Diarmuid," Victor's voice cuts through the patter of raindrops, his tone carrying an edge of command that bristles against my already frayed patience.

I turn to face him, schooling my features into a mask of neutrality. "Victor," I reply, my voice steady despite the storm raging within. "A fitting day for a funeral, wouldn't you say?"

His eyes, sharp and calculating, meet mine. "Indeed. The heavens themselves seem to mourn the passing of an O'Sullivan."

I nod, turning away to hide the flash of anger that I know flitters across my face. This is neither the time nor the place for his rub, reminding me that he suspects I am the one who put Andrew in the ground and will one day return for him.

A few more prayers are said over the grave before we all start to leave and find shelter from the harsh rain.

The dim lighting of our usual spot casts shadows across the table where Lorcan, Ronan, and I sit, nursing our drinks in silence. We have all removed our suit jackets and ties. Our shirts are a bit damp, but the heat in the bar will soon warm us up, and the brandies will heat our blood. It's a private corner we sit in at The Church bar, the usual spot we always meet. Today, it is our refuge from the chaos of our lives outside.

Movement close by has us all turning around. Wolf stumbles towards us, his gait uneven, a clear testament to how much he must have drunk already. I glance at Lorcan, and his features tighten. The two women trailing Wolf struggle to match his erratic pace, their expressions a mix of resignation and discomfort.

He opens his arms wide, and a smile that shows his teeth stretches across his face.

"Meet my escorts: $4000 and $3000," he slurs as he gestures to the women.

Lorcan raises an eyebrow, his response dry. "Clever names."

Wolf's grin widens, unfazed by the sarcasm. "I'm being realistic. It doesn't matter what I call them because all I will ever see is what they cost."

Ronan leans forward, his tone laced with disbelief. "I didn't think a man in your line of work would need to pay for prostitutes."

Wolf's laughter is loud, drawing glances from nearby tables; a wave of my hand has them looking away. If they keep looking, I'll have them removed from the premises. "They are not for me, my beautiful cousins. They are yours. My wares. I brought them for you."

"How generous." Lorcan's reply is as sharp as a knife.

Wolf's expression shifts, the drunken facade slipping momentarily to reveal the cold businessman beneath. "I'm not being generous. I am paying you."

Ronan's eyes narrow. "This is a business transaction?"

I remain silent, observing the weight of the moment settling in. Wolf's actions are a grim reminder of the world we inhabit, where everything has a price, and everyone is a commodity. The air between us feels charged, heavy with unspoken questions and the harsh realities of our choices, of my choices. If he knew I took his father's life, what would he do?

"They killed my fucking father!" Wolf's sudden outburst slices through the murmur of the bar like a gunshot, silencing conversations mid-sentence. His pain, raw and unfiltered, hangs heavily in the air, a stark contrast to the drunken haze that had clouded his actions moments before.

Reacting swiftly, I snap my fingers to catch the attention of a nearby waitress. When she approaches, I give her a pointed look and a quick, discreet nod toward the women accompanying Wolf. "Take them to the main part of the bar," I instruct quietly, "and close the door behind you." She nods, understanding, and guides them away with practiced ease, leaving us in a bubble of sudden privacy.

As the door shuts with a soft click, Lorcan guides Wolf to a vacant seat at our table.

"Perhaps she should bring some coffee and water in, yes?" Lorcan suggests eyeing Wolf with a mix of concern and caution.

Wolf's reaction is immediate, a mix of anger and defiance. "Fuck, no! We are drinking tonight."

The weight of the moment is suffocating as I wait to see what Wolf will do. His gaze lands on me, and the drunkenness seems to dissolve. Around us, the bar slowly returns to its usual buzz, the patrons resuming their conversations, their laughter a distant echo against the backdrop.

Wolf finally looks away from me. "I'm drinking; I don't give a fuck what you are doing."

Ronan attempts to intervene. "I don't think there is room in you for more alcohol," he says.

But Wolf is beyond reasoning, his grief morphing into a bitter resolve. "No, no, no. It's never enough. It won't be enough. Not until the son of a bitch that killed my father is dead. Not just dead. Mutilated. Did you hear what they did to him, Ronan? Did you hear?"

I nod, my expression somber. "We heard, Wolf. I'm sorry for your loss," I say, the words slipping out in a tone that suggests empathy. Inside, I'm amused, finding a dark humor in the situation. The irony of apologizing for a deed I did isn't lost on me.

Wolf's reaction is swift, his pain translating into anger. "I don't want to hear another fucking person say that to me. Look, we made a deal, right? With those cult motherfuckers?"

Lorcan glances around nervously. "Wolf, keep your voice down," he hisses, the tension in his voice betraying his concern for our secrecy.

I lean back casually. "The door is closed, Lorcan. This room is soundproof." My reassurance is meant to ease the tension but also to assert control over the situation. Wolf needs to calm down.

Wolf snaps at Lorcan, "Yeah, shut the fuck up, Lorcan." His dismissiveness sparks a snort of amusement from Ronan.

Wolf continues. "We made a deal with those cult motherfuckers that we would do what they say as long as they help us with our shit. Well, we need help. We need every secret asshole they have to get in on this."

I exchange a glance with Lorcan, unable to resist a jab. "Did you hear that, Lorcan? Secret assholes." My words are laced with sarcasm, a light jab in the otherwise tense atmosphere.

Lorcan, unamused, fires back dryly. "I don't play for that team."

Wolf's eyes, burning with a mix of grief and determination, lock onto each of us in turn. "I'm serious, guys. Look, we are family. The four of us in this room. Fuck Victor. Fuck the Kings. Fuck all of them. We need to take care of our own. Whatever we need to do to get to the guy who did this, we will do it."

Ronan's voice is steady, and his decision is immediate. "I'm in."

Lorcan's shock is palpable. "What?"

Ronan doesn't falter. "I know that you are nervous about your upcoming election, dear brother, but Wolf is right. An O'Sullivan has been maimed and murdered. If we don't make an example of the person who did this, it can happen to any of us. I'll do most of the dirty work, but can you pull some strings if I need it?"

Lorcan, after a moment's hesitation, nods. "Yeah. I can do that."

All eyes turn to me, the final piece of this precarious puzzle. The weight of their stares is a tangible thing, pressing down with the gravity of the situation laid bare before us. Wolf's sex trade, Ronan's legal enterprises, Lorcan's government ties—each plays a critical role in the fabric of our syndicate. But I'm the assassin who will be expected to kill the person who took Andrew's life. Funny how this is all coming full circle, and being part of this circle gives me full control.

"I'm in," I say, my voice steady, betraying none of the turmoil churning within.

As we sit in the dim light of our secluded meeting place, a pact is forged. The waitress arrives at our table with fresh brandy for everyone. I hadn't even noticed anyone ordering the drinks. But Lorcan sits beside the small buzzer that goes directly to the bar. He must have been the one.

He doesn't appear happy but raises his glass. When the door closes, we all do the same. "To justice," he declares.

"To killing the motherfucker who killed my father," Wolf chimes in. We all click glasses, and I have no idea how I'm going to pull this off.

When Kings Rise

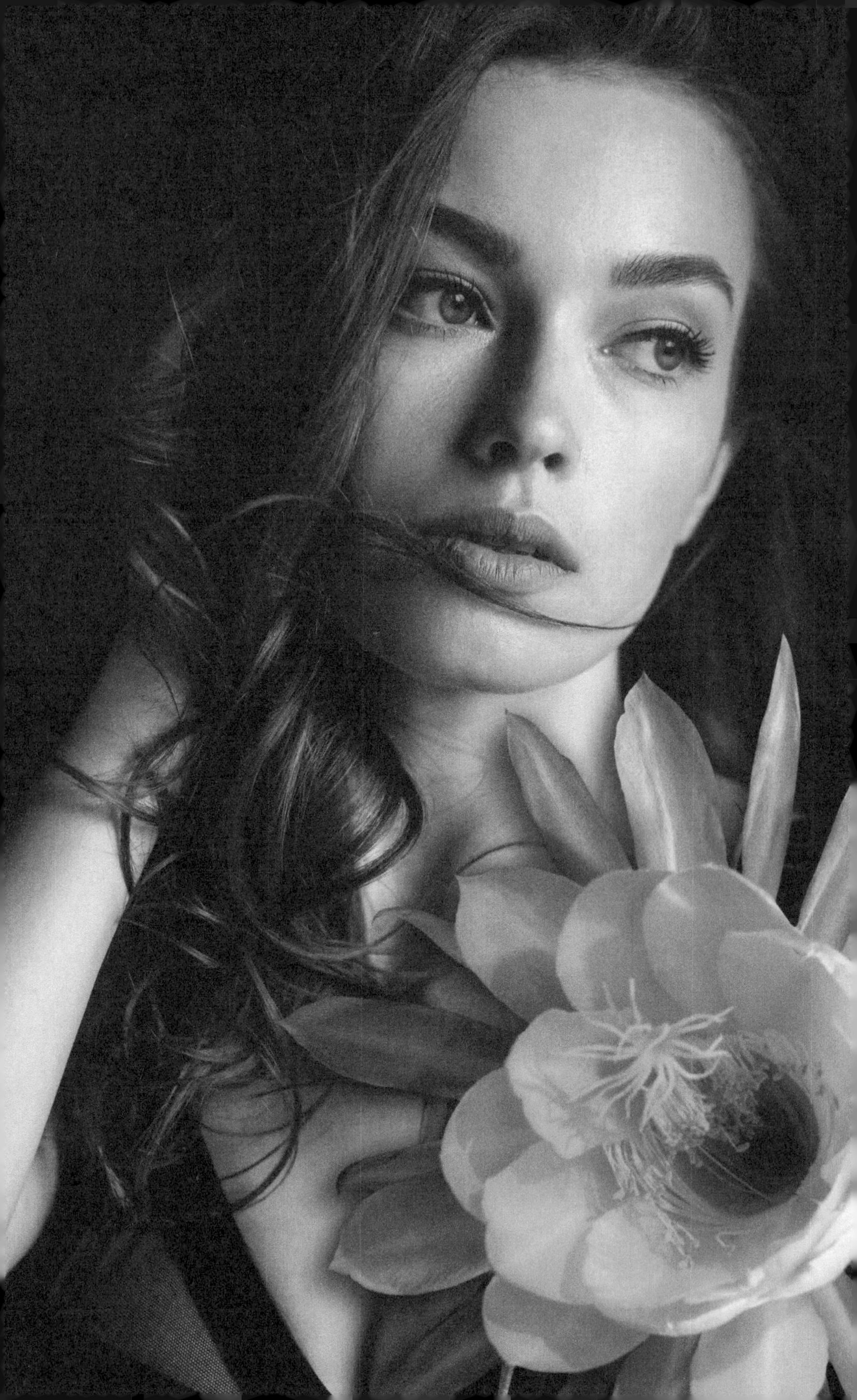

CHAPTER ELEVEN

Selene

THE SOFT GLOW of the evening light filters across my vanity mirror, casting a warm hue on the array of makeup and hair products scattered in front of me. I'm seated comfortably, the chair's plush cushion a small comfort as I unroll the curlers from my hair, each lock falling into place with a gentle bounce. My reflection stares back at me, a mix of anticipation and nerves for tonight's work event. It's not every day you get to represent your department at such a prestigious gathering.

Just as I secure the last curl into place, a knock at the door has me pausing, glancing at the clock. Who could it be at this hour? I stand, smoothing out my dress before making my way across the main room of my apartment. The hardwood floor feels cool under my bare feet, a sharp contrast to the warmth of the room heated by the late afternoon sun.

Opening the door, I'm greeted by the familiar face of my grandfather, his eyes carrying a mix of apology and concern.

"I'm so sorry, a chroí," he starts, his voice carrying the soft lilt. "I know you are getting ready for your work event, but you have a visitor."

I tilt my head, puzzled. "Who is it?"

He steps aside, revealing a figure. Niamh stands there, her teeth chattering, eyes wide with what appears to be fear or shock, and a tight grip on something in her hand.

"Selene! I need to talk to you!" Her voice is urgent.

I nod to my grandfather, mouthing a silent thank you for bringing Niamh here. "I'll take it from here," I assure him, my voice steady despite the flurry of questions swirling in my mind.

Gently, I place my hands on Niamh's shoulders, guiding her into the warmth of my apartment. The door shuts with a soft click behind us. The contrast between the cozy interior and the crisp air outside makes me aware of the tremble in Niamh's frame.

"What's going on?" I ask, leading her toward the sofa. My heart races, not just from the disruption, but from concern for my friend standing before me, who is visibly distressed.

"I didn't know where else to go, and I know we are not exactly friends, but we could be friends. I mean, in different circumstances—"

"Shush for a moment. We need to get you warm. Take off your clothes," I interrupt her. All I can do is wonder what has happened. And why has she come to me.

Niamh looks around my apartment for the first time and hesitates.

"Oh, for Peter's sake. I'll turn around. Take off those clothes, and wrap yourself in the blanket from the couch. Really. How can you possibly still be shy with me?" We have seen each other naked and even heard each other's cries of pleasure.

Niamh's cheeks pinken, but she nods, and I turn around to give her some time to get changed. I hear the wet clothes hit the floor and then it sounds like she settles herself onto the couch.

"Okay," she whispers, her voice still trembling.

She's wrapped herself in the thick blanket, and more color floods her face, but her shivers are still there, still visible in her hands that tighten on the blanket.

I sit down beside her. "Now, how did you find me?" I ask.

"It was an ordeal," she says.

"All right?"

She nods and pauses as if trying to gather her thoughts. "I had my phone, but I don't have it anymore. I mean, oh God, I am so cold."

I want answers but I don't press any further. Instead, I get up and move to the kitchen to put on the kettle. I get down a cup and make her a steaming cup of tea. She accepts it and wraps her cold fingers around the mug.

I wait as she takes a few sips. But there is a wildness in her gaze, as if she's afraid.

"You're safe here, Niamh."

I can't help but feel a twinge of something—pity, perhaps. Her presence here, in my home, is a testament to her desperation.

"I found your parents' address online. Most addresses are online if you've lived in one place for a long while," she confesses, her voice steadier now with the warmth of the tea seeping into her bones.

The mention of my parents sends a jolt through me. "You went to my parents?!" I can't keep the shock from my voice, the thought of Niamh encountering them, of all people.

"And I'm so sorry that I did. I just assumed that you were living at home like Amira and me. I didn't realize that your parents are—"

"—assholes," I finish for her, a bitter laugh escaping me. It's a harsh word but fitting. Their estrangement is a wound that's never fully healed.

"YES. I'm so glad you said it," she agrees, a flicker of a smile gracing her lips.

"My parents gave you this address?" The very idea that they'd help, even slightly, is surprising.

"Yes, but it was a trip getting here. I was jogging, and then I needed to find you. It has been raining for hours. I tripped, and my phone went into a gutter, but I still have this!" Niamh raises her fist, clutching something tightly.

"And what is this?" I ask.

Niamh opens her hand to reveal a business card with a phone number on it.

"A guy's phone number, and I shouldn't have it." She looks down at the card with a frown.

"Okay?" I have no idea where this is going. Does she no longer want to marry Diarmuid? Is this what it's all about?

Niamh releases a long sigh. "I wanted to find out more about Andrew O'Sullivan, Diarmuid's uncle who was murdered."

I nod. "I heard about that."

"So, I went to the village where his body was found."

"I bumped into Rian, a podcaster who likes to look into unsolved murders. He was looking into the case of Andrew O'Sullivan. It wasn't just Andrew's body found at the burial site; they found a woman's, too."

This surprises me; I hadn't heard anything about a second body. "Do you know who the woman is?" I ask.

Niamh shakes her head and takes another sip. "That's why Rian gave me his number. He said if I was interested in what happened, I could ring him."

"Are you going to ring him?" I ask.

Would I ring him? Would I want to know more? I'm not sure.

"I don't know," Niamh answers honestly before sipping her tea.

I checked the time; we don't have much left before the event.

"So, what are you planning to do to get ready for tonight?" I ask, watching her closely. Niamh's reaction is immediate, a mix of confusion and dawning realization.

When Kings Rise

"What do you mean? What's tonight?" Her voice is tinged with genuine puzzlement, the weight of her earlier ordeals clouding her memory.

I can't suppress a slight smile at her baffled expression. "The annual Diners of Influence event at the Hand of Kings mansion," I remind her, emphasizing the importance of the evening. This event is not just any social gathering; it's a cornerstone of our community's calendar.

Niamh's face drains of color, panic rising in her eyes like a storm surge. The reminder seems to hit her with the force of a physical blow.

Seeing her distress, I step in, a surprising sense of protectiveness washing over me. I've felt protective of her since meeting her; it's nice to take care of someone else and forget my worries for even just a moment. "Don't worry. I've got it," I assure her, my voice gentler than I would have expected. It's a strange sensation. But, I recognize it for what it is: I want a friend.

I lead her to the bedroom, my mind already racing through the logistics of preparing us both for the evening. "Take a shower," I instruct firmly, pointing towards the bathroom.

Niamh nods, still wrapped in the blanket, and enters the bathroom, closing the door after her.

While she's in the shower, I turn my attention to the wardrobe, laying out dresses on the bed. Each piece is beautiful and sexy, designed to make an impression. I've laid out the last dress when Niamh emerges from the bathroom wrapped in a towel, her hair damp and her expression wary. She surveys the dresses I've chosen, a flicker of admiration in her eyes quickly overshadowed by concern. "I'm worried about showing off my shoulders," she admits, her voice small. "My mother always called them manly."

I meet her gaze; all the dresses are sleeveless. "They're not manly; they're strong. Men don't own strength." It's a declaration, a challenge to the insecurities that have been unfairly thrust upon her. It seems I'm not the only one with an asshole for a mother.

As we dress, the room transforms into a whirlwind of fabric, shoes, and accessories. I help Niamh with her jewelry, hair, and makeup, each step bringing us closer to the image of sophistication and power we aim to project.

"Why are you helping me?" Niamh asks, her voice laced with wonder as I apply her makeup with careful strokes.

"I'm not concerned about the results of this competition between us," I reply truthfully. This night, this event, transcends our personal battles. There's something greater at stake, a realization that's slowly dawning on both of us.

Niamh's next question catches me off guard, a piercing look in her eyes. "All of us are here because we have something to lose. What do you have to lose?"

For a moment, I'm speechless, the question striking at the heart of my own fears and doubts. "I have no idea," I confess, the admission more revealing than I intended. It's a moment of raw honesty.

The two of us are now ready for the event, and we both look at ourselves in the full-length mirror.

I still see the shadow of fear in her gaze.

"The woman….from the grave?" I start.

Niamh nods. "I've been thinking about her, too."

I smooth down my navy dress before turning away from the mirror. "She probably has a family out there. Someone waiting for her to come home." I frown; that must be horrible.

"Someone who misses her." Niamh continues as she turns to me.

There is a brief silence between us. It's wrong that no one knows who she is or that she is dead. Why was it kept out of the spotlight? But with Rian's help, we might be able to get answers.

I nod at Niamh, a confirmation of what we should do. "Then, let's get her home."

Niamh smiles. "I agree." Before we leave, she picks up the card with Rian's number on it and stuffs it into the pocket of her dress.

When Kings Rise

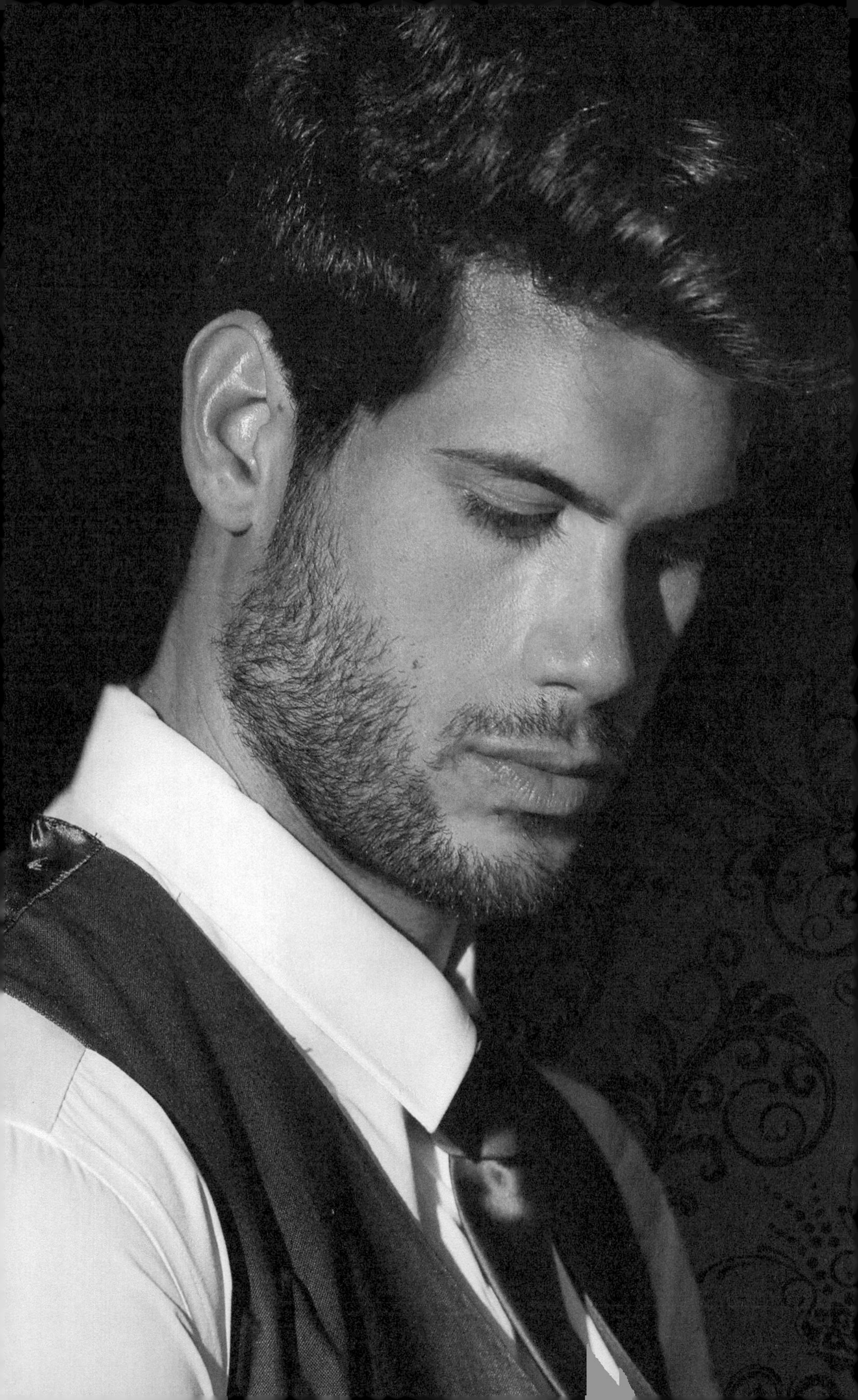

CHAPTER TWELVE

Diarmuid

HANDS OF THE KING EDICT SIX

Kings are made to lead our world, but they must be guided. One Hand shall place the Kings in their places. One Hand should make Kings. One Hand should destroy Kings.

THE RAIN HAS ceased its relentless assault on the world outside, leaving behind a serene yet sodden landscape. As I step onto the Hand of Kings manse's grounds, the aftermath of the storm is immediately apparent. Every surface, every leaf and blade of grass, is sheathed in a heavy coat of raindrops, like nature's own jewelry. The lights from the manse cast their glow onto the gardens and lawns, turning the water droplets into shimmering diamonds.

I hand my coat to the doorman, but he doesn't make eye contact. In fact, he barely acknowledges my presence beyond the necessities of his role. I suppress a smirk, finding a twisted amusement in their obedience.

The grandeur of the manse never fails to impress, a testament to the power and prestige of its occupants. But tonight, it feels different.

As I step into the warmth of the place, the murmur of conversation and the subtle strains of music greet me. The air is thick with anticipation, every guest playing their part in the night's proceedings. I scan the room, my eyes adjusting to the transition from the dimly lit gardens to the brightly illuminated interior. Here, in this den of influence and intrigue, every smile hides a motive, and every handshake is a calculated move on the grand chessboard.

I navigate through the crowd, acutely aware of the space that seems to open up around me. It's as if my reputation precedes me. I don't mind. Let them whisper, let them speculate.

Good. The last thing I need is drama tonight. As I move through the room, the precise orderliness of everything, from the perfectly pressed uniforms of the staff to the hushed, almost non-existent conversations among the guests, signals Victor's presence.

Michael, Victor's Page, finds me among the throng. His greeting is formal, almost excessively so, but that's to be expected given his position. His is the lowest rank in the Hand of Kings, yet tonight, he bears a message of supposed compassion. "Victor will not be attending the event, but he wanted to extend his condolences," he informs me, his voice steady, betraying no hint of the personal sentiment behind the words—if there even is any.

Victor is rubbing salt in the wound. Finding a chink in my armor. I smile at Michael. "Tell Victor I thank him for his condolences," I say as Michael nods and departs.

As more guests arrive, the dynamic shifts subtly. My brothers, ever predictable, make their customary beeline for the open bar, their nods in my direction serving as our only form of acknowledgment. The crowd

When Kings Rise

is a mix of actors, politicians, religious figures, and even a NASCAR owner/driver. It's a testament to the event's reach and the diverse interests it attracts. The air buzzes with the undercurrent of networking and deal-making.

Then, amid the sea of faces, I spot Niamh and Selene arriving together. Seeing them together shows me they are bonding. I'm not sure how wise that is.

As I watch them make their way through the gathering, the real work of the night begins. Beyond the handshakes, the polite smiles, and the arranged table settings lie the real battleground.

And so, I ready myself to join the fray, my eyes always searching, always assessing. Tonight, like so many nights before, is a game of chess played on a grand scale. But as I've learned, in this game, every piece has the potential to be a king—or a pawn.

"You ladies look beautiful," I say as Niamh and Selene approach me, two stunning brides. I take each of their hands and place a kiss there that lingers longer than it should. Both blush, and I hope they are thinking about the pleasure I gave them. Both of them were equally delicious.

The moment is shattered when Amira appears. She's barely dressed, her black tight-fitted dress more fitting for risqué entertainment than a high-end gathering. Her lips are painted red, and she reaches up on the tip of her toes and presses her lips to mine. The reaction from the crowd is mixed—a few gasps, a smattering of applause. Amira smiles under the spotlight and gives a shy smile to the crowd.

"Where is the rest of your dress?" Selene's comment is biting, and I suppress a smile at her obvious jealousy. But she is just as stunning as Amira.

Amira grips my hand as if I have chosen her. The thought of pushing up the short black dress and fucking her right here and right now sends waves of pulsing want to my lower regions.

I don't release Amira's hand as the promise in her eyes has my mind reeling.

But the moment is fleeting. Isaac Waryn, the priest with ties to Brien Cahill's departure to the United States, appears at my elbow, pulling me back from the brink of scandalous indulgence.

"Diarmuid, how are you?" he asks, but it's as if he wants to say more. The fact he's even approaching me in the presence of the hand of the kings makes me annoyed. I release Amira's hand.

"Very well, and yourself?" I ask.

"I'm good." He answers and glances around.

"If you can excuse me." I glance at my three brides as my reason not to talk to him.

He seems to understand that this isn't the time or the place. "Could you give me directions to the main dining room?"

I'm ready to tell him where it is when Selene steps forward. "I can show you, Father."

He smiles at her, and she walks away with an eagerness that I think is brought on by her annoyance with Amira's earlier display of affection.

When Selene departs, this leaves Niamh standing alone. She's a vision of elegance and nervous anticipation. She looks incredible, her beauty a stark contrast to the raw, unbridled allure of Amira. And yet, it's Amira who consumes my thoughts, her audacity and the glint in her gaze promising me so much more than …fun, and it ignites a fire within me.

Excusing myself, I take Amira's hand, leading her away from the prying eyes and whispered judgments. Together, we step out into the night, leaving it all behind. The cool air of the evening envelops us, a welcome respite from the intensity of the manse's interior. In this moment, with Amira by my side, I will be uninterrupted in exploring the depths of desire and defiance that she so effortlessly showcases.

"Where are we going?" Amira asks. I wait for shyness to soak into her gaze but only excitement and want is there.

I have the same want, and I hold out my hand, and she easily slips hers into mine.

"I believe the last time we were together, we got sidetracked," I say, thinking about how the maid had struck her.

She nods. "You took care of it." She stops walking and smiles up at me.

I reach out and cup her face. "You are mine to take care of."

She inhales a quick breath that makes her breast strain against the soft, silken fabric of her flimsy dress.

I lead Amira toward a secluded sanctuary known only to a few. The fountains, now silent and drained for the season, offer a hidden alcove of privacy.

The farthest fountain, hidden from the view of the manse's windows, is where I take Amira. Here, the oversized vases that once adorned the walkway are absent, leaving behind square slabs of marble that serve as pedestals. With care, I remove my hand from Amira's and take off my coat. I spread my jacket over the cold marble, a makeshift bed.

Amira bites her lip as she glances down at the jacket, and as if sensing my intentions, she lies down and spreads her legs. The dress rides up higher, revealing tanned thighs.

The cold is forgotten as we both yearn for the same thing. I kneel between her legs, and her gaze remains transfixed on me. Like she doesn't want to miss a moment. Dipping my hand and sliding it up her thigh and slipping one finger under the black panties she wears, I don't stop until the warm folds give me access, and she moans loudly. I move my finger in and out before pushing in two fingers; when I remove them, I place my fingers in my mouth. I want to take my time with her, but our disappearance will not go unnoticed.

She spreads her legs further, the dress now riding at her midriff as I unbuckle my belt and push my trousers and boxers down far enough to release my raging cock. I need this release so badly, and Amira seems to be the only one willing. I will have all three, but tonight, it will be Amira.

The minute I bend down to place my cock at Amira's opening, her hands find my shoulders, and she's pulling me to her with a greed I can easily match. The minute I slide my cock in, her folds stretch around me, and I lower myself. Her fingers curl around my shoulders; her eyes widen at the sudden intrusion between her legs.

I can't be gentle or slow, and I'm not. Amira lies still under me, her eyes tightly shut as I start to push my full length into her. Her teeth clamp down on her lip as if she is stopping herself from screaming out. I don't want her to scream and draw attention to us. I pound faster and harder; her eyes snap open, and before she can make a sound, I press my lips to hers and grip her hips, demanding her body to open up further for me.

I don't stop until I'm fully in, and then I fuck her like I want to. She cries into my mouth, and it's delicious as I take her virginity under the fountain. The faster I go, her cries turn to whimpers before they morph into moans that have her opening her eyes and looking into mine. With our mouths barely touching each other's, we breathe and pant as I fuck her until I release my seed into her body. She cries out only seconds later, her own release washing over my cock that's still wrapped in her warm folds. The muscles tighten and clench around me like they are demanding every last drop of my cum, and I'm happy to oblige.

Afterward, we make our way to the main dining room. We arrive just before Victor, the man whose presence dictates the rhythm of the night. His voice is a familiar drone that I scarcely register. My focus is elsewhere, lost between Amira and Michael's earlier words that Victor wasn't coming. Was that message only for me? What was Victor playing at?

As we walk, I glance at Amira. On our way back, rain has started to fall, and one raindrop traces a slow, deliberate path between Amira's breasts. It's a distraction until we are ushered to our table.

Niamh, one of my brides, has her attention fixed on her plate, a deliberate attempt to remain unseen, unnoticed. Selene, however, offers a stark contrast. She isn't eating, her plate untouched, her focus not on the food but on me. Her eyes hold a tempest, fury, and accusation woven together in a silent rebuke that speaks louder than words ever could.

She's furious.

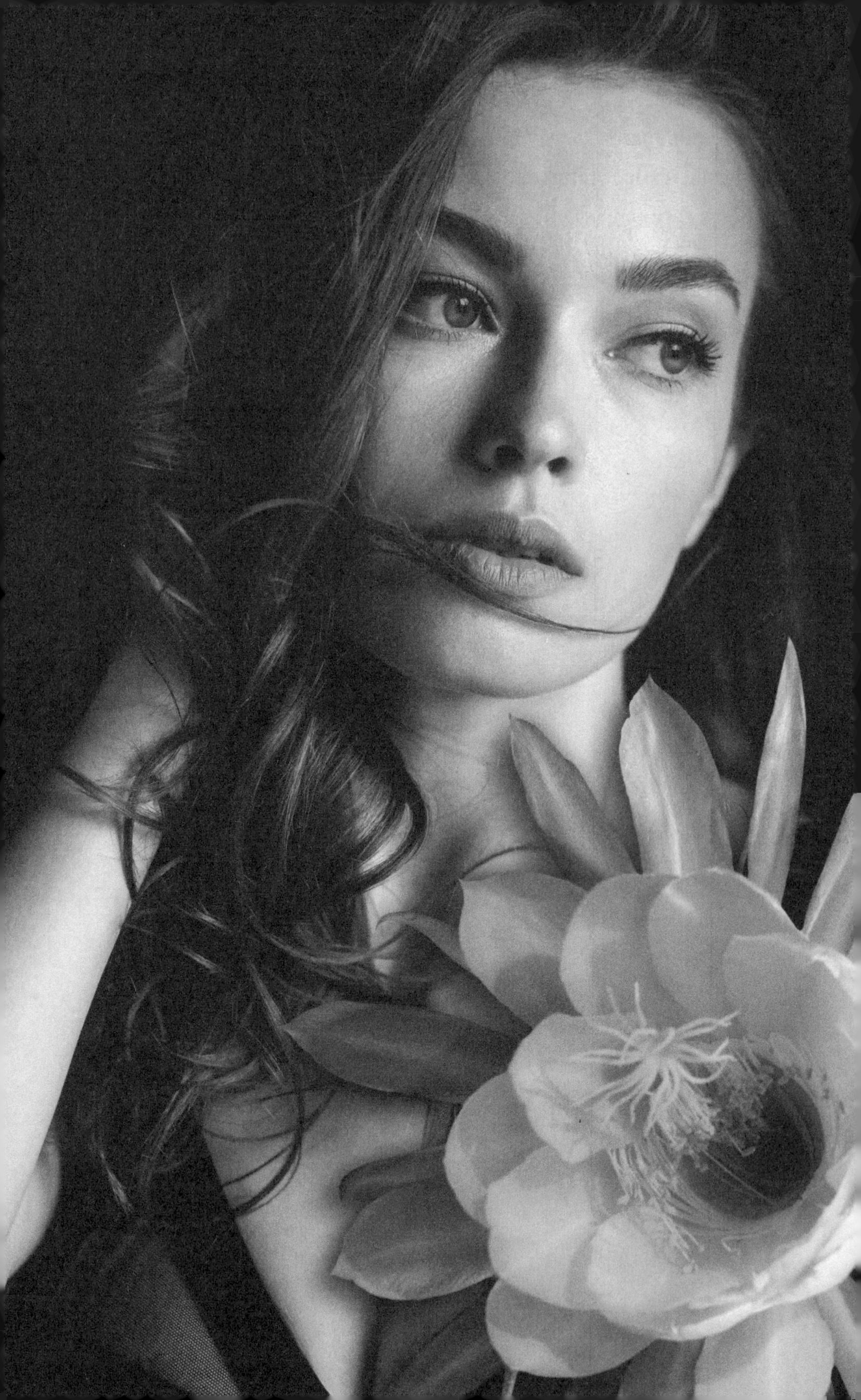

CHAPTER THIRTEEN

Selene

AS I GUIDE Isaac Waryn toward the main dining room, a sense of guilt tugs at the corners of my heart. I couldn't help but feel a pang of regret for leaving Niamh behind. It was never my intention. In my mind, I assumed she would know she was welcome to join us. It seems I assumed too much.

Next time, I need to just tell her to come. The last thing I want is to create distance between us over a misunderstanding. I would hate for Niamh to think I had intentionally left her with Amira.

Walking beside the priest, a habit from countless similar events, prompts me to almost offer him my arm. But the memory of his clerical collar stops me mid-gesture. Sure enough, Father Waryn continues with his hands clasped behind his back, maintaining a respectful distance. I smile at a few people who are curiously looking our way. I know my dress is very fitting here. The rich blue ballgown is the exact same color as my eyes.

"I'd imagine that you feel embarrassed about what we just witnessed back there," Isaac says while looking over the crowd also, and nodding at people whom I have no idea who they are.

He's referring to the earlier incident—an uncomfortable moment that I wish could be erased from memory. "You imagine correctly," I admit, the words heavy with apology. "I'm so sorry for her behavior, Father. She wasn't raised right."

Isaac glances at me. "I may be a man of the cloth, but I wasn't born in a monastery. New love is a strong emotion that can make us do all sorts of foolish things."

His words catch me off guard, sparking a defensive reflex. "I beg your pardon, Father, but they are not in love." My tone is sharper than intended, a reaction to the assumption that doesn't fit the reality. Amira is cruel and trying to grab attention. None of us know Diarmuid, so whatever any of us are feeling, it isn't love.

Isaac's next question halts me in my tracks. "Then, if I may ask, what is the situation between you?"

"Father?" My voice betrays a flicker of confusion, mingled with apprehension.

Taking a moment, I let my gaze wander through the ornate hallway, appreciating the brief bit of privacy before we reach the bustling main dining room. The silence here is a stark contrast to the lively chatter that awaits us, and yet, my heart races with a nervous energy. The thought of engaging in conversation with a priest, above all people, makes me want to run.

"I'm afraid I don't understand your question, Father," I finally say, attempting to mask my discomfort with confusion. Isaac's gaze meets mine.

"Or you are afraid to answer it," he counters gently.

"Maybe," I concede, the word barely a whisper. Admitting even this much feels like standing on the edge of a precipice, unsure of the fall.

Isaac's tone softens. "Whatever your relationship with that man,

When Kings Rise

please let me give you a warning." It's not his tone or his body language that sends off all the alarm bells in my body; it's his gaze. He's afraid.

"He has a job. This job is important for his employers. It is a job that no one else can do quite as well or as…eagerly." His choice of words sends a chill down my spine, bringing to mind the O'Sullivan mafia ties.

My mind races, trying to piece together Isaac's cryptic message. Before I can form a response, he continues, "There are bad people in this world, my child. The worst kind performs the worst sins imaginable, and they do it for pay. You see, there was this child—"

His revelation is abruptly cut off by the sound of footsteps. Someone emerges from the dining room, passing us with a glance before disappearing down the corridor.

"A child?" I'm clasping the priest's arm.

He appears uncomfortable all of a sudden. "I've said too much." The priest glances into the bustling dining room.

"You've said nothing." I want to know what he is saying. "Did Diarmuid hurt a child?" Revulsion tightens my core.

The priest shakes his head, but something doesn't feel right. "But he hurts people?" I prod quietly.

The priest doesn't answer, but he doesn't have to. I see it in his gaze.

Oh god. My mind trips and races over his earlier words. *"He has a job; this job is important to his employers."* If he's part of the mafia and hurts people, does that make him a hitman?

The realization crashes into me with the force of thunder: Diarmuid is a hitman. My mind races, trying to reconcile the man I thought I knew with the stark reality Isaac's words have painted.

As we approach the grand doors of the main dining room, Isaac's gesture for silence—fingers pressed to his lips—halts any questions I might have had. The bustling energy of the room engulfs us. I find my seat beside Niamh, scanning the room for familiar faces. The absence of Amira and Diarmuid doesn't surprise me.

"Where are they?" I whisper to Niamh, trying to sound casual.

"Diarmuid led Amira outside," she replies, her tone indifferent.

I lean closer to Niamh, my voice low but urgent. "Don't go off alone with Diarmuid. Stay beside me for the rest of the night, all right?" The protective instinct in me flares to life, a fierce need to protect her from the man we are both promised to.

I watch as Diarmuid and Amira enter the dining hall. Both of them are a bit disheveled looking. Disgust makes my stomach turn.

He deserves Amira. I can't bring myself to eat, my appetite stolen by the revelation of Diarmuid's true nature. My gaze fixates on him, a silent accusation. Disgust bubbles within me.

A part of me knows I should act pleasant; after all, I was groomed for him. My parents' warnings echo in my mind, foretelling dire consequences if I fail to secure my place by his side. Yet, at this moment, their threats feel distant.

My resolve hardens, a defiant flame burning away any lingering doubts. I refuse to bind my fate to a killer's.

After pushing my food around my plate and listening to distant babble and Amira's loud giggles, I excuse myself from the table.

"I need to use the ladies' room," I say as I stand. I don't look at Diarmuid or Amira but focus on Niamh. She rises straight away. "Me, too."

Niamh follows me upstairs, where there are no guests. I know there is a bathroom up here that we used before when we were requested to go to gatherings with Diarmuid.

It's our little haven, a brief respite from the calculated smiles and watchful eyes. As we close the door behind us, the clamor of the party below fades to a distant murmur, and I take a deep breath, bracing myself for the conversation ahead.

"I found out something about Diarmuid," I begin, my voice steadier than I feel. The weight of the secret presses against my chest. "He's a hitman." Saying it out loud causes my stomach to tighten.

Niamh's reaction is immediate. She covers her mouth with her hand and turns away. "The day he took me to church, he stopped at a post box and had me get an envelope out of it. When I gave it to him, he said it was from Victor, and the name on it was the person he was commanded to kill."

Oh God. Revulsion pours through me, but along with it comes fear. *It's true.* He's a hitman and for Victor.

"What are we going to do?" she asks, her voice tinged with a vulnerability I've rarely heard from her. The question hangs in the air. I have no idea. It's not like we can walk away from Diarmuid.

"Is that why you were digging into Andrew's death?" It made such sense. Niamh didn't seem like the snooping type. Her shaken state when she arrived at my apartment, her desperation to find out more was now adding up.

Before she can answer me, the door swings open. Amira saunters in, her laughter cutting through the tension like a knife. "Oh, shoot. I thought that I would be the only one up here," she says, a smirk playing on her lips.

"Coming up to pick the grass blades out of your ass?" I retort, unable to keep the bitterness from my voice. It's a petty jab, but I'm not in the mood for her games.

"Oh wow, slut shaming, are we?" Amira fires back, her tone mocking. It's clear she's not the least bit bothered by my comment.

"I don't care about you having an active sex life, but don't you think we should be making a good impression at an event like this?" I counter.

"Check the dew on the marble outside. My ass made a great impression," she quips, unfazed.

"Whatever. You can have him, Amira. We would much rather just be dropped so we can go on with our lives," I say finally.

As Amira laughs off my words, I turn to Niamh, seeing my own resolve reflected in her eyes.

Amira's laughter fills the space between the sterile walls of the bathroom, her amusement making me tighten my fists. "You really are morons," she says, her voice laced with disdain.

"You don't understand..." Niamh says, her face is unnaturally pale even under the makeup I had applied not long ago.

Amira cuts her off. "I understand. What? Do you think that if Diarmuid rejects you, you just get to go on with your merry lives?"

"Well, I know that I would have to figure out a Plan B, but basically, yeah," Niamh says.

"Why couldn't we?" I challenge.

Amira's response is a mix of arrogance and pity. "Oh my God, I should just let you go. I would win him, and you guys can figure it out yourselves. Luckily for you, I am already winning, so I don't mind being nice." Her words drip with condescension. I want to tell Niamh to forget it. We don't need to listen to Amira's words. She's just on a high from having Diarmuid.

"Once you become a Bride, you belong to them. Even if you get rejected, you still belong to them. Understand?" Amira's tone shifts, the gravity of her statement hanging heavily in the air.

Niamh's confusion mirrors my own. "What do you mean?" she asks, her voice small.

The implications of Amira's words are chilling. The idea of being forever bound, with no true escape, is a cage I hadn't envisioned. I had naively assumed that rejection would be a release, a chance to reclaim my life and start anew. I knew I'd have to face my parents' wrath, but never being free of this world sends a cold dread that settles in my stomach.

This is a fate I can't accept. My mind races, desperate for a solution, a way out.

As Amira stands before us, a smirk playing on her lips, I realize that our understanding of the situation has been naïve at best. How does Amira know all this, and we don't?

When Kings Rise

Amira's words cut through the air, each one landing with the weight of a verdict. "Diarmuid is a Duke, right? He is supposed to become a King?"

"In the mafia?" Niamh asks, before glancing at me.

Amira's laugh is humorless, sharp. "No, dumbass. In the Hand of Kings. You really have no idea what you have gotten yourself into. They are like the Illuminati, but they never fell. Hundreds of years of building power, and you are a Bride of a future King. If you fail, no other King or Duke is going to want you, so you get passed to the Marquesses, then the Earls, and so on. Fail enough times, and Wolf will get you."

"Who's Wolf?" The question escapes my lips before I can stop it, a reflection of my growing horror. Each word that leaves Amira's mouth keeps getting worse and worse.

Amira's answer sends a shiver down my spine. "Diarmuid's cousin. He operates the O'Sullivan sex trade."

The room spins as the gravity of our situation becomes painfully clear. Niamh and I exchange a look of shock, our shared fear unspoken but palpable. The world we thought we knew, the dangers we believed we understood, pale in comparison to the nightmare Amira unveils.

She has got to be lying. But why would she?

"You can try to run, but you won't get far. Your passport will magically stop working. Your family will be stalked. Your money will disappear." Amira's voice is cold, matter-of-fact.

"Face it, ladies. Your options are on Diarmuid's arm or someone's whore." The finality in Amira's statement is a death knell. She doesn't seem to care how we are taking this as she turns to the mirror and fixes her hair. She smiles at herself.

Panic wells up within me, a tide of desperation and fear threatening to drown my resolve. The thought of being passed down the hierarchy of power like a pawn in a sick game is unbearable. And Wolf... the mere mention of his name and his vile trade sends a wave of nausea crashing over me.

Beside me, Niamh's eyes are wide with the realization of what Amira is saying. The revelation of the Hand of Kings, of the real power and darkness behind Diarmuid's position, casts a shadow over any hope of escape.

It can't be like this. Amira spins and gives us one final look. "Best of luck." She leaves the bathroom, and I can't seem to find my footing.

Something ignites in me, a spark of defiance. The fear is overwhelming, yes, but the thought of succumbing without a fight is intolerable. I glance at Niamh, seeing my own determination mirrored in her eyes. No matter how dire the situation, we can't let fear dictate our fate.

"We'll find a way," I whisper, more to myself than to her. "We have to." With every fiber of my being, I vow to fight, to seek a sliver of hope in this darkness—for Niamh, for myself, for our very souls.

When Kings Rise

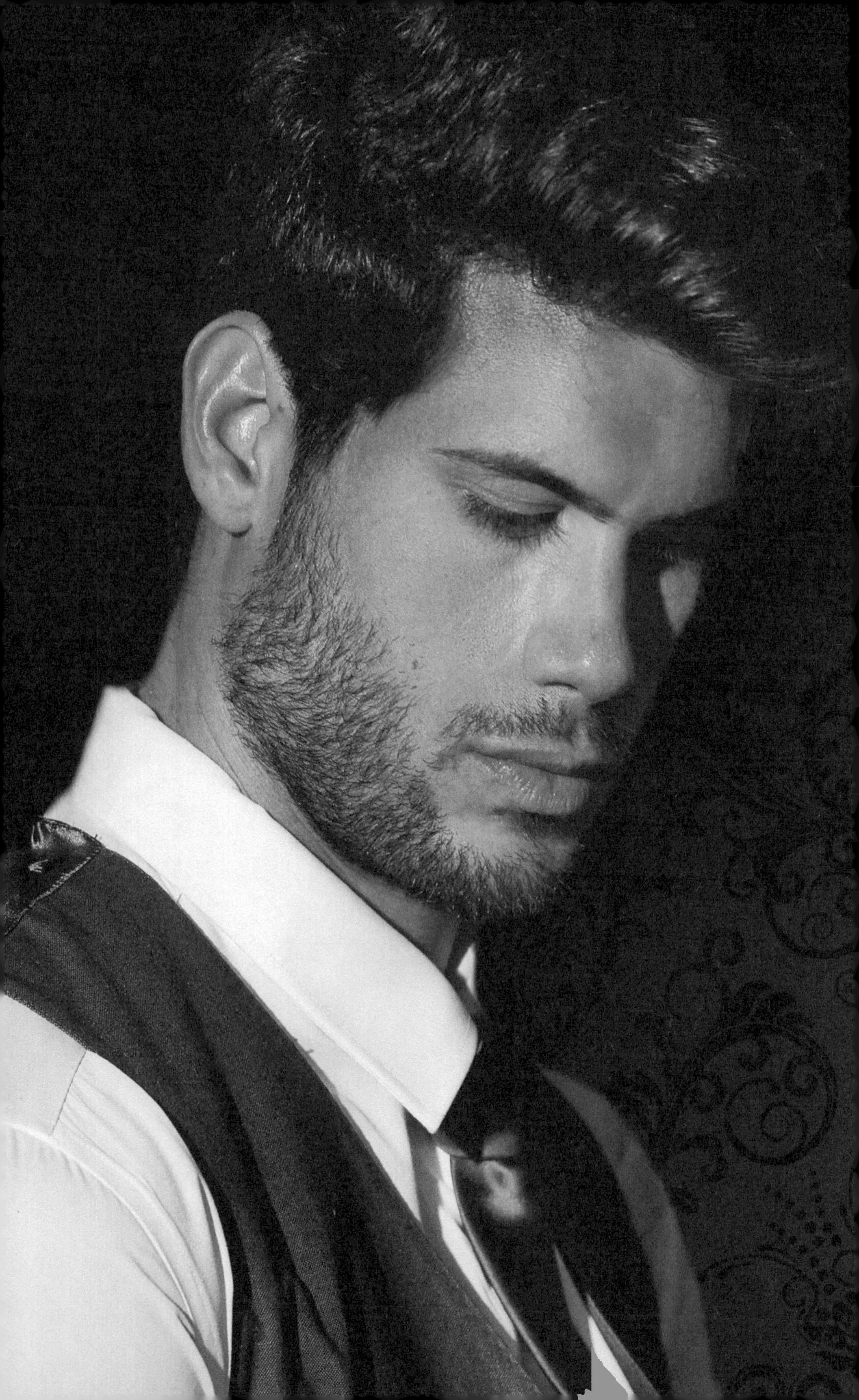

CHAPTER FOURTEEN

Diarmuid

THE CHANDELIERS CAST a warm, luxurious glow over the long dining table. I sit, somewhat stiffly, in my designated seat of honor, surrounded by the echoing laughter and the clinking of fine china. This dinner is Victor's doing, a chance to parade the prestige of our order before eyes hungry for the slightest hint of weakness. But to me, it feels like a gilded cage, each course a rib in the frame.

The starter arrives, a tiny marvel in the bowl of a spoon, crafted by a French chef whose name escaped me as quickly as he had introduced himself.

Next, the scallops, a sea-kissed treasure from the cold waters of Bedford, Massachusetts, are placed before us. They're seared to perfection, a testament to the journey they've undergone to grace our plates. Yet, as the flavors unfold on my tongue, I can't help but crave something as simple as a pot of stew.

The main course is presented with a piece of A5 Wagyu beef so small it almost seems lost on the expansive porcelain. The chef from

Kobe, who had tenderly prepared it, speaks of the beef as if it were a piece of fine art.

As the courses parade before us, each more elaborate than the last, my thoughts drift to the real reason behind this grand display. I've watched other Dukes undergo this bridal ritual, a test to see how potential Consorts fit in with high society.

Amira shifts her chair subtly yet decidedly closer to mine, her movement smooth and deliberate. She has drawn the attention of nearly everyone in the room. It isn't hard to see why. Her dress, if one could call it that, openly displays her cleavage. The small black spaghetti straps are tiny on her tanned shoulders, and I know when she stands, the dress will barely cover her. The fabric clings to her form in a way that leaves little to the imagination. Plunging necklines and daring slits were its signature, making it a piece more suited for a sultry night out than a fine dining experience. The smirk on her face and joy in her eyes suggested she was fully aware of the effect she had, reveling in the attention. As much as I tried to focus on the culinary artistry before us, I can't ignore how sexy she is.

"Are you enjoying the meal?" She coos up at me, her hand slowly moving under the table to rest on my leg. I thought after having her, my need would die down, but now it starts to grow again.

"It's delicious," I declare.

Even though I want her body, I can't help but feel a sense of distrust with Amira. She is someone who seems to crave the spotlight with such intensity. Was it a result of neglect, a plea for attention unmet by her parental figures, or simply a facet of her personality?

I glance at Niamh. She barely makes a ripple in the social currents of the dinner. Her modesty and poise suggest years of discipline. In her is a quiet strength, a resolute spirit that doesn't need the limelight to affirm its worth. Maybe she would be better suited to me.

I shift my leg away from Amira's touch, and straightaway, a frown appears on her face.

My gaze then drifts to Selene. She elegantly navigates the raspberry champagne sorbet. She must sense me watching her as her gaze clashes with mine. In that glance, I see not just her earlier anger or irritation but a challenge, an invitation to delve deeper than the surface. It's an anger that isn't raw or uncontrolled but calculated, a reflection of a mind as sharp as it is beautiful.

She drops her gaze from mine, not in shyness, but as if she has seen enough.

So, have I had enough? As the last course is served, people retire to the lounge. The ladies leave to refresh themselves, but not before I notice one last look from Selene. She intrigues me.

The lounge I enter is quiet. I select a glass of whiskey from a tray of assorted spirits.

I take my drink and step out onto the balcony, where I'm greeted by the cool evening air, a stark contrast to the heated atmosphere of the dining hall. The whiskey burns a path of warmth as I sip, watching the driveway stretch out before the main door.

Lorcan's presence is announced not by the sound of his footsteps but by the shared understanding of needing a moment away from the festivities. He leans against the balcony railing beside me, his gaze fixed on the darkness beyond.

"Twelve courses." He finally breaks the silence, his voice carrying a mix of awe and incredulity.

"Yes," I reply.

"Twelve fucking courses," he repeats, with a chuckle that borders on disbelief.

I lift my drink in a silent toast to his observation. "I'm having my thirteenth now," I say, a smile tugging at the corner of my mouth.

Lorcan laughs.

"I noticed that you had an extra appetizer," Lorcan says, his voice casual, as if he's commenting on the weather.

"I won't talk about this," I respond firmly. What I do with my Brides is private, and my business alone.

"Come on, you know that Ronan or I will be next. I just want a little bit of information," Lorcan persists, but his 'just' feels heavier than it sounds.

"My moments with them are mine," I say again.

I can almost sense Wolf approaching before he steps out onto the balcony and joins me and Lorcan. His steps are heavy. He's already been drinking, a fact that surprises none of us. It's become part of his persona, a shield as much as a weakness.

"I need a gun," he slurs slightly.

"Wolf, this is not the place," I respond instantly.

His words become reckless, teetering on the edge of madness. "Well, if anyone tells the authorities, I'll just kill them. People are allowed to do that in Ireland. No one gets punished if you kill someone in Ireland," he proclaims, a twisted smile playing on his lips, oblivious to the gravity of his own words.

Lorcan's response is immediate, his voice laced with anger and fear. "Shut the fuck up, man. Get a hold of yourself! The fucking cardinal is over there." His eyes dart towards the dignitary, a silent plea for Wolf to recognize the danger of his rantings.

Wolf, however, seems lost in his own vendetta, his voice rising. "Perfect. I can ask his forgiveness after I kill my father's murderer."

It's then that Ronan appears, his question simple yet loaded with concern. "What the hell is going on here?"

"Wolf is trying to drag our entire business down," I explain.

Ronan nods, understanding flashing in his eyes. "We need to get him away from everyone before someone tells Victor."

Together, we box around Wolf and try to guide him back to the house. But he's being awkward, pushing against us. I shove him; he wobbles but rights himself. The guests' curious gazes feel like spotlights on us, but a glare from me is enough to make them avert their eyes, a silent command they dare not disobey.

As I help Wolf up the stairs, I hear the front door open. Looking down, I catch a glimpse of Niamh exiting, her silhouette graceful and determined, with Selene following close behind her.

Their departure is a silent alarm. Selene's exit, with Niamh in tow, is not just an escape; it's a statement, they have bonded. I don't have time to follow them, not with Wolf in such a volatile state.

I focus back on the task at hand, guiding Wolf away from prying eyes and ears before he says something that will get him killed.

But Wolf notices my two Brides leaving too. His finger, unsteady yet determined, points directly at Selene, his words slurred but clear: "Get out while you can, love."

She nods, a silent acknowledgment of a warning perhaps long expected, and leaves without a word.

Wolf's attention, however, swiftly shifts. His gaze moves, heavy with alcohol, and he points upwards. "But you, you can stay." His words float up to Amira, who leans over the railing, curiosity etched into her features.

Anger crashes through me, and in a moment of decision, my hands release their grip on Wolf, a calculated risk. His body, unprepared for the sudden absence of support, sways and then crashes against the banister, the impact sharp and sudden. The combination of the blow and the alcohol coursing through his veins proves too much, and his lights go out, his body slumping to the ground in a heap of silence.

"Alcohol thins the blood, Diarmuid. A hit like that could kill him." Lorcan glances around to see if anyone saw, his disapproval of my actions evident in his tone.

I don't give a fuck. He deserved it.

"Then, half of all our problems are fucking solved," I retort.

Loran and Ronan pick up Wolf and carry him upstairs to a guest room, the effort obviously draining with each step my brothers take. Wolf isn't small by any means. But I enjoy the view of him slumped over, his feet trailing along the wooden floor. At least he's quiet.

I'm aware that Amira follows closely behind.

"Can you open the door?" Lorcan asks me, glancing over his shoulder. I don't move, so Amira does as my brother asks.

"Thank you, Amira," he says as he and Ronan get Wolf into the room.

Wolf groans loudly from the bed. Amira steps in again. "I'll get a washcloth," she says and disappears into the adjoining bathroom.

The tension in the room thickens as my brothers glance at me. "She is not taking care of him."

"I barely take care of myself; I'm not about to do it for him." Ronan's retort is quick, laced with his own brand of humor and resignation.

"Diarmuid, what is the point of having spare women if you can't get them to mop up a drunk?" It's Lorcan's comment, though, that breaks the strained peace, his words cutting deeper than he probably intends.

The disrespect in his tone ignites something within me. Without fully processing the decision, I find myself pinning Lorcan against the wall, my anger finding a physical outlet. The threat in my eyes is as clear as the words unspoken between us.

Ronan steps in before things escalate further. "There is no need for that," he says, pushing us apart.

Amira returns, her arrival marked by the practical items in her hands—a washcloth, a water pitcher, and a glass. Her posture, one hip jutted out, an eyebrow lifted, speaks volumes, her words cutting through the tense silence. "Either you three have become really close and cuddly, or I missed a fight."

Her comment, light yet pointed, draws no response from us. She moves to the side table, setting down her washcloth and pitcher.

She places the cloth on Wolf's forehead, and he groans again. When none of us move, Amira glances at me. "I can watch over him until he wakes."

"No," I say immediately.

"She's right. Let her watch over him so we can return before people start to notice," Ronan says. If it was Lorcan after his smart-ass comment about Amira, I would be saying no again, but I glance at Amira, who nods at me.

"Go," she says.

With a final nod, we file out of the room. The reminder of Selene and Niamh leaving has me walking down the stairs.

"Come, we can have a drink," Lorcan calls from behind me.

"I have somewhere I need to be," I tell him, my mind already steps ahead of my body.

I find myself outside of Selene's house. The knowledge of where each of my Brides lives is something I found out before they even met me. Selene's home, a converted garage beside her grandparents' house, stands quiet under the night sky.

Knocking on her door, I'm met not with the warm welcome one might hope for but with annoyance. Her expression is clear; her willingness to entertain me, markedly less so.

"May I come in?" I ask as she folds her arms across her chest. Her dress from earlier has been swapped for denim jeans and a loose-fitting sweater.

"No." Selene barks.

I raise my brow. "You know you can't stop me from coming in." I remind her of my power here.

"If I can't stop you from doing whatever you want, then don't fucking ask. Selene fires back, her voice sharp enough to stir the quiet of the night, catching the attention of a neighbor. A light flashes in the dark, and I take another step toward her, but she doesn't back away from the door.

It's clear she doesn't want me in her home, but I'm not giving up. "Let's take a drive, then." I offer an olive branch.

Her expression morphs, showing her discomfort at the thought of being alone with me.

I grit my teeth, not one to bow to the whim of a woman. But I want to know what has made Selene so upset.

"A brief walk down the street. A public place."

She unfolds her arms and lets out a heavy huff. "Fine, wait here." She closes the door in my face. She doesn't keep me waiting for long before she appears with a heavy pair of brown outdoor boots and a heavy cream jacket thrown over her sweater.

"I see you and Niamh have bonded," I start.

"She's very nice." Selene weighs each word.

"And Amira?" I ask.

Selene glances at me now, her gaze intelligent and yet concealed. "You seem to like her."

I hide a smile. Is that jealousy I detect? "I like all three of you."

"Why are you here?" she asks. From meeting her the first three times, my impression of her was that she was always controlled—but not tonight. I wonder what transpired. But I had pegged her as the troublemaker.

"To find out why two of my Brides left so abruptly."

Selene starts walking again. "The night was almost over, and I was tired. So, you came all this way to find out why I left?"

I came for more than one reason. I wanted to get away from Wolf and my brothers' disapproving looks.

I nod.

"I'm not a dumb, spoiled, rich girl. I'm educated. I can tell when people are lying, and you do nothing but lie," she accuses, her voice a mixture of anger and disappointment. Her abrupt words make me stop walking this time.

"I have to be deceitful in my business," I answer honestly. We have so many secrets to carry. Sometimes, the lies can get tangled with the truth.

"What business?" she probes, but her voice isn't as sharp anymore.

"Imports," I answer.

"Liar."

"Be careful, sweet Selene," I whisper and give her a warning glance.

"I know what you are." Her voice is heavy, too heavy, and I don't like it.

I immediately go with humor. "Fuck, are you going to call me a vampire?" My attempt to deflect with humor sounds feeble even to my own ears.

"How do you know about that movie?" She counters, a smirk playing at the edge of her words.

"How does an intellectual such as yourself know about that movie?" I retort.

"You are redirecting the conversation," she accuses, pinpointing my tactic with ease before tightening her jacket around her perfect frame.

"You are easy to redirect," I shoot back before I run my thumb along my lips, lips that she glances at for a moment. I swear her blue eyes soften before they ignite again with something close to fear.

"You are a killer," she states, a declaration so raw that it strikes a chord inside me that hasn't been pulled since I was a kid.

"Am I now?" I reply, with a raised brow, attempting to regain control of the conversation, and steering us back into a more humorous footing.

"Stop it! I know what you've done. And who you've done it to," she insists, her voice a mix of anger, fear, and a daring kind of courage.

"You have entered a dangerous world, Selene," I warn.

"So, you've done it. You've killed people. Men, women, and

children," she states, a cold summation of my sins laid out in the open, each word a weight, each accusation a mirror reflecting a version of myself I've fought to keep hidden, not just from the world, but from myself.

The discipline of masking my emotions has been a cornerstone of my existence since childhood, a necessary armor forged from the unforgiving punishments from Victor and Andrew. I have perfected the steady stare and a firm lip that has seen me through countless situations. Yet, Selene's accusation causes me to pause. It's a microsecond of hesitation. To any onlooker, it would have been invisible, insignificant. But Selene's eyes widen.

Her reaction is immediate; her breath falters, and then she sprints back toward the safety of her home. Her steps are panicked, and it takes me a moment to respond.

I chase after her. As I run, my mind races, trying to piece together the hows and whys of her accusation. She's made it clear she knows about the killings, the lives I've taken, even mentioning children—a detail so specific, so damning, that it sends a chill through me.

How could Selene know? Her inference, her mention of children, implies she believes Brien Cahill is among my victims.

I push my body harder, and the need to understand and clarify become my sole focus.

She's made it nearly to her home, running alongside the building, when my arm circles her waist, and she's airborne before I swing her and pin her to the wall. She's breathing heavily and, straightaway, she starts to fight me.

As I cover her mouth with one of my hands, anger races through me. "The world isn't fair. Sometimes your parents are shitty, and you have a shitty life. Sometimes, people die who shouldn't have died. Sometimes, a father drives up a gambling debt, so his son gets killed for it." My emotions become erratic; I've never lost control like this, and it isn't fair, but I'm not working to make the world fair. I am working to get what is mine.

Tears fall from her eyes and soak my hand. She stops fighting me, and I remove my hand from her mouth. She licks her tears before she glares at me. "You are a monster."

Her response cuts deeper than any knife I've ever wielded. I agree with her.

I press my body against hers. "I am a monster with clearances, allowances, and ultimate freedom. I could tear into you right now, and nothing would happen to me."

My anger turns to something darker as I look down at her wet lips. Raising a hand, I run my thumb along her bottom lip. She swivels her head quickly away from me. Like my touch disgusts her, I grip her face, forcing her to look at me. "This kind of monster needs to be pleased, or no one is safe."

Her heavy breath fans across my face. My movements are abrupt and almost frantic as I press my lips to hers. She doesn't kiss me back, but that doesn't hinder me from taking what I want.

My tongue shoots out and runs across her lips, forcing them to part. When my hand trails across her breasts and in between her legs, she speaks.

"Please don't." Her plea is soft.

But the darkness she has ignited in me doesn't allow me to stop. "Are you going to stop me?" I ask.

Her gaze darts around the space. "Physically, I can't." She grits her teeth.

"Did you not swear obedience when you became one of my Brides?" I dip my head in and kiss her again before trailing kisses up to her earlobe. "Who told you that I was a hitman?"

When I look back in her eyes, I see her resolve as strong as before. I hate that she is protecting someone.

"Tell me, and I'll stop."

"If I tell you, will you kill the person?" She questions.

I grin. "Most likely."

"Then I won't tell you." I hate the thought that she could care for someone else. She must know she's mine; if she doesn't, I will show her.

I run my hand back between her legs, and she tightens her thighs together as if she could stop me.

"You like my touch; I remember how wet you were for me." I pop the first button of her jeans and then pull down the zipper while holding her hands above her head.

"Let's see if you are as repulsed as you are acting?" I dip my finger inside her, and she's wet. I grin in victory.

She tries to wiggle, but my tight hold on her keeps her back firmly against the wall.

"I think my troublemaker likes this," I whisper in her ear again before I push my finger deeper.

"So, tell me, who you were talking to?" I insert a second finger, and she gasps, her core tightening around me. She's fucking perfect. As I watch her gaze transform from hate to pleasure, I almost don't want her to tell me who told her. I want to make her come right here and now on my fingers.

I use my thumb to rub her clit as I continue to fuck her with my fingers. I want nothing more than to bend her over and take what is mine, but I'll wait for the right moment for that. For now, she will learn who her master is.

"Give me a name, and I'll stop," I whisper, moving my fingers harder and faster inside her while my thumb circles her clit. She's shuddering, her gaze glazed over.

"No," she whispers as her eyes flutter closed.

I love that she won't give in; I love what I'm doing to her.

She groans, and her core tightens. I work harder on her clit, knowing she's close to coming, and when moisture fills my palm, dripping down from my fingers, I know I have gotten what I wanted.

She comes hard on my hand, and when her body stops shaking, I extract my fingers and lick each one slowly. Her mouth is open as she watches me. "I can keep coming back and doing this until you tell me."

A spark of something flashes in her eye. Maybe she wants me to come back every night.

Tears still stain her face, and she wipes her cheeks with the back of her hand.

Looking at her and into her eyes is a declaration that I am indeed a monster, and she has no intention of telling me.

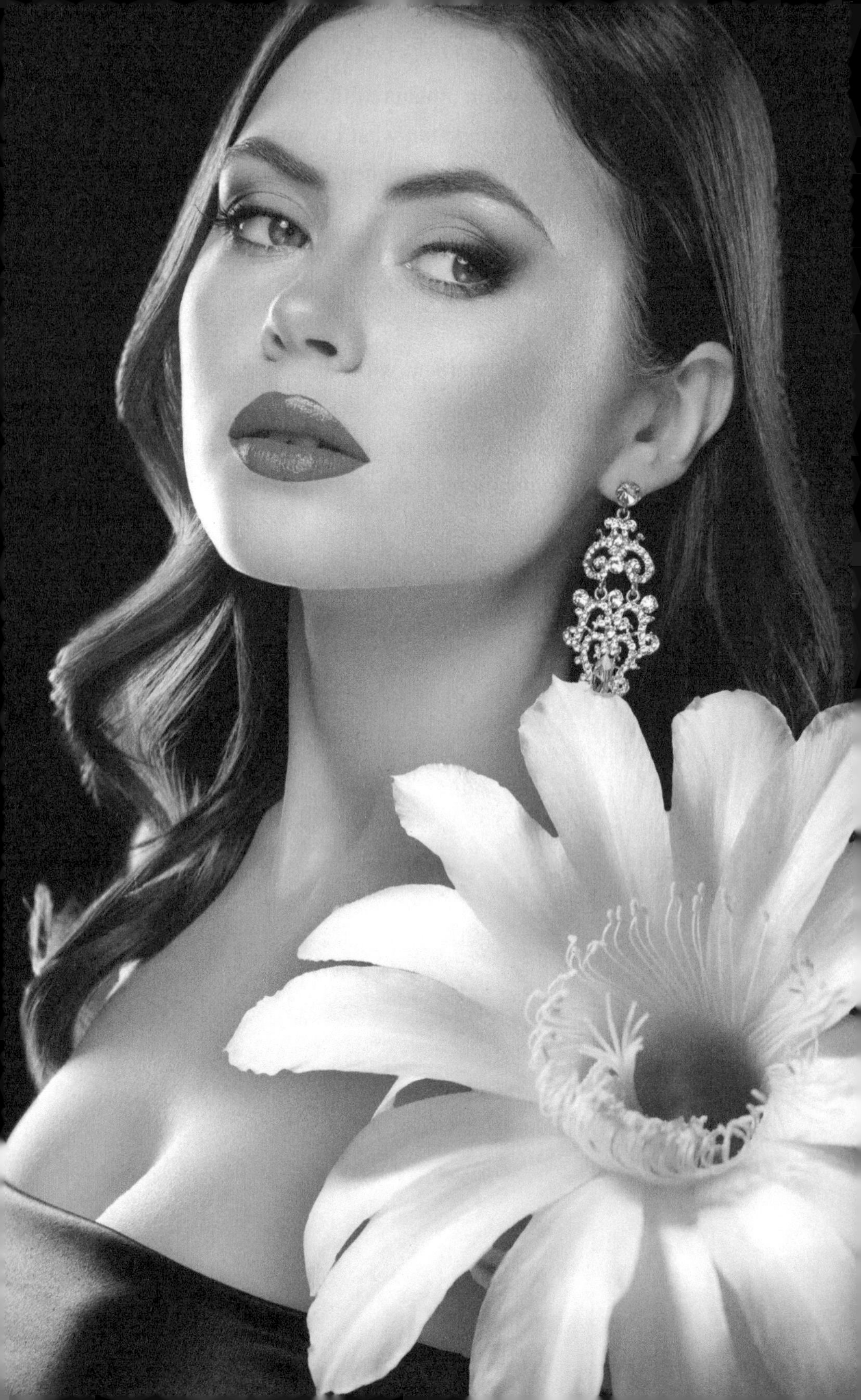

CHAPTER FIFTEEN

Amira

I DIDN'T WANT Diarmuid to go. I would have liked to spend more time with him. But I'm trying to show him my range—how I can play whatever part he needs—a role that changes between lover and now caretaker of his family's troublesome relatives. Babysitting Wolf will show Diarmuid that I can be a good wife.

Wolf is still asleep as I sit on the side of the bed and examine his face. I can't help but notice the unmistakable family resemblance—the sharpness of his jawbone and the distinctive shape of his brow. They are the same features that Diarmuid and his brothers have. All are extremely handsome.

The light catches the hair on Wolf's cheeks and head, revealing a surprising red tint. A ginger, I muse.

Every so often, Wolf stirs, each time awakening with a parched throat and a confused look in his eyes. He reaches out for a glass of water, his hand shaking slightly. To him, I'm a stranger—a face without a name. And why should he recognize me? My interactions with Wolf have been fleeting, and the more I've learned about him, the more I've felt a creeping sense of unease.

He is just as deadly as his cousins, maybe more so because Wolf is unpredictable, based on the rumors I've heard. The most disturbing whispers, the ones that send shivers down my spine, are his involvement in the family's sex trafficking ring. I had shared this knowledge with Niamh and Selene with a smile, but inside, my stomach soured.

He starts to cough again, and I sit and watch him struggle to catch his breath. I've heard of people choking on their own vomit. If he does throw up, I won't be able to clean it. Knowing that I need to impress Diarmuid, I rise and walk around the bed, picking up the glass and bringing it to Wolf's lips.

"You're okay; just take a sip." I try out a smile as his gaze focuses on me. He does as I command. The water seems to lodge itself in his throat, and he's sputtering again. A memory thick and hard slices through my mind.

My mother bent over a small bucket as she gasped for air, but there was no forgiveness for all the alcohol she had indulged in. Her hand had reached out to me for help, and I remember standing in the hallway watching her, hoping her last breath would be stolen on the kitchen floor.

Unfortunately, it wasn't.

Wolf gasps and takes in a few lungfuls of air. His brows furrow. He won't remember this tomorrow, so I have nothing to fear. But just in case, I offer words of encouragement. "There you go. You are doing great." The pitcher on the bedside table is empty, and I pick it up and leave to refill it. I'm contemplating using the tap water in the bathroom, but even as a child, I was scolded if I drank from the bathroom taps. The system in the attic wasn't safe for drinking, and it fed into the bathrooms. The thought of going downstairs and bumping into my mother, though, made me choose to take my chances with the tap water.

I take one final glance at Wolf, who has his eyes closed again, before I slip out of the room. I don't have to walk far before a figure in a crisp, white uniform catches my eye—a maid, moving with purpose toward me. She glances at the empty pitcher in my hand, and without a word, she reaches out for it. I'm not used to people helping me, but it's something I could get used to. So when she takes the empty pitcher from my hands with a nod, I release it. She disappears out of sight. I'm not sure if I should wait or return to the room, but the chirp of my phone from the pocket of my dress distracts me. I fish it out of the small pocket along the thigh of my dress. The one thing I always ask for is for my garments to have pockets. The screen lights up with a message that instantly sends a shiver down my spine. It's from my mother.

"Where are you?" The text is brief, but I can hear her voice in it. Her voice is edged with that familiar blend of worry and disapproval. My fingers hover over the keyboard, my mind racing. How do I explain, yet again, that I am attending the most important yearly social event in our city? An event that, for better or worse, could shape my future and that of our family? She knows this, but with all the alcohol fueling her body, she's very forgetful.

I type out a response, reminding her of where I am.

No sooner have I sent the message than a reply comes through, and with it, a knot forms in my stomach.

"Whoring again." Her words lash through the phone. She's always been angry at me, but even more so since the loss of my brothers—a void that nothing can fill—has left her grappling with a depression so deep it colors every word she says to me. The absence of my youngest brother, Michael, who we only hear from through sporadic letters, adds to the constant fear that one day, those letters will stop coming.

But I shouldn't have to be her punching bag. That's all I am to her.

"No. Securing our family's future, Tess." I use her name for extra emphasis.

"Your water." I glance up at the maid, who has returned with the water, and I take it from her before making my way back to Wolf.

I slip the phone back into my pocket, a sense of resolve hardening within me. Today, proving myself to Diarmuid is more important than fighting with my mother.

With a deep breath, I lift my chin and step forward, ready to face whatever is behind this door. But my ringing phone has me pausing.

"Can I hold that for you?" The maid is still standing in the hallway. I want to tell her to leave, but instead, I hand her the pitcher and answer my mother's persistent rings.

"I'm busy, Mother," I say.

"Have you so easily forgotten about your brothers?" Her words are slurred and filled with pain. This week marks a painful anniversary—the death of Dominic, my older brother. The memory is a sharp, constant ache, a reminder of the price paid by those forced to serve the Hand of Kings. Dominic and Kevin, both lost to a cause they had no choice in, their futures snuffed out prematurely. Dominic's death, in particular—gunned down during a police raid—haunts our family, a wound that never truly heals. But I feel that pain, too.

The timing of the anniversary only serves to heighten my mother's volatility, a fact I'm painfully reminded of as she continues to spew her poison down the phone.

"Of course, you have forgotten. All Amira cares about is Amira."

I turn away from the maid and hiss into the phone. "I haven't forgotten. But you seem to forget you still have a daughter." I'm braver with the distance between us. If I were home, she would surely strike me. I don't want to go back to that house, to the suffocating atmosphere of sorrow and resentment.

"You're not much of a daughter." Her words lash out, and I end the call. The pain of her words is too much.

I turn to find the maid watching me. I slip the phone back into my pocket. I reach for the pitcher, but not before the maid's gaze meets mine, a flicker of concern in her eyes. It's a kindness, perhaps, but in that moment, it feels like pity, and something within me recoils.

"What are you looking at?" The words snap from my lips, sharper than I intended. It's a defense mechanism, an instinctual cover for the pain that's threatening to spill over. The maid, taken aback, merely hands me the pitcher and moves past me, her momentary concern replaced by a professional detachment.

I retreat to the guest room, only to find Wolf sitting up in bed, looking woozy but alert. He finishes a glass of water as I enter. When Wolf's gaze meets mine, I recognize something in his deep gray eyes. Like me, he carries his own pain. A shadow seems to cling to him, visible in the weariness of his eyes and the careful way he holds himself.

"Where is Diarmuid?" are his first words.

I place the pitcher on his bedside table. "I'm not sure," I answer honestly. "I think he's with Lorcan and Ronan."

He huffs at that. "I'm sure they are discussing what to do with me."

I hesitate, not sure how much to pry, but any opportunity to get to know Diarmuid better is one I will grab with both hands.

"They seemed very concerned when you fell." I lie remembering how Diarmuid had released Wolf in anger, allowing his cousin to fall and bang his head.

"They don't care about me." His words are bitter, and he takes another drink. "They would be glad if I were buried alongside my father."

"I doubt that is true," I respond.

He snorts. "You don't know my cousins." He waves his hand in the air. "You will soon enough, and then you will be running."

"He's been very kind to me," I say and think of how we shared a moment out in the garden. How he took my virginity from me and only me. He never touched Niamh or Selene, and why would he? They don't compare to me.

"Kind until he gets what he wants," Wolf says, watching me closely now.

He already got what he wanted, my mind sings. I shake the thought away.

A silence slips over the room for a moment.

"Who were you talking to on the phone? It seemed heated." Wolf sits up even further in the bed. He grows more alert by the minute.

I'm wondering how much I should share with him. "My mother," I answer.

Wolf doesn't pry, but I need to talk; I never talk about her. "It's our brother's anniversary—well, two of my brothers'—death. Michael... well... we aren't sure about him, to be honest." I find myself drifting to the end of the bed.

"I understand loss like that. I buried my father," Wolf says, with an almost vulnerable look in his gaze, but he doesn't have pity in his eyes, and that is what keeps me talking. I can't stand pity.

"I'm sorry about your father. It's a burden."

He nods in response.

"My brothers served the Hand of the Kings, also," I say and fold my hands onto my lap.

"Oh, I may have known them. What is your second name, Amira?"

The fact he knows my first, surprises me. My last name isn't a secret. "Reardon."

Wolf's eyes light up as if he recognizes the name, but he shakes his head. "I can't say I worked with them, but then again, I've worked with a lot of people, being a Duke." He offers a lazy smile.

"You're a Duke yourself," I tease, a smile playing at my lips as I think to myself that if things don't work out with Diarmuid, maybe, just maybe, being close to Wolf might save me from being stuck at home with my mother. He didn't seem as bad as people made him out to be. In fact, I could see myself liking Wolf.

He returns the smile with a wry twist of his lips. Wolf suddenly pushes to his feet, a bit unsteady but determined. "I have to get to work," he announces.

"Your work can wait for another day," I protest half-heartedly, intrigued by what could possibly demand his attention so urgently. There's a part of me that knows all too well the nature of his duties, yet being in the room with him makes me wonder how much is really true.

He pauses, considering something. "Why don't you come see my office?" He suggests a challenge in his eyes. I hesitate, aware of the reasons why such an action would be frowned upon. "It probably isn't appropriate for a Bride to be alone with another man," I murmur.

Wolf's response is immediate, confident. "I have nothing to hide from Diarmuid," he states, his tone leaving no room for argument.

The invitation hangs between us. I stand and consider his offer, the knowledge of what he does, the intrigue of the unknown, and the sheer boldness of his invitation stirring a reckless desire inside me.

The transition from the luxurious, suffocating atmosphere of the event to the starkness of the old school building is jarring. We hadn't far to go, just across the courtyard, before we arrived at what Wolf says is his office.

I'm reconsidering my decision, glancing back at the mansion with all its glistening lights. But as Wolf steps into the foyer, I find myself following him. The chill of the stone walls contrasts sharply with the warmth we've left behind, making me very aware of how scantily I'm dressed. The building, despite its seemingly abandoned exterior, exudes an air of careful maintenance. The wooden floors gleam under the steady glow of well-maintained lighting, free of the cobwebs one might expect in such a place. A calendar on the wall, its pages fresh and current, seems oddly out of place in the otherwise timeless space.

I can't help but feel a mixture of confusion and curiosity as we walk through the entrance. "This is not what I expected," I admit, my voice echoing slightly in the open space.

Wolf glances at me, a hint of amusement in his eyes. "The O'Sullivans have always been good at hiding in plain sight," he explains, his tone casual but carrying an undercurrent of seriousness. "Ever since they joined forces with the Hand of Kings, secrecy has become a cornerstone of their operations. Most of my dealings are through contacts within the Hand. Diarmuid's side of the business, given his involvement in the illegal arms trade, tends to work more closely with the O'Sullivan network."

The revelation of Diarmuid's activities, while shocking, doesn't surprise me as much as it should. Yet, hearing it spoken aloud by Wolf brings new clarity. Maybe Wolf has more power than Diarmuid if he works so closely with the Hand of the Kings.

As we move beyond the foyer, the facade of the old school continues to unravel, revealing its true purpose. The presence of regular offices makes this place almost seem mundane.

That is, until we step into Wolf's office. It's such a large, yet inviting room. The warmth radiates from a fire that has been kept lit for a while now. Off to the left is paned glass, hiding another room, but I can't make it out. The chill from walking across the courtyard starts to ease.

"There are things I know that could upset the ruling governments of most of the world powers," he states, a grim seriousness etching his features as he sits behind an old and oversized desk. He points at the chair in front of his desk for me to take.

My curiosity piques. "What kind of things?" I ask, not sure I'm ready for the answer.

"Disgusting, sinful things, Amira," he replies, his gaze fixed on a point somewhere beyond my head.

The gravity of his words sends a shiver down my spine, a realization of the depth and darkness of the secrets Wolf harbors. It's a reminder of the world I'm stepping into and also a reminder that I shouldn't be here.

He rises, and I'm surprised when he holds a hand up for me to stay seated. "Wait here."

I sit in the warm room and wonder where he has gone. It's not long before I get up, unease rippling through my system.

After what feels like an eternity, Wolf re-enters his expression unreadable. "What's going on?" I ask.

"I'm going to allow you to see me work," he answers, a cryptic smile playing at the edges of his lips.

He guides me to another door, this one secure with both analog and digital locks. Once through, we enter a room that stands in stark contrast to everything I've seen thus far.

"The sex training room," he states calmly, like that statement alone shouldn't have me running back to the safety of the manor.

He directs me to a chair in the corner of the room and tells me to wait. "I have this feeling you would like to see my work."

I can't answer. I'm curious what a sex training room is. Maybe I could learn something to use on Diarmuid to gain the upper hand. I know Wolf won't touch me—it's forbidden—so I nod and wait to see what he wants to show me.

"Have you seen anyone else arriving here?" he asks all of a sudden. I shake my head. "No."

"At least a dozen people have seen you. This place has lots of secrets and a lot of protection, too. It has to."

The far door opens before I can respond or decipher what his words could mean. A young woman enters.

Wolf smiles kindly at the woman and signals her to approach him with two fingers.

She does with a bowed head. The dressing gown she wears covers her body, but I don't think she wears anything underneath it.

"The Dukes and Kings of the Hand of the Kings prefer virgins. But I know a woman has greater value if she has some experience." I sit up straighter.

He leads the woman to a chair, but before she sits down, she strips off her gown. Beneath it, she is naked like I thought.

Wolf stands behind her, his gaze never leaving mine. He runs both his hands down across her breasts and squeezes her nipples lightly. The woman's face is covered with her long, loose hair, and I want to see her eyes. But I don't dare speak as Wolf continues to play with her breasts. His strokes aren't gentle, and redness starts to grow along the skin from his constant grabbing. He's watching me as he rolls her nipples between his fingers and pulls them outward. I lick my lips, and he grins before he moves around her and kneels at her feet.

"It is my job to teach women how to be the best lovers possible before I marry them off or sell them." He states and spreads the woman's legs, revealing her private area. I'm surprised to see she has no pubic hair. I'm wondering if it's something Wolf prefers, or if the woman did it herself.

I have so many questions, like why wasn't I trained? Or what does he mean to sell them off? But the words are cut off as he runs his large hands up her thighs and spreads her legs as far as the chair will allow. He's left enough room so I can see as he slips a finger inside the woman, and she groans in pleasure. My stomach dances with butterflies. I'm back to looking at the woman's face, but I can't see with all the hair hanging loose.

"I want to see her eyes." I find the words.

He smiles with delight and reaches up, grabbing a bunch of her hair to yank her head back. Her oval face has creamy, flawless skin, her lips are slightly parted, and her bottom lip is large. Her white teeth are stark against her pink lips. She's very pretty. Would Diarmuid prefer her more than me? An unyielding anger has me wanting to hurt her, but I stay in my seat. Wolf slips another finger inside the woman. This time, his movements are harsh, and the woman hisses in what sounds like pain, and that makes me smile. He doesn't think she is prettier than me. He wouldn't dare touch me like that; it reminds me of my worth. I'm so much more than these whores.

When I meet Wolf's gaze, he smiles back at me. "Does this turn you on?" he asks as he continues to plow his fingers into the woman. He keeps a tight grip on her hair, never allowing it to cover her face, keeping the curtain back so I can watch the show.

"Yes," I whisper.

He nods. "There is no shame in that. You can learn by watching, but you can touch yourself if you wish."

I want to touch myself, but I also know the complications of my actions.

Wolf, as if noticing my hesitation, rises to his feet; the woman instantly pulls her legs together, and her annoying hair falls down around her face. Wolf gathers her hair again, and I'm surprised at how thoughtful he's being for me. He scoops it back in a low ponytail and uses his fingers as a tie. While looking at me, he unzips his pants, and his large cock springs free. He's still looking at me when he directs the woman's head to his cock by pulling her hair. She looks up at him with large eyes, but he isn't paying her attention. No, he's paying me attention.

"Circle your mouth and suck." He instructs her she reaches up, and I notice the shake in her hand as she places the tip of his cock in her mouth.

"Open wider," he commands, and she does.

He groans. "Watch those teeth; no man wants to feel your teeth." His words are gritted, and as he teaches the woman how to give him a blowjob, I grow wetter. The sounds of the woman's saliva and the sucking sounds along his cock has me squirming in my seat. I've never watched porn, never mind watching someone perform a sexual act in front of me. Not to this degree, anyway. With Selene and Niamh on our first night at the Hands of Kings' mansion, that was brief. Not like this at all.

Wolf continues to watch me, and his features morph into an ecstasy that I want a part in. He pushes his cock further down the woman's throat, and she's gagging, but he doesn't stop. His hand loosens around her hand and spreads out on the back of her head, restricting her from pulling away from his cock. With each ragged breath from him and each choking noise from her, I become braver.

I let my hand fall between my legs and gain access easily to my core beneath my short dress. I'm swollen with need and touch myself.

Wolf smiles. "Good girl. You can be my student, one I will never touch, but you can watch and learn."

The idea makes me wetter. I could learn and enjoy myself. The woman's face is puce as Wolf's movements grow frantic while he fucks her mouth with no mercy. My own fingers move quicker over my swollen bud. I want to come, but some part of me doesn't want this to end. The woman's hand reaches up and touches Wolf's torso as if she's getting ready to push him away, but something makes her drop her hands.

Wolf's groans fill the room, and when he slams into the woman's mouth, spit and his cum pours out of the corners, but he doesn't remove his cock. It's too much for me, and I come, sitting in the chair as the woman slaps his stomach, finally losing control as she tries to break free. Tears pour down her face. But Wolf isn't letting her go. Not even close.

When Kings Rise

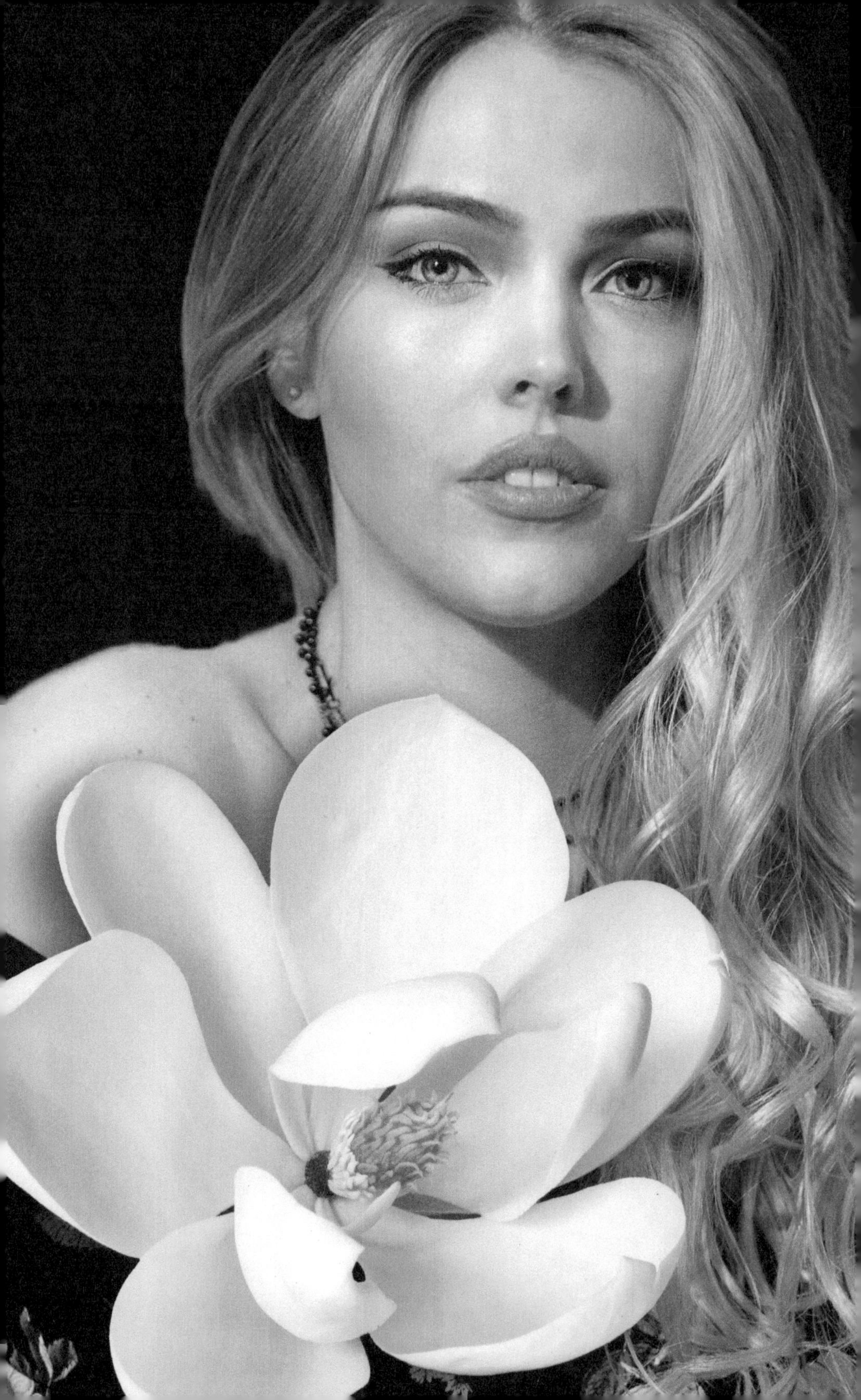

CHAPTER SIXTEEN

Niamh

AS THE HAND of Kings' car pulls up to my family's stately home, the silence of the night wraps around me like a cloak. I expect anger, worry, perhaps a lecture waiting at the doorstep. But there's none. Instead, when my father shuffles to the door, wrapped in his bathrobe, his face lights up with a smile so warm it feels out of place in the cool night air.

"Ah, Niamh, you're back," he says, his voice laced with a giddiness that makes my stomach churn. I know immediately what he's thinking—his assumptions about Diarmuid and me, about what my absence signifies. I bite my lip, wondering how his expression might change if he knew the truth. That my night had been spent in an entirely different company than what he'd approve of.

"Goodnight, Dad," I manage, stepping past him before the questions start.

"Oh, goodnight, love." His cheeriness makes my shoulders tip closer to my chest. No father should be happy when their daughter comes home so late, but the fact that he isn't asking questions makes me climb the stairs quickly.

As I make my way to my room, the sight of Ella's door catches my eye. My heart aches with the need to see her. My hand hovers over the wood, longing to feel the warmth of her presence, to hear her sleepy voice tell me everything is okay. But I stop. She's been run ragged by Mom's insistence on perfection between school and ballet, and she deserves her rest. With a heavy heart, I retreat.

Once inside my room, the mirror catches my reflection. Selene's choice in the dress I'm still wearing shines under the moonlight filtering through the windows. She has an eye for fashion that could make anyone feel beautiful, even when their world is quietly crumbling. I promise myself to thank her, to let her know her efforts didn't go unnoticed.

But as I peel off the layers of silk and lace, my thoughts drift back to Ella. Panic rises within me, a familiar, unwelcome guest. My father's assumptions about Diarmuid couldn't be further from the truth. Amira, with her perfect timing and sharper instincts, had swept in before I had the chance. Not that I was playing the same game. My stakes are different; they're personal. They're Ella. She needs me to be strong to succeed where it truly matters.

My mind races, and the sudden ring of my old phone slices through the silence of my room like a beacon in the dark. I race to the bedside table and glance at the screen. It's Selene. My heart skips a beat, part hope, part dread. What does she want? Could this be about Amira? Or something else entirely?

I brace myself and answer. Having a backup phone with my original number was my mother's doing. She insists we always have a backup. I keep it charged but barely use the older model.

"Ay, Niamh! I'm sorry to disturb you. I just want to know the name of that guy you met in the woods. Not just the name. The number. I need the number," she rushes out, her words tumbling over each other.

My brow furrows in confusion, the fatigue from the night's events making it hard to keep up. "Selene? I don't understand. Why—" I begin, but she cuts me off, her urgency palpable even through the phone.

"—we agreed to help this woman and we are. We need to. Someone is looking for her," she explains, her voice a mix of determination and worry.

"Selene, it's getting late," I protest weakly, hoping to push this conversation to morning so I don't wake up Ella; I walk to my bedroom door and make sure it's closed tight.

But Selene is relentless. "Text him, then. Tell him we will be there tomorrow morning," she insists, leaving no room for argument.

I sigh, running a hand through my hair, the events of the night catching up to me all at once. With a nod to myself, even though she can't see it, I agree. "Okay, Selene. I'll do it now. We'll sort this out."

I end the call, the room falls silent once more, and I scoop the card out of my dress pocket. I still need to replace my new phone, another thing I'll add to my list of things to do. I quickly type out a message to Rian. Despite the exhaustion tugging at my limbs, a sense of purpose steadies my heart. We had made a promise to help and help we would.

The morning light creeps through my curtains, painting my room with the soft hues of dawn. I wake with a flutter of excitement in my chest, a rare feeling these days, spurred by the thought of spending a few precious moments with Ella before the day fully begins. But as I pass by her room, my heart sinks. Her bed is neatly made, empty, the lingering scent of her perfume the only sign she was ever there. She must have left early for class, another reminder of the space growing between us.

I return to my room and get dressed quickly. Rian had agreed to meet me and Selene this morning at his apartment. After slipping on a pair of jeans and a cream-colored sweater, I tie my sandy hair up with a hair tie and wash up in the bathroom before I enter the kitchen.

Breakfast is a quiet affair, my thoughts still lingering on Ella and the day ahead. I break the silence. "I have to go off this morning," I say, breaking a croissant in half.

My mother gives my breakfast a disapproving glare. She normally only allows me to have fruit in the mornings, but since I'm not under her strict eating rules, I enjoy the pastry in front of me. I have no idea why we have so much if we can't eat it.

I'm expecting a load of questions. Instead, my announcement is met with an outpouring of pride and excitement. Their eyes shine, not with the joy of my accomplishments or happiness, but with the reflection of their own desires. They see me not as their daughter but as a key to unlocking the life of luxury and ease they've always craved, a life they believe marrying into royalty can provide. The weight of their expectations sits heavily on my shoulders, a crown of thorns disguised as gold. My appetite dwindles as my father reaches across and takes my mother's hand, a silent message that things are going their way.

"We have a guest at the door," The maid informs us all. I rise quickly, gather my old phone, and stuff it into my pocket before anyone notices and wonders what happened to my new slick phone.

My father rises, too. "Best meet the man in question." My father is dressed in a suit and proudly pushes back his shoulders.

I can't help but stifle a laugh at the sheer disappointment etched across my father's face. His dreams of Kings and grandeur were momentarily shattered by the arrival of a friend, not a suitor.

"Father, this is Selene McNamara," I say.

He covers his disappointment quickly and takes Selene's hand, giving it a firm shake. Before he asks any questions, I grab a coat off the hook and link my arm with Selene's. "I'll be back later." We race from the house and walk down the driveway out onto the road.

"Your father seems nice," Selene says, glancing back over her shoulder.

When Kings Rise

"He's still watching," I reply without looking back.

"Yes." Selene frowns, and I tug her to the left, out of sight.

"He thought Diarmuid was coming to get me this morning." I offer up the explanation to her that she hadn't asked for.

At the mention of Diarmuid's name, Selene tenses.

We walk the rest of the way to the bus stop in silence. Only a few people are waiting, and we arrive just in time as the bus pulls up. Selene gets two tickets for Sandyford, and we find a seat.

The bus ride to Sandyford is filled with an uneasy silence, broken only by my inquiry into Selene's unusual determination.

"Did something else happen?" I whisper, not wanting anyone else to hear our conversation.

Selene's features tighten. "No. I just want to help."

She isn't telling me everything and seems unwilling to talk this morning. So, I leave her alone and glance out the window. As the cityscape blurs past the window, I realize that perhaps, for Selene, this is more than just solving a crime; maybe it's a way to take control in a world where we truly have none.

When we finally arrive at Rian's place, I'm not sure what I was expecting, but it certainly isn't this. The building itself is one of those new, nondescript blocks that seem to have sprung up overnight to cater to the city's ever-desperate demand for living space. It's clear from the get-go that Rian's apartment, like many others here, was designed with functionality in mind over comfort, intended for those willing to compromise space for a place to call home.

Stepping inside, the contrast is startling. Every inch of Rian's studio apartment is consumed by his work. The walls are plastered with photographs, notes, and maps, all connected by a spiderweb of strings that trace patterns only he could decipher. Timelines stretch across the walls, and stacks of books and papers clutter every available surface, creating a chaos that's both bewildering and strangely ordered. A single corner stands out in stark contrast—clean and carefully arranged, the dedicated space for his video podcasts, a slice of normalcy in a room swallowed by obsession.

Selene's voice cuts through my initial shock, her words tinged with dark humor. "Feels like we've stepped into the den of a serial killer, doesn't it?" she comments, and despite the gravity of our visit, I can't help but let out a laugh. The tension in the room lightens, just a fraction, as Rian turns to greet us with an energy and warmth that's immediately disarming. He pushes his glasses up on his nose as he smiles warmly at us.

I can't help but smile back; his open and friendly approach makes me trust him. I glance at Selene. She's more guarded.

"We wanted to talk about Andrew O'Sullivan," Selene starts. Rian isn't put off by the instant jump to why we are here.

The moment the name is mentioned, Rian springs into action, adding another string to his complex web of information—the physical connection of string to pin, linking Andrew O'Sullivan to the woman.

"Do we know who the woman is?" Selene's question weighs heavier than anything else in the room; that's why we are here, after all.

Rian speaks with a confidence that's both reassuring and concerning. "No, not yet. But the best lead we have is through the medical examiner's office," he asserts, his eyes scanning his network of clues as if they might reveal a new path at any moment.

Selene's interest piques at this, her mind already racing ahead to the logistics. "Do you have any fake IDs or something that could get us in?" she asks, her voice a blend of hope and practicality.

Rian shakes his head, a rueful smile playing on his lips. "No, nothing like that. But, if the body is being claimed by a possible family member, there might be a way to gather enough information to aid our investigation without needing to sneak in." He moves to a small fridge and extracts three frubes, offering one to me and the other to Selene. I decline again, just like I did the day I met him, and he doesn't seem put off as Selene declines, too, with the curl of her nose. I'm not sure how he can eat when we are talking about a dead body.

"Why does milk turn into yogurt when you take it to a museum?" Rian reads the joke before he rips off the top of the plastic.

"I don't know," Selene says.

"I'm intrigued," I say.

Rian smiles. "Because it turns into cultured milk."

As he drinks his yogurt, I try to get us back on track.

"You do realize that getting involved with this could be dangerous for you, right?" I warn him, hoping he understands the gravity of our situation. I'm saying it for me and Selene also.

"I don't care about the danger," he says, his focus unwavering. It's clear that the pursuit of truth, the unraveling of this mystery, outweighs any personal risk in his mind.

"Okay, so you can go in then?" I ask.

"Oh no, they know who I am; I could never get in. You two will be going in by yourselves," Rian states matter-of-factly.

His words settle over us. I want to protest that this is getting out of hand.

I look over at Selene, ready to protest, ready to turn and leave. We can forget we ever started down this path. We can remain ignorant.

We can remain safe.

But something gleams in Selene's eyes. Determination. A grim kind of resolution. Her gaze flickers to mine, and when she speaks, I find myself nodding despite my fears.

"We're in."

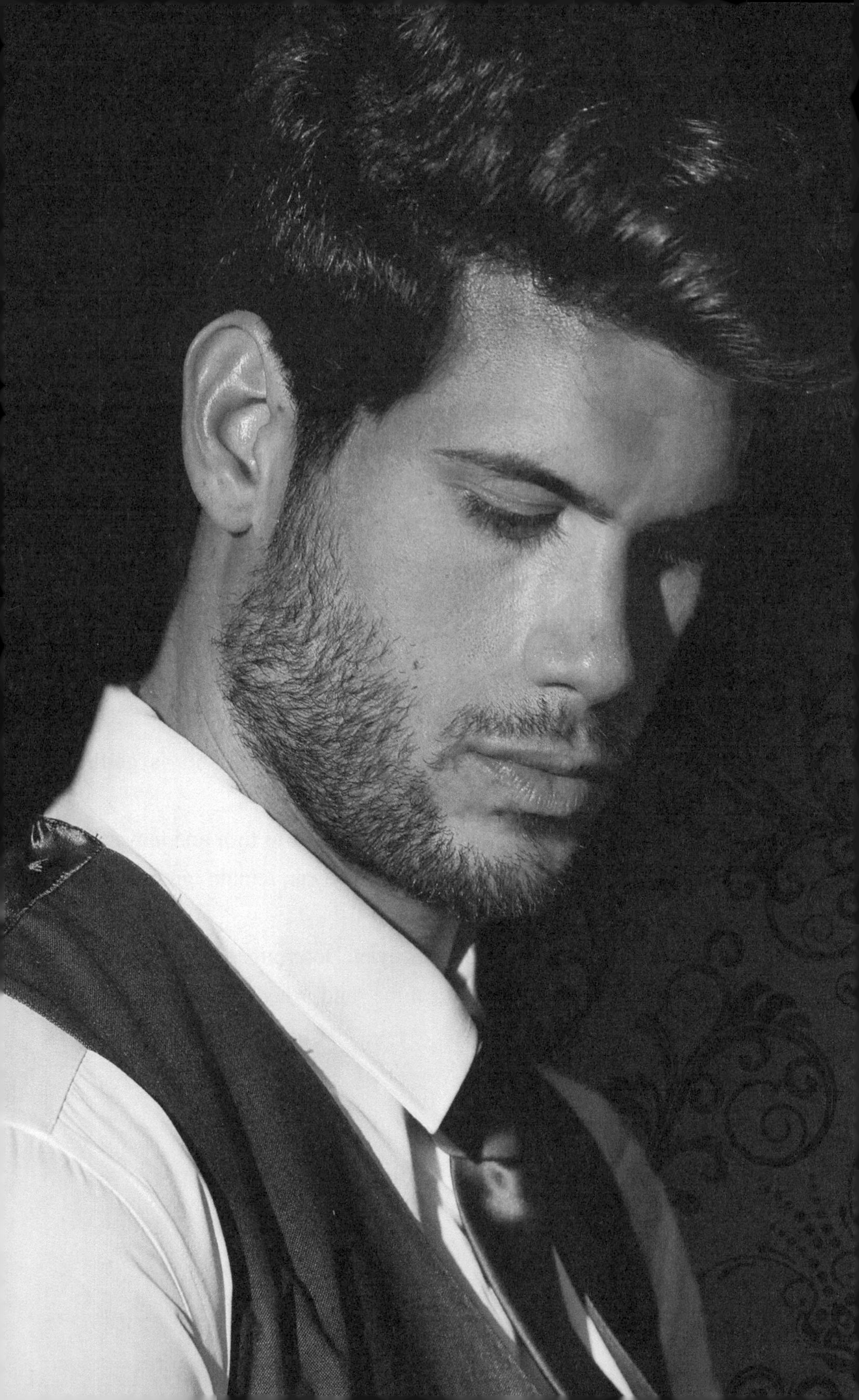

CHAPTER SEVENTEEN

Diarmaid

AS I WALK through the hallowed halls of St. Gertrude's Church, the air vibrates with the harmonious singing of the choir. Their voices lift and soar, reverberating against the ancient stone walls. I tread lightly between the altars, my steps measured, my heart attuned to the sanctity of this place. Pausing at the transept, where the architecture of the church forms the solemn shape of the cross, I feel a moment of reverence, and my mind sings that we should all burst into flames for our sins. I turn away, hating how much control the church—or, more precisely, Victor—has over us all.

Beyond the reach of the choir's celestial sounds lie the hidden chancel, shrouded in thick, velvet curtains. Behind the far right curtain, shielded from view, I can hear whispers that get swallowed up amidst the choir's song. In this moment, I'm reminded of tales of secret societies, of hidden truths and ancient mysteries—a feeling akin to stepping into a scene from *Angels & Demons* or *The Da Vinci Code*. The intrigue that surrounds these hidden chapels stirs a sense of anticipation within me, hinting at the depths of the organization's reach, perhaps as far-reaching as the tunnels beneath Newgrange.

I'm not alone in the church; the choir members' voices are not the only presence here. Floor sweepers, altar polishers, and light duster all move through the space with a purpose. Their tasks are seemingly mundane yet vital to the preservation of this sacred environment. Victor, ever the strategist, leaves nothing to chance.

Slipping through the heavy velvet curtain, I enter the concealed chancel. The space closes behind me with a hush, the fabric falling into place. My eyes adjust to the dim light, and the figures before me come into sharp focus. Lorcan, Ronan, Wolf, Victor, and there, like a shadow from my past, Oisin Cormick. Oisin, the killer who guided my hands to become what they are today. The sight of him, especially next to Victor, sends a shiver through me, unearthing memories long buried.

A memory flashes before my eyes, unbidden yet vivid. I'm eleven, witnessing the execution of a woman for the first time. Until then, death had been a distant concept, one I executed without personal connection, and always men. But her scream—it pierced the air. She was a prisoner; her crimes justified her fate in the eyes of our laws, but the reality of her death shook me to my core. She had killed her husband and brother all in the name of money. As she screamed for death, I had turned away, unable to watch, a moment of weakness that didn't go unnoticed. Cormick had seen, and he had reported it.

The aftermath of that day is a blur of pain and reprimand. Two figures loomed large in my punishment: Victor, with his cold disappointment, and Uncle Andrew, a man whose anger was both swift and brutal: punches, kicks, the unforgiving hardness of walls against my back. The air was filled with shouts, the sound of objects breaking, thrown in a rage meant to discipline, to harden. Those moments shaped me, molded me into the weapon I am, yet left scars no one can see, scars that ache when the past surfaces.

When Kings Rise

As I stand among them now, the weight of those long-ago punishments heavy in the air around me, I remember the lessons disguised as torture. Victor's method was always precise and calculated to teach endurance and control. The memory of the candle's flame flickers in my mind, its heat a ghost on my skin. I was made to hold my hand just above it, the fire close enough to singe, to warn of the pain that comes with failure. Each instinctive recoil, each flinch away from the heat, was met with the sharp lash of Victor's whip across my back. The scars that web my skin are hidden relics of those lessons, marks of a past I carry with me. They are a part of me no one has seen, secrets kept even from my Brides.

Despite the turmoil within, I manage to compose myself, taking a seat calmly beside Lorcan. The contrast between the tranquility of my outward appearance and the storm of my memories is stark. Around us, the conversation shifts, a momentary distraction from the path of reflection.

"The choir sings ever so beautifully," Oisin remarks.

Victor, however, is quick to critique, his ear finding fault where others find beauty. "They are flat," he declares, his voice devoid of warmth.

Victor stands at the helm of our gathering, his authority unchallenged, his command absolute. Previous meetings with my brothers, filled with the casual back-and-forth of familial bonds and rivalry, now serve as stark contrasts to this moment. Here, in Victor's presence, no one dares interrupt. No one dares to contradict. The respect—or perhaps fear—he commands is palpable, suffocating. Despite the rage simmering within me, a desire to smash his head against the cold marble, I remain still.

Oisin's presence complicates matters further. The man molded me, but he's retired, so I have no idea what he's doing here.

Victor's voice cuts through the silence, drawing my focus.

"The investigation into Andrew O'Sullivan's murder has revealed no more leads. The Gardai have been ineffective in their search. Whether due to incompetence or corruption, they are no further on." He glances at each of us.

"We need answers, actions. This can't be ignored."

Victor places his hands behind his back, tilting his head up and looking at the ceiling before giving each of us a sharp look. "That is why I've pulled Oisin out of retirement."

Everything in me freezes. This is the last action I thought Victor would take, especially since he suspects me.

The very idea that the man from whom I learned everything is now an active player changes everything.

Victor glances at Oisin. "Failure is not an option."

I try to keep my features relaxed, but my heart races in my chest. I glance at Wolf, who is delighted with this news.

I can't afford even the tiniest slip up. Selene's earlier accusations about me being a hitman crash heavily against my skull. No matter if the information is something she pieced together or was told, I need to make sure my Brides are in line at all times. One tiny slip up from here on out could cost me my head.

"You will all work closely with Oisin. Whatever he needs from you, you will grant." We all nod, but Victor's gaze lingers on me.

I hate him with so much passion. I will have my moment to destroy him, but for right now, I need to think of a way that he doesn't destroy me first.

When Kings Rise

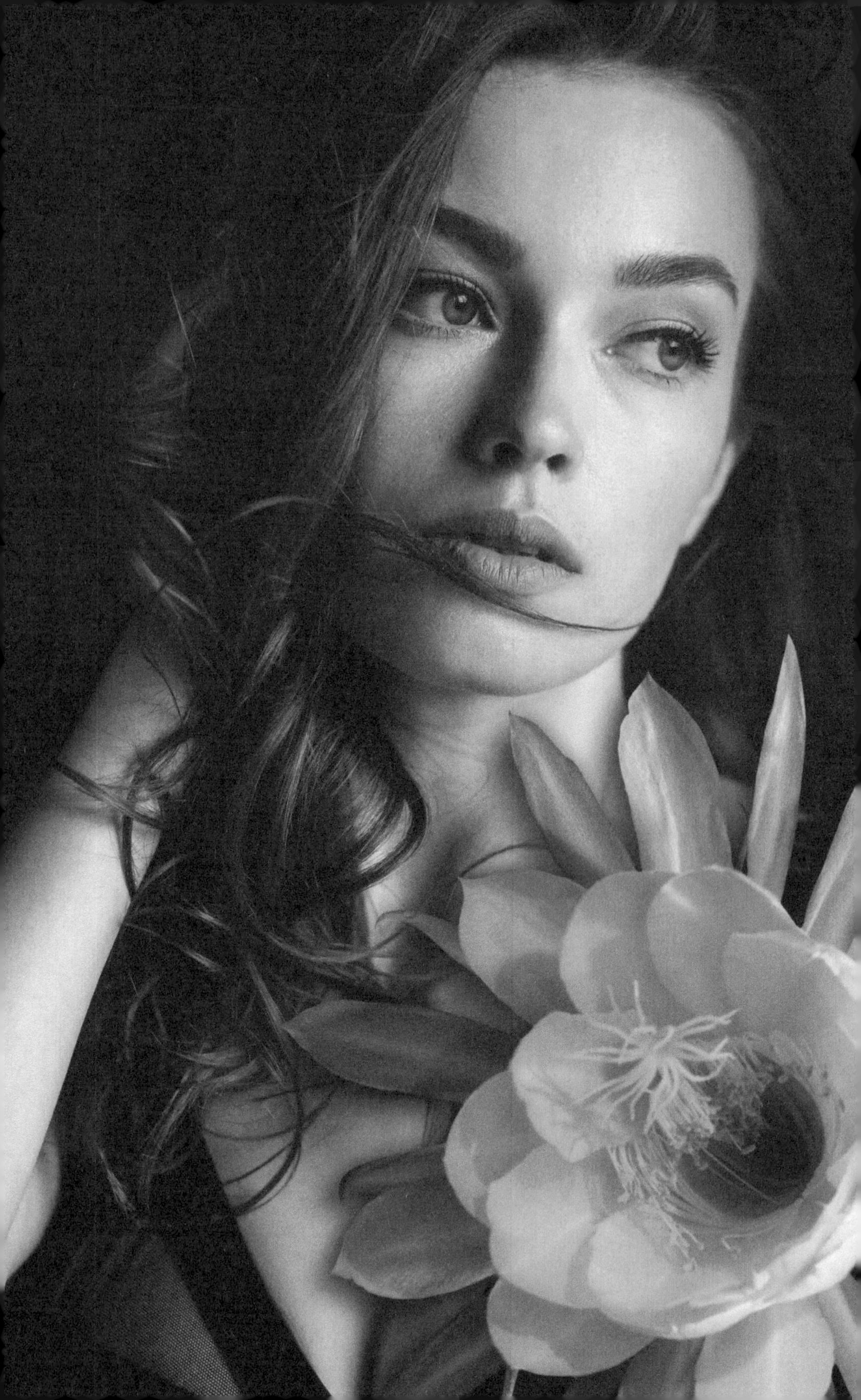

CHAPTER EIGHTEEN

Selene

THE RED BRICK building towers over us, its grandeur reminding me more of a castle than the coroner's office. Despite its apparent age, I know it's a fairly new construction. The sweeping driveway leading up to it is deserted except for us.

I adjust my stride to match Niamh's erratic pacing. She bumps into me yet again, a little too hard this time, causing me to stumble slightly. I glance at her, puzzled by her inability to walk in a straight line despite being perfectly sober. This may be all too much for her, especially since Rian shared more information with us that he was holding onto.

The girl who was found on top of Andrew's grave has brown curly hair. It's such a small detail, but for both of us, it's starting to paint a picture, especially since we have a photo of a missing girl who doesn't appear to even be eighteen.

"Niamh, are you okay?" I ask, keeping my voice low despite the emptiness around us. It feels like the walls have ears, and the last thing we need is unwanted attention.

She gives me a tight-lipped smile, her eyes darting around nervously. "Just, I want this done and over with."

I can't help but agree. It was the mushroom pickers who had found the body. They told Rian the details of what they found, and we have one more final detail. She was wearing a tweed jacket on the twenty-second day of September.

As we reach the grand front door, Niamh straightens up. She rings the bell, and the sound echoes, seemingly swallowed by the building's vast interior. Moments later, the door swings open to reveal a stern-faced woman in a coroner's uniform.

"Can I help you?" she asks, her tone as cold as the air that flows out from the building. All the dead people are in there. I swallow, ready to speak, but Niamh seems to have found her nerve.

Niamh steps forward, and I admire her courage as she speaks. "Yes, we believe we may be related to a woman who was found on September 22nd. We were hoping to... to claim her, if possible."

The woman's expression softens slightly, but she maintains her professional demeanor. "I see. Do you have any proof of your relationship? Anything at all that can help us verify your claim?"

Niamh and I exchange a glance. This is the part we hadn't fully prepared for.

"We... we have this," I say, pulling out a photograph from my jacket pocket. My fingers are sweaty around the image. I find it hard to look at the young, smiling girl that Rian had somehow dug up from the internet. He must have spent hours comparing the small details he had gathered to find a girl who matched this description and who had been declared missing in this area on September 22nd.

The coroner studies the photograph, then us, her gaze lingering a little too long on our faces. "Do you have any ID?" She finally says.

Shit. Of course, she would ask for ID. I curse Rian for not thinking of this; I curse myself, too.

"We got here as quick as we could; I'm sorry, I don't." I hang my head and take a large lungful of air.

"She always wore her tweed jacket around this time of year. It was her favorite." I hope sharing this knowledge will prove that we know her.

The assistant still watches me. I don't know what she sees on our faces, but all of a sudden, she's hesitant.

No.

Niamh chimes in, her voice stronger than I've heard it today. "We got a letter from her two years ago, but nothing since. We've been so worried. Please, we just need to know if that's her."

The assistant nods, seemingly swayed by Niamh. "Let me check the records for the clothing description." She steps away, leaving us to wait at the reception desk.

The wait stretches out like a long, dark alleyway in front of us. Niamh doesn't speak, and I don't try to engage in conversation. If I do, I might end up convincing us to leave.

The assistant returns.

"Normally I need ID, but there's something about your request... Follow me."

As we step into the building, I can't help but feel like we've just crossed an invisible threshold. The interior is just as imposing as the exterior, with high ceilings and long, echoing corridors. My heart races with a mixture of fear and anticipation. What are we about to discover? And more importantly, are we ready for the truth that awaits us?

The coroner's assistant leads us down a long, sterile hallway, her steps echoing off the walls, adding a somber rhythm to our procession. The air is thick with a blend of disinfectant and something else, something I can't pinpoint. Niamh's hand brushes against mine. Her features are strained, and she does appear to be a grieving sister. I think she is imagining her own sister on a slab. I know they are close.

My heart pounds in my chest, a frantic drumbeat as we draw closer to the room, and the reality that I'm going to see a dead body that isn't painted for the Gods or clothed for the living sets in. It will be raw, white, and appear very much dead.

We're shown into a small, stark room. The assistant pauses at the door, her face unreadable. "I'll need a moment to prepare," she says before leaving us alone with our thoughts and our fears.

Niamh turns to me, her face pale under the harsh fluorescent lights. "Do you think we're doing the right thing?" She whispers, her voice barely carrying across the room.

I clutch the photograph tighter, the edges wrinkling under the pressure. I glance down at the picture. The girl has the kind of ordinary face that blends into the crowd.

"We're too far in to doubt ourselves now," I reply, my voice steadier than I feel. "Remember, Megan ran away because of that boyfriend right before graduation. That's our story."

Niamh nods, biting her lip, the anxiety clear in her eyes. But there's also resolve there, a determination that mirrors my own.

When the assistant returns, she doesn't waste time on pleasantries. "Follow me," is all she says, leading us to another room, this one with a somber purpose. The air is colder here, the kind of cold that seeps into your bones.

She stops by a table that's covered with a sheet, her expression softening just a tad. "We're not allowed to do this usually, but something about your story..." She trails off, leaving the sentence hanging in the cold air.

Niamh squeezes my hand, her presence a comforting anchor in the storm of my thoughts. I'm thinking we could just peek under the sheet and leave. I glance at Niamh, who's watching the door.

I want to know what lies under that sheet, but it also terrifies me.

"The body... she was found wearing a tweed jacket," she starts, and for a moment, the world seems to stop spinning.

The revelation hits us like a physical blow. As the assistant prepares to reveal the body, I brace myself, not just for the possibility of recognizing the girl from the photo but for the consequences of today.

The assistant, with a solemn nod, carefully pulls back the sheet. My heart races, but I force myself to look. Niamh, on the other hand, can't bear it. She turns away instantly, burying her head in my shoulder. Her reaction, while genuine in its horror, fits perfectly with what the assistant would expect.

The girl on the table is indeed young, her features peaceful yet hauntingly still, marred only by the stark, unnatural bruising around her neck. The discoloration stands out, a silent testament to the struggle that marked her final moments. It's jarring, and my calm façade begins to crack under the weight of this visible violence.

"The coroner is ruling it as suicide," the assistant says, her voice steady but lacking conviction.

I can't help myself; my gaze flicks to the assistant, searching her face for any sign of doubt, and I find it. There's a hesitation in her eyes, a flicker of uncertainty.

I need her to talk. We have come this far. "She would never have taken her life," I say. My gaze wavers with guilt at seeing this girl's body. A part of me knows we have no right to see her in such a state.

She hesitates, glancing at the door. "It's just that... the bruising, the positioning. It doesn't sit right with me," she confesses in a hushed tone. "But I'm not the coroner. I just assist."

"But you've seen things that don't add up," I press, my resolve hardening. "Things that might suggest... something else?"

She looks torn, caught between her professional duty and the truth she must suspect. "I can't say for sure. There are just... doubts. The angle of the bruising, the lack of other injuries... It's as if she was...," she trails off, unable to finish the thought.

I nod, but Niamh squeezes my shoulder as if to tell me we have to go. The bruising around her neck speaks of a struggle.

The sheet is placed back over the girl, and the assistant steps away.

Niamh lifts her head from my shoulder, her eyes red but determined. "What do we do now?" She whispers, her voice barely audible.

"We find the truth," I reply, my voice laced with a newfound determination. "For her and for all the Megans out there whose stories don't end as neatly as they're told."

The assistant returns to us, chewing her lip. Lowering her voice to a whisper. "There was skin found beneath her fingernails," she divulges, glancing nervously around as if the walls themselves might be listening. "The coroner tested it and claimed it matched her own DNA. Said it wasn't from an assailant."

I process her words, the implication sending a cold shiver down my spine. The absence of scratch marks on her body—it doesn't add up. She hadn't been clawing at herself in a frenzied attempt to escape some inner torment. No, she had been fighting for her life, grappling with a very real and external threat.

"She didn't scratch herself," I mutter, more to myself than to Niamh or the assistant. "She scratched the person who killed her." The words taste bitter in my mouth, a vile truth hidden beneath layers of convenient lies.

Niamh's grip on my arm tightens. A wave of nausea washes over me, the room spinning as the reality of what we're suggesting settles in. I clamp a hand over my mouth, fighting back the urge to vomit right there on the coroner's pristine floor.

The assistant mistakes my physical reaction for emotional turmoil, which, in a way, isn't entirely wrong. She quickly offers me a tissue, her eyes filled with a mix of sympathy and concern. "I'm sorry; this must be incredibly hard for you," she says, her voice gentle.

Gathering my composure, I take the tissue, using it more as a prop to cover my moment of weakness. "I... I can't be sure this is my sister," I stammer, playing back into our narrative with a thread of truth.

The assistant nods, understanding—or believing she understands—our plight. "We have a composite sketch," she offers. "An investigator tried to piece together what she might have looked like before... before she ended up here."

She retrieves the sketch from a folder and hands it over. The drawing depicts a young woman, vibrant and full of life, a stark contrast to the cold, silent form on the table. It's a glimpse into what could have been, a life cut tragically short.

Niamh and I lean in, studying the sketch. It's generic enough to be anyone, and yet, in the lines and shadows, I see the faces of every missing person, every unsolved case.

"Thank you," I say softly, handing back the sketch. "This... This helps, but I don't think it's our sister."

As we put distance between ourselves and the building that now seems more like a facade for darker truths, the reality of our situation settles in. The coroner's too-quick judgment and the ignored evidence, it all points to something sinister—a network of power and silence, possibly the mafia, a cult, or a terrifying mix of both.

"Are you okay?" I ask Niamh beside me.

"I can't stop thinking...what if that was Ella." Her lip wobbles.

I link my arm with hers. "It's not." I remind her.

She nods, but the heaviness in her gaze doesn't lift.

I take out my cell and call Rian. I relay everything we have learned.

"What's next is the most tedious job you'll ever do: figuring out who that is in the sketch," he says, a heavy sigh punctuating his words.

After ending the call, I turn to Niamh, who looks as if she's carrying the weight of the world on her shoulders. "All we have to do," I begin, my voice steady despite the turmoil within, "is to steadily chip away at this case. We give Diarmuid no reason to suspect what we've been up to."

Niamh nods, her eyes meeting mine with a renewed spark of determination. "One step at a time," she agrees, a semblance of a plan forming between us. "We stay under the radar, gather what we can, and build our case. For her," she adds, a vow to the girl in the sketch and all those like her who have been silenced too soon.

When Kings Rise

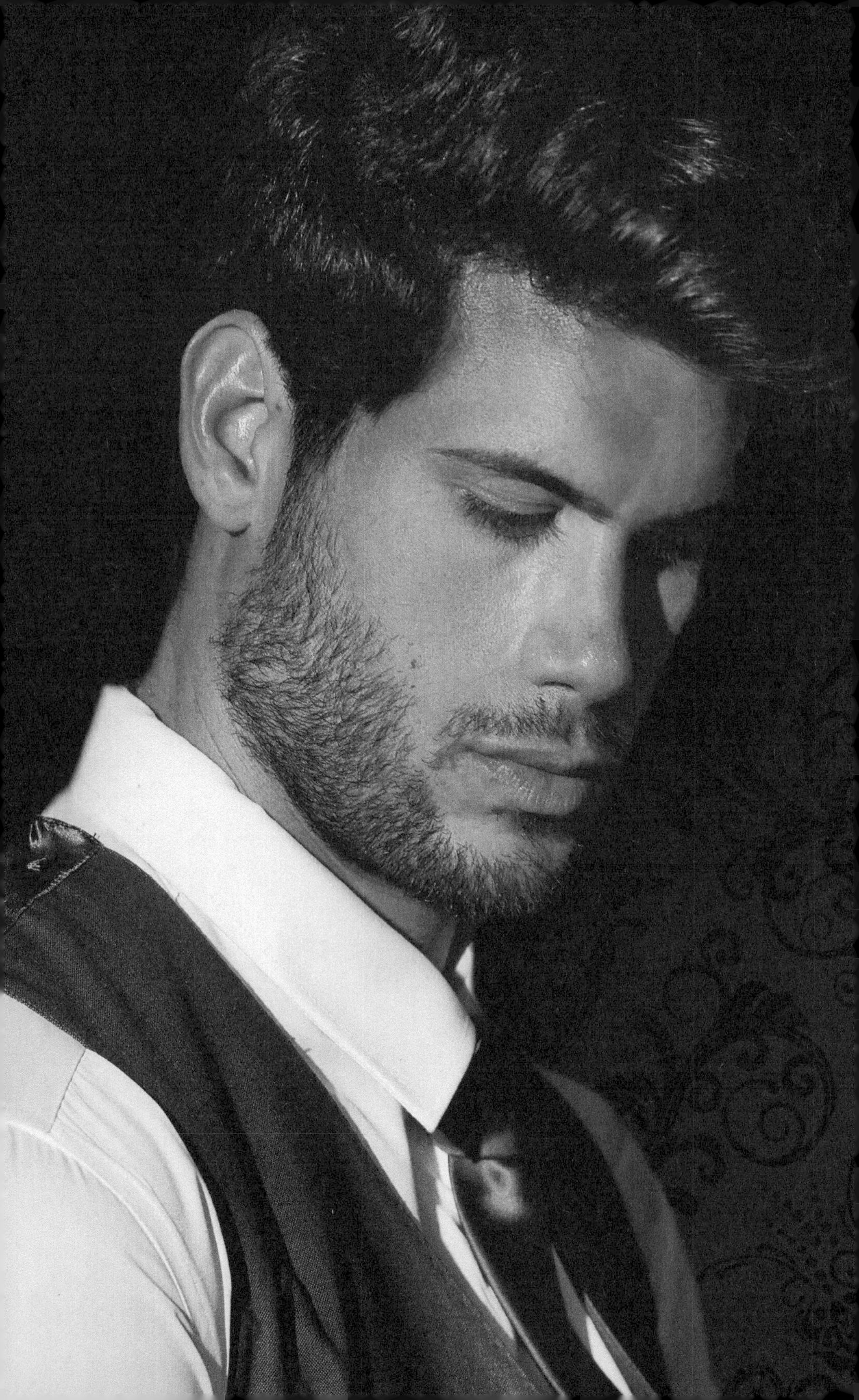

CHAPTER NINETEEN

Diarmuid

TONIGHT, I WANT to get to know my Brides on a more personal basis. I've ordered them to be waiting for me in the main bedroom. I shake off all the earlier tension, all the secrets I carry, and shed it all like a second skin at the door. Tonight will just be about fun.

As I enter the room, Selene and Niamh are already there. Both are wearing their gowns over the lingerie I selected for tonight, their gazes shielded, and I know they have bonded. I wonder if Selene shared what happened between us the last time, along with the knowledge she had acquired about me being a hitman. I remember her talking to the priest, Isaac, and he may have planted seeds in her head.

I offer a polite smile to try and break some of the tension. Niamh offers one back; Selene doesn't. She's still mad. I internally smile. I'd pegged her as a troublemaker from the second I laid eyes on her, and I'm correct. But by taking what I wanted from her the other night, I hoped it showed her she cannot fight this.

Amira walks in, her gaze flitting around the room. It skims past

Selene and Niamh in disgust before landing on the table that I had brought in here by the staff. Light refreshments line the table.

"It's wonderful to see you all here," I say, my voice steady. "Please take a seat." I pull out each chair for them. Selene and Niamh sit together, and Amira sits beside me. She raises her chin high like she's seated where she should be.

I've spent time with each of them, and yet tonight, I have the urge for more. "Let's make this evening count," I propose, raising my glass in a toast.

Each of them picks up their glasses and takes a drink. Amira's dressing gown is slightly open, showing me a tantalizing flash of her flesh. She places her hand on my thigh with a coy smile on her painted-red lips.

My cock grows instantly, but I don't just want one tonight. I want to explore all three.

No one has touched the soft pastries, but I'm not interested in them either. They are just a way to relax my Brides.

Amira's hand trails up my chest until she touches my cheek very boldly. I don't stop her as she turns my face to hers and presses a kiss to my lips. Her tongue flicks into my mouth, and I suck it, causing a groan to slip from between her lips.

I turn my face away to find Niamh watching me with curiosity. That's all I need. I rise, not trying to adjust the erection prodding through my navy trousers. I walk to Niamh and hold out a hand. She glances at Selene as if for permission, and that annoys me, but she places her small hand into mine, and I pull her from her seat.

"You're very beautiful," I say.

Her brown eyes widen with surprise, and I reach back, unclipping her long sandy hair, letting it fall down her back. As I take off her dressing gown, I address Selene and Amira. "You are all beautiful." And so they are. Each one is as stunning as the other.

Niamh's dressing gown pools at her feet, and she steps away, showing me the sexy black lingerie that's sculpted to her body.

Her breasts are perfect, and the flimsy see-through fabric doesn't hide her hardened nipples. I reach up and brush my hand across the hard tips. Her tongue flicks out, and she wets her lips. I smile and lead her to the large bed.

"You can join us," I say to Amira and Selene.

Amira doesn't need me to say it twice; she's already peeled off her dressing gown and disposed of her underpants before she kneels in the center of the bed. Her boldness makes me reach out to her, and I press a kiss to her lips that she sinks into. I turn my head to Niamh and kiss her soft lips. Her kisses are more reserved than Amira's, but I sense a curiosity in how she moves her lips across mine.

Selene finally stands at the foot of the bed. "Remove your gown," I order.

Fire sparks in her eyes, but she does as I say. Her breasts are the largest, and her wide blue eyes reveal so much her mouth doesn't say. She doesn't trust me. It's clear. I hold out my hand to her, and she takes it as I lead her up onto the bed. With my three Brides surrounding me, I take my time and run my hands along their perfectly sculpted bodies.

Amira runs her hands along my back and grips my suit jacket, pulling it off. I allow her as I kiss Selene. I brush my tongue along her lips, demanding they part. When they don't, I run my fingers between her thighs, and she groans, giving me access to her mouth. Does she remember I finger fucked her and licked my fingers after? She tastes sweet, just like the wine she sipped on at the table. I turn my head and find Niamh with wide brown eyes watching me. Gripping the back of her neck, I drag her face to mine and kiss her gently on the lips.

Amira is still demanding my attention, her arms circling my waist. I've already had her, but I turn, and she immediately plows her tongue into my mouth, her movements daring and sending my temperature skyrocketing. When her fingers start to work on the buttons of my shirt, I grip both her hands and stop her. That's one thing I won't allow. I won't remove my shirt.

She frowns, but I push her back until she's lying on the bed.

"Join her." I command Selene and Niamh. Once again, Selene is the last to follow my command. I will definitely have to revisit her home and teach her some manners. But for now, I'll enjoy them. As they lie on their backs, I get off the bed and remove my trousers and boxers. My cock is rock hard, and the thought of fucking the three of them has me climbing back up on the bed.

"Spread your legs," I say to all of them.

I taste Amira first while running my hands along Niamh's thighs. I work my way up until I push past her panties and find her folds pushing them apart. My tongue sinks into Amira's pussy, and her hands find my hair and push my head down greedily.

I work a finger inside Niamh, and she arches higher as if her body demands more. I push a second finger in as Amira grinds her pussy into my face. When I rise, my face is wet. I shift on the bed, swapping my hand over so I can still finger Niamh and taste Selene. Selene watches me, and I grin as I push her panties aside and lick along the outside of her folds. She doesn't groan until my tongue flicks across her clit. Only then does she give in.

Movement at my side has me licking Selene heavier as Amira's sweet mouth captures my cock. She spits on it before she starts to suck. I push my fingers deeper inside Niamh and consider placing a third in her, but Selene's sweet pussy demands my attention. Amira's lips touch the base of my cock, and then she's taking all of me in her mouth. Her suction is perfect, and if she keeps going, I'll end up coming.

Niamh arches again, her core tightening around my fingers, and I remove them and rise from in between Selene's legs. Amira releases my cock and quickly moves behind me as I rise to my knees. Her tongue laps my balls, and I close my eyes for a moment at the glorious sensation.

Niamh's gaze is bewildered, and when I reach down and yank her body to mine, she gives a little startled cry. Pushing her legs as far as they will go, I place the head of my cock at her opening. She tenses almost immediately. Leaning over her, I don't enter her body but press a wet kiss to her mouth, my hands moving along her swan-like neck. She rolls her head in pleasure, her hips rising as she seeks my cock. Amira continues to suck my balls, and I glance at Selene, who watches it all, her guard slowly coming down, lust shining heavily in her eyes.

I push the tip of my cock into Niamh and tighten my hold on her neck. Pain mixed with pleasure is an experience everyone should have. Niamh reaches up and covers her hands with mine, but I don't release as I push fully inside her. I start to tighten my grip, not hard enough to cut off her air supply, but a threat hangs there that I suspect will heighten her orgasm when she comes.

The moment my cock is fully inside Niamh with my hands gently encircling her neck, something shifts in her gaze. Her reaction is immediate; she's clawing at me in a searing panic, her eyes wild. Her legs kick under me. It takes me a moment to retract my hands, and from her neck, but she doesn't calm down.

She's scurrying away from me; her legs slam together as she fights for air. Amira continues to lick my balls as if not noticing Niamh's distress.

"Stop," I order.

Selene has reached Niamh and grabs her arm. "Take a breath." Tears pour down Niamh's face. I have no fucking idea what is happening, but a wave of protectiveness washes over me, mingling with guilt. This was not what I intended.

I slowly reach for Niamh, but she recoils from my touch, and I won't have that. I grip her face, making her look at me. "You're okay," I say.

"Take a deep breath, in through your nose, out through your mouth," Selene instructs.

Niamh starts to calm down, but fear is still etched into her features.

As I watch Selene soothe Niamh, I'm struck by the depth of my own feelings for these women, particularly for Niamh. It's a realization that comes with a weight of responsibility I hadn't fully acknowledged until now. I had sought to create an environment where barriers could be lowered, not realizing that in doing so, I might inadvertently expose wounds I knew nothing about.

What has frightened Niamh so much? "I would never hurt you," I reassure her.

It's then I notice Amira standing at the side of the bed, her dressing gown back on her body and a look of pure displeasure on her face.

Niamh continues to huddle in a corner of the bed. She looks more like a scared animal than the vibrant person I know her to be. Selene holds her and rubs her hair; it calms Niamh down. I scoot off the bed and get dressed.

"Let's get dressed," I suggest gently, my voice aiming to bridge the gap between us. "It'll make everything feel a bit more normal, right?"

Selene rolls her eyes, but there's a glimmer of understanding there. "Because clothes make everything better," she says, sarcasm lacing her tone, yet she moves to do as I suggest.

I kneel beside Niamh, offering a smile that I hope conveys sincerity. "You're safe here," I tell her, my voice soft but firm. "Nothing is going to happen to you." It's a promise, an oath I intend to keep.

Selene, pulling on her dressing gown, snorts. "Well, at least until Diarmuid picks his Consort. Then all bets are off."

I shake my head, meeting Niamh's gaze. "That's not true. My priority is to keep you all safe, always." It's a declaration made not just to Niamh, but to Selene as well, a vow against the fears that haunt their thoughts.

Amira enters the line of my sight. Her expression has hardened. "Safe? How can anyone be safe here?" Her voice is laced with skepticism, her eyes scanning the room and landing on Niamh with a disquieting intensity.

"Amira," I warn, not wanting to see Niamh worked up again.

My tone is sharp, sharp enough to cut through Amira's pouting.

A smile grows on her face. "Maybe Niamh can go home so Selene and I can pleasure you."

I stand, my decision clear. "Amira, you need to leave," I command.

She laughs, a sound devoid of any humor. "Make me."

She pushes too hard. Selene has returned to Niamh and helps her cover herself with her dressing gown. I reach across and grip Amira's arm, removing her from the room.

Once in the corridor, her actions caught me off guard. Her lips are on mine, her tears mingling with a kiss that's more desperation than desire. "I'm not afraid," she whispers between sobs, her words punctuated by the press of her body against mine. "Do whatever you want to me. Nothing can make me afraid."

"Amira," I start, my voice barely a whisper. "It's not about fear. It's about respect, about choices. I want you to feel safe, not because you think there's nothing left to lose, but because you truly are safe."

The air between us crackles with an intensity that's hard to breathe around. Amira's eyes are wild, her emotions refusing to be calmed. "I am not doing anything until Niamh is okay," I state firmly, my decision unyielding. It's a line drawn in the sand, a declaration of where my priorities lie.

Amira's reaction is instantaneous, a fury unleashed. She screams, a sound that echoes off the walls, filled with pain and anger and something else—desperation. Her nails find my cheek, a sharp, stinging sensation that's quickly followed by the warm trickle of blood.

The commotion draws the manse's workers, curious and concerned, to our floor. Their presence only adds to the chaos, their eyes wide with shock and confusion. "Disappear!" I bellow, a command laced with a threat that I hate to make. "Or else." I don't finish my threat.

Amira continues to fight against me. "I'm the best choice!" she insists, her voice cracking under the weight of her conviction. "You're wasting time, our time!" Her declaration is a plea, a demand, a vision of the future she's convinced herself of—one where competition and choice don't exist.

She lashes out again, catching me off guard for a second time. Her nails slice my skin once again, and I grip her arms, stopping her madness. Picking her up, I march down the stairs barefoot. She's still shouting.

"It should be just us." She's lost her fucking mind.

I don't stop until I reach the front door. I use one hand and open it, clearing the last steps until I deposit her onto the driveway.

"What are you doing?" She shouts as I turn and jog up the steps. I give her a stern look before I lock the door behind me. My face burns, and I touch my cheek. Fresh blood is sprinkled along my fingers.

I march up the stairs with a determination to find out what the fuck caused Niamh to freak out so badly.

When Kings Rise

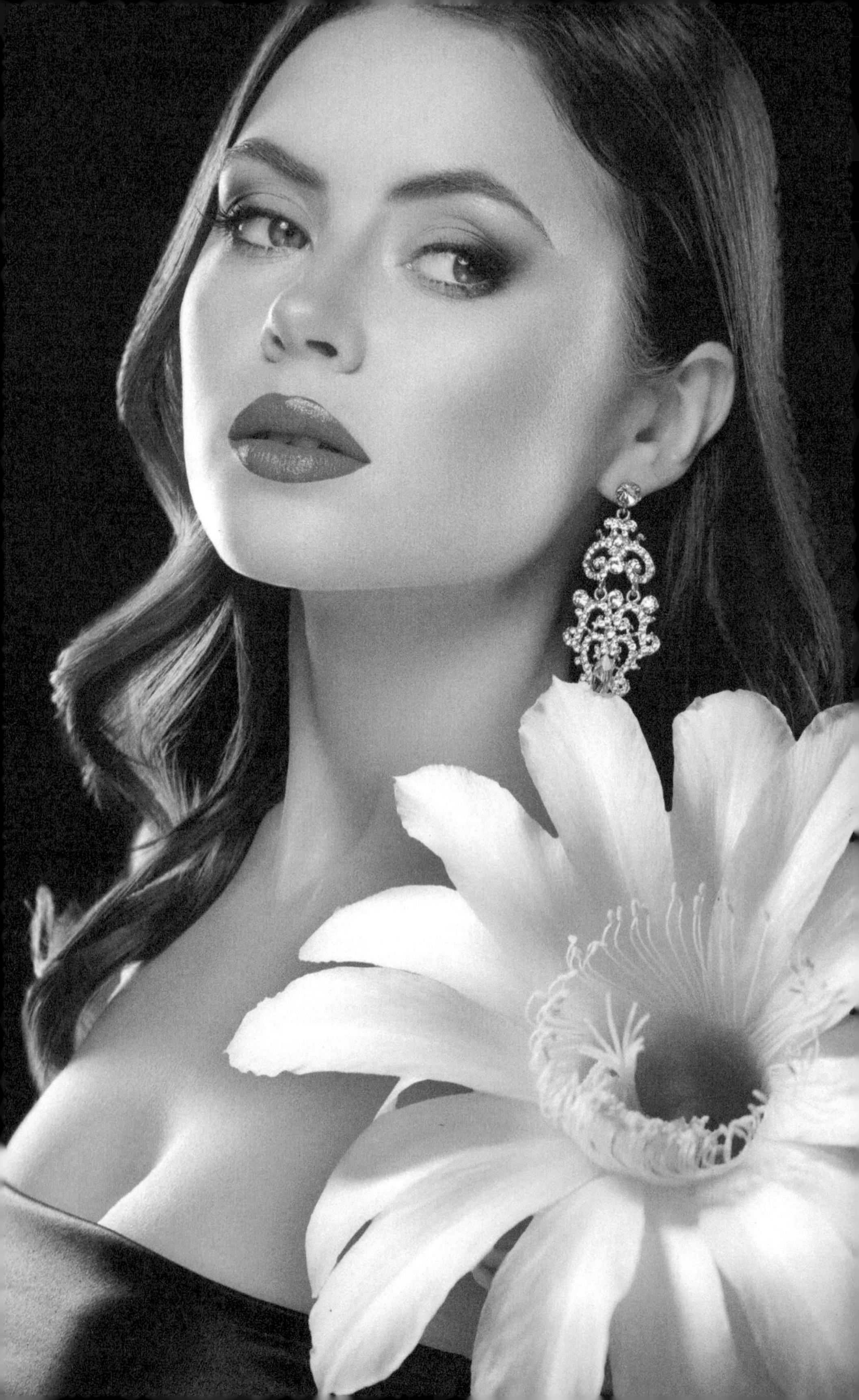

CHAPTER TWENTY

Amira

R UINED... I'VE RUINED everything. The realization hits me as I pick myself up from the concrete driveway, the cold seeping through the thin fabric of my dressing gown. I can't go home, yet the thought of fleeing to somewhere—anywhere—makes me realize I've nowhere else to go.

What have I done? Why did I lose my mind like that? Questions whirl in my head as the coldness of the driveway presses against my skin. My brain registers the chill, yet I can't bring myself to move, to react. I'm paralyzed, not just by the cold, but by the flood of memories and emotions that choose this moment to overwhelm me as I stare up at the mansion.

Dominic died this week, six years ago. I was just thirteen, barely stepping into my teenage years, when I lost my brother. My mother couldn't bear it. She couldn't bear most things, leaving me to grieve alone. Oh, God. How is she going to react to this news? The thought of facing her, of adding this failure to the mountain of disappointments that already defines our relationship, is unbearable.

I search the windows of all the rooms, trying to remember which one on the third floor we were in. Is Diarmuid up there with Selene and Niamh laughing at me? Did Niamh act like that for attention? Of course, she did; she was such a bitch to force my hand. Angry tears pour down my cheeks.

"Now, this is a view I rather enjoy." The words drip with unwelcome amusement and insinuation.

Wolf stands there in the driveway, his gaze not meeting my eyes but wandering in a manner that makes my skin crawl, focusing on parts of me as if I'm nothing more than an object for his viewing pleasure.

With Wolf's hungry gaze on me, I make no attempt to shield myself.

"I hoped I'd run into you again," he says, and the memory of watching him train the last girl assaults my memory, along with the things I learned watching him. I had used the same techniques on Diarmuid, hoping to gain the upper hand.

I don't speak, and Wolf takes a step closer.

"Speak to me." There is nothing soft in his words.

"I need to go home," is all I manage, my tone flat, devoid of the turmoil that rages inside me.

"Do you have Diarmuid's permission to leave?"

His question sparks a bitter smile that doesn't reach my eyes. "One could say that," I murmur, the irony not lost on me.

He looks at me for another minute. "I'll drop you home."

I don't exactly want to be alone with Wolf, but standing in the driveway with the image of Diarmuid and the girls laughing at me has me following him around the back of the mansion where his car waits.

The drive home is a silent journey, punctuated only by the occasional flick of Wolf's lighter and the soft exhale of smoke. The car's interior is filled with the sharp, acrid scent. Wolf doesn't fire any questions my way. Either he doesn't care, or his mind is somewhere else. It's a relief, in a way, to not have to explain, to not have to relive the humiliation and pain through recounting.

Outside, the city moves past us in a blur of lights and shadows. Streetlamps cast shadows on the pavement, fleeting glimpses of people living their lives—a couple laughing on a street corner, a group of friends sharing a late-night snack—moments of normalcy that seem so alien to me now. The world goes on, indifferent to the upheaval in my own life.

As we turn down the long driveway to my house, a sense of unreality washes over me. Wolf, without having asked, knew where to take me. Under different circumstances, I might have found that alarming, questioned how he knew where I lived. But tonight, I'm beyond caring, beyond questioning. I'm just a shell.

When the car comes to a stop, I don't thank Wolf. I don't say goodbye. I simply get out and walk toward my front door. Wolf is gone before I even close the door behind me.

The moment I step inside, the house embraces me with an eerie silence, save for the faint hum of light emanating from the kitchen. My body is a tight coil of tension as I tiptoe through the foyer. I want to go to my room, but the state of the entry to the kitchen stops me cold.

The trash can lies on its side, contents spilled. The sugar bowl, once a fixture on the counter, now lies shattered against the wall, its contents strewn about in a chaotic spray of white against the darkened tile. The scene is one of absolute destruction.

Despite my urge to keep moving, I'm rooted to the spot. And there, amid the devastation, sits my mother. Her presence is almost ghostly, her head resting on the countertop, her eyes closed.

For a suspended moment, I entertain the thought that she's finally drank herself to death. But then her eyes blink, and the harsh light of reality washes away the brief illusion of peace. Her face contorts.

"Amira!" Her scream sounds like a dying animal. It shatters the coldness inside me. I need to run to get to my room. But my legs won't cooperate.

"Get in here!" She roars again. I close my eyes briefly, praying for a respite from my mother's madness. If I don't do as she says, what will she do?

I take a small step into the chaos of the kitchen. I don't ask what happened.

When I come into full view, her face pinches in anger. "I would trade you for your brothers," she spits out, a confession so cruel it seems almost unholy. "I ask God every day to do this." Her words are venom, designed to wound, to break.

Her next words strike hard. "Whore," she hisses, a label meant to degrade, to diminish. It's a blow aimed not just at who I am but at the very essence of my being.

In that moment, something within me shifts. The pain of her words is real, but it ignites a spark of defiance. "You don't have to worry about me being a whore anymore," I respond, my voice steady despite the chaos inside. "I'm not a Bride anymore." It's a declaration, not just of my status but of my refusal to be caged by her judgments, her expectations.

"So, not only are you a whore, but you are a bad whore." Her voice drips with disdain.

My hands tighten into fists. "I am only a whore because you and Da fucked up our lives," I shoot back, my voice laced with a venom born of years of suppressed anger and hurt. "You're the reason why my virginity was given away. You're the reason why Da is out with the hooker of the week. You're the reason why my brothers are dead." I'm smiling as each word drips from my lips. I laugh at her as her face turns white.

She's still for a moment, but it's only a short moment before she launches herself from the seat, and my back slams heavily into the kitchen floor, taking the air from my lungs.

"You dirty bitch." She grabs the rubbish around us, stuffing it into my mouth, cutting off any air that tries to find its way into my starved lungs.

Before I can react, she drags me off the floor with a strength she shouldn't be able to wield. She turns the tap of the sink on, and cold water violently splashes across my face before she rams a bar of soap into my mouth.

"You will wash your mouth out, you sinful, sinful girl."

I choke on the acid taste of the soap. I push her away, and she stumbles, the soap dropping to the floor. She's ready to launch herself at me again when I scoop up ice-cold water and aim for her face. The shock has her screeching, but she's already grabbed a pan caked with grease and swings it, missing my face, but it slams against my shoulder. I cry out as I tumble to the ground.

"Get up!" she roars.

I'm stunned for a moment, but when her fingers curl around a cutting knife, I get to my feet. A sharp edge grazes my arm, drawing blood—a stark red line. The pain is sharp, and I try to wrestle the knife out of her hands. My bare foot slams down on her bare feet, and she cries in pain, her grip on the knife loosening. I scurry to the ground and pick it up, rising quickly to my feet and bringing the blade to her neck.

"You're a sorry excuse for a mother." Tears burn my face. My heart hammers in my chest. "I want you to die!!!" I roar into her face. "Die, you bitch." I scream again, pushing the blade. Something in me snaps as blood makes a trail down her neck. I drop the knife, the clang loud at our feet.

As I stagger away, the taste of blood coppery in my mouth and my wounds stinging painfully, I know that this is a turning point. I've lost everything. This is not just about survival; it's about forging a new path, one where my parents no longer dictate the direction of my future.

I take the first step up the stairs when a heavy weight lands on my back. I spin as my mother's fist connects with my jaw. She puts all her hate behind the thump, and I'm dazed for a moment. I shuffle up two more steps and spin, kicking her in the face. She falls down the three steps, and for a moment, her still frame makes a hysterical laugh bubble up my throat. But when she raises her head, I know she will kill me. I claw at the stairs, scrambling to get to the safety of my room.

"Get back here!" Her screams are right behind me.

Fear wraps its dark fingers around me as I dart through my bedroom door. My heart races, pounding against my chest with the same ferocity that my mother slams her body against the door just as I get it closed and turn the key. She pounds against my bedroom door. I can hear the wood complaining, threatening to give way under her relentless assault. If she gets in here, she will kill me. The thought flashes through my mind, clear and terrifying.

My jaw aches, and I'm dazed for a moment, but I scramble around my room. My eyes scan the familiar space for anything that might aid my escape.

Tennis shoes—I gather them up along with a sweater and a pair of jeans. I slip everything on in a heartbeat. Panic courses through me as another crash sounds at my door.

Next, cash. I hastily shoved it into my pocket from a drawer I'd always hoped would remain a secret. It's not much, but it's all I have.

The window is my only exit, and it looms in front of me like a beacon of hope. I wrench it open, the night air slapping my face, sobering me with its chill. I'm halfway through, one leg dangling out into the void, when the inevitable crash sounds behind me. The door has given in. My mother's fury has me scrambling, but I don't look back. I can't.

The fall from the window is brief, a momentary flight that ends with my feet hitting the ground hard. I stumble, but I don't fall. Adrenaline is a miraculous thing, lending me the strength and speed I didn't know I possessed. As I run, my mother's rage-filled screams chase me, a haunting soundtrack that follows behind long after I've escaped into the night.

But for now, I run. I run from a house that was never a home, from a woman who was never truly a mother. Each step is an alchemy of liberation and terror. The night is cold as I race away from my mother and into the unknown.

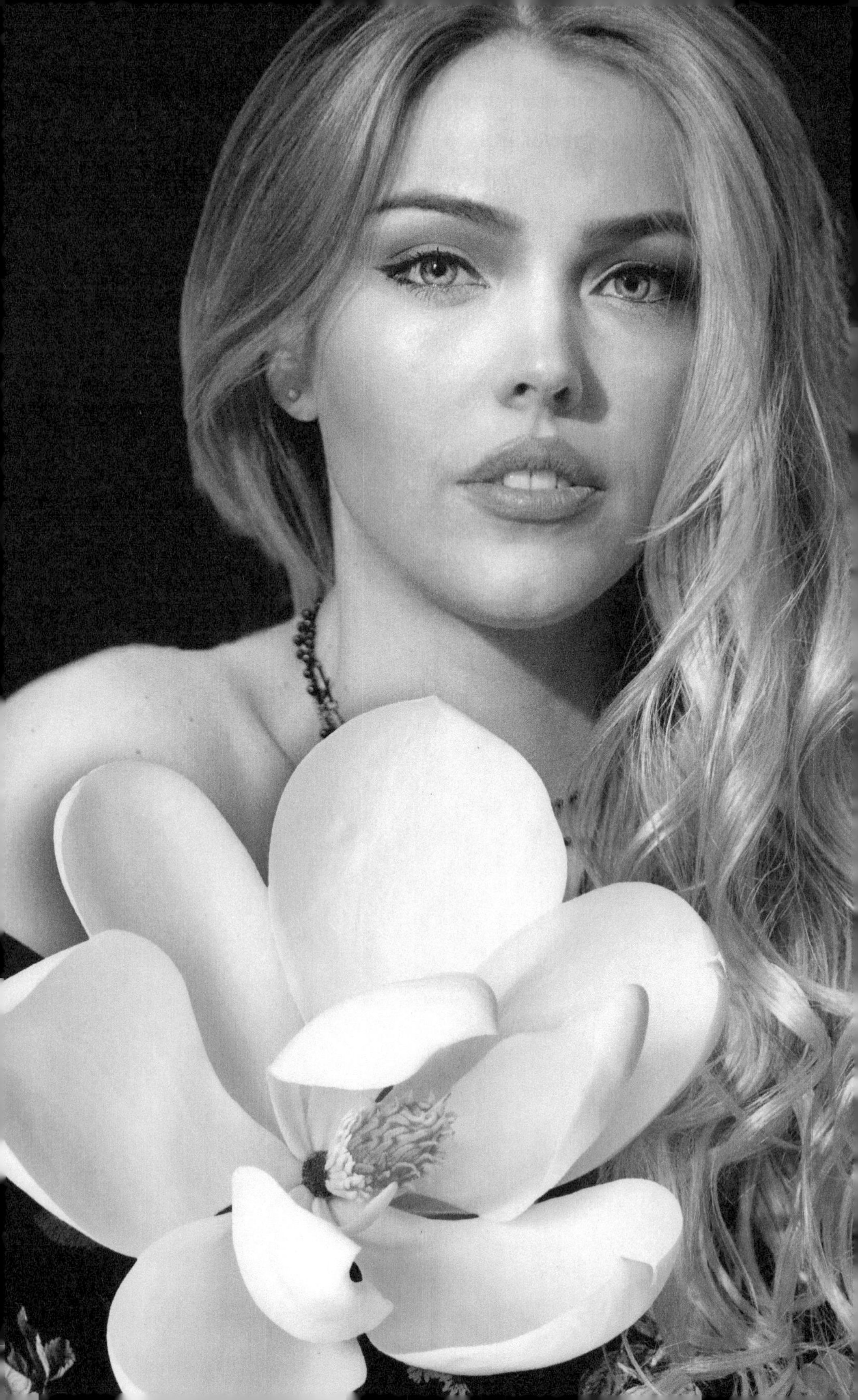

CHAPTER TWENTY-ONE

Niamh

I'M NESTLED HALFWAY between the stage and the back doors of the Gaiety Theater. To my left is my mother, her attention riveted upon the unfolding spectacle. My father occupies the space on her other side. Both are engrossed, their gazes never straying from the stage. Around us, a sea of formally dressed spectators share in this silent performance.

The dancers, with their vibrant costumes and energetic movements, bring the story of the four seasons to life. Winter's chill has been banished, making way for Spring's vivacity. It's a transformation I've witnessed year after year, yet it never ceases to stir something within me. The stage is awash with colors.

I search for one performer among the many. Ella, my sister, the one person I can pinpoint in a crowd of a million without fail. Tonight, though, her role makes her stand out even more. Not a flower, nor a bird. No, Ella is Bacchante—a character with a name, a story, a presence that is undeniably her own.

A swell of pride rises in me. She embodies Bacchante with such conviction that, for a moment, I forget she's playing a role. To me, she's the very essence of Spring itself—wild, joyful, and unrestrained.

Intermission breaks the enchantment of the performance with the stark reality of a crowded space. The stage, just moments ago alive with the vivid storytelling of dancers, now lies hidden behind the heavy curtains, awaiting the second act. Dancers in yellow cloaks mark the transition, their movements a whirlwind of color and grace, guiding their fellow performers offstage in a final, fleeting tableau before the curtains draw close.

The shift in the auditorium is immediate. A collective exhale fills the air, the sound of hundreds of people rising, stretching limbs stiffened by the long sit, engaging in whispered conversations, or navigating the aisles toward the restrooms or concession stands.

I watch as my parents stand, my father taking the lead as he always does from his preferred spot at the end of the row. They merge into the stream of people, my mother's voice barely audible over the hum of the crowd, discussing Ella's performance—every movement, every leap, every turn scrutinized. The pride in her achievement is tempered by a relentless pursuit of perfection. Even in this moment of triumph, the conversation centers on what could be better, on the imperfections only a trained eye could catch. It's an all-too-familiar pattern, their high expectations always casting a long shadow.

I find myself alone, my parents' attention anchored to Ella's performance. Their absence by my side is a familiar scenario on nights like these, where my sister's talent steals all their attention. Oddly enough, I like this time alone.

Navigating the throng of intermission-goers, I push my way out of the theater and onto the sidewalk. The transition from the artificially lit interior to the outside world is startling. The grand arc of the theater's entrance frames my exit as I step into the crisp embrace of the autumn air. The freshness of it is nice after the warmth of the theater and the intoxicating scents of perfumes and aftershaves.

Outside, the city is alive—the distant hum of traffic blends with the closer sounds of conversations and footsteps on pavement. Streetlights cast a golden glow.

My moment of solitude is cut short by an usher who approaches me as I attempt to re-enter the theater for the second act.

"Niamh Connolly?" He asks, an odd formality in his tone.

"Yes."

He nods. "I am here to lead you back to your seat." He holds out his arm for me to go ahead.

"I already know where my seat is," I say.

"Let me have the honor." He smiles softly.

As I follow the usher up the stairs and down a hallway tinged with the muffled sounds of an audience settling back into their seats, a familiar anticipation builds within me. The red curtains that mark the entrance to the private boxes loom ahead, but it's the sight of one particular box, distinguished by its door, that signals this is no ordinary seating upgrade.

The usher opens the door, and I peek inside. Diarmuid is there, his presence commanding even in his silence.

"Sit next to me," he says without even turning around.

The opulence of the private box is immediately apparent, not just in its furnishings but also in the presence of bodyguards, discreetly hidden behind curtains at both the front and back.

"Leave us." Diarmuid's terse command has them departing, and it leaves us alone.

As I take my seat beside him, my attention is involuntarily drawn to Diarmuid. Dressed for the occasion, his appearance transcends the usual definitions of formal attire. There's an undeniable elegance to him, a refinement that accentuates his presence. The realization that he's both familiar and entirely enigmatic brings a blush to my cheeks. Here is a man who has changed the course of my life, and yet, the gravity of our situation feels all the more real in this secluded setting.

Selene's warnings echo in my mind that Diarmuid is dangerous. That he survives, and indeed, thrives in his world is a testament to his strength and perhaps, to aspects of his character that are better left unexplored. The severity of his deeds, as hinted by Selene, keeps rising to the forefront of my mind.

As the curtains of the box draw back and the performance resumes with the vibrant depiction of Summer, Diarmuid breaks the silence. His voice is calm amidst the storm of my thoughts.

"I wanted to make sure you are okay after the other night," he starts.

My body tenses at that question. The night he put his hands around my neck, and all I could think about was that poor girl in the morgue, with marks around her neck, too. I don't think Diarmuid killed her, but it had pulled me under a dark current of fear and panic that I wasn't able to escape that night, not even when he had returned and questioned me. I had no answer for him then, and I have none I can give him now.

"You're kind," I say, a simple truth.

"You are one of my Brides, a role that carries with it duties and responsibilities. It's my job to take care of you," he answers simply.

"What about Amira?" I ask, remembering she hadn't returned after he had escorted her from the room.

"Amira is my concern, not yours."

Maybe he's right. I'm not overly fond of her. I take another peek at Diarmuid. "How did you know I was here, and how did you get tickets?"

He looks at me, and a slow smile crosses his lips.

"The Kings own the private box," he states.

This doesn't surprise me as much as it should. Their reach and influence, it seems, extend even into the cultural heart of the city.

"And how I knew where you were... I had you followed." He continues with a matter-of-fact tone.

When Kings Rise

However, the fact that Diarmuid had me followed here sends a chill down my spine. It's one thing to be under the protective gaze of a powerful organization, quite another to be shadowed without my knowledge.

"Why am I being followed?"

Diarmuid doesn't answer straightaway, as if he is weighing his words. "The organization has its own unseen dangers; this level of protection and surveillance is necessary."

That doesn't exactly answer my question. His lack of detail shows me he doesn't fully trust me. But we don't know each other that well. The only time we are together is with Selene and Amira.

"Why are you here, at this ballet?" he asks, shifting the conversation. I allow the turn of questioning, knowing I'm not going to get any more out of him.

"My sister Ella is in the play. She has the role of Bacchante. She's only sixteen, so her role is very significant for someone her age."

"It sounds like your sister has a bright future ahead of her," he states.

"I guess."

"The role of Bacchante at sixteen doesn't guarantee success in the world of ballet?"

The moment stretches between us, charged with an energy I can't quite name. Diarmuid's gaze is intent, probing, as if he's trying to read the very essence of my thoughts. It's disconcerting, and yet, I find myself unable to look away.

"Well, the first woman to play Bacchante, Marie Petipa, ended up dying of impulsive insanity," I share, a bit of trivia slipping out in an attempt to lighten the mood or maybe to impress him with my knowledge.

He chuckles dryly, the sound echoing slightly in the spacious box. "You must be a very supportive sister to know so much about ballet."

The comment stings, though I know he doesn't mean it to. I shift uncomfortably in my seat. "This used to be my world," I confess, a hint of nostalgia coloring my words.

"You hated it," Diarmuid observes, more a statement than a question.

I nod, the admission slipping out easier than I expected. "I did. But my mother... She wants a prima ballerina in the family."

"Your sister isn't a prima ballerina now?" he probes, his interest piqued.

"No, to be a prima ballerina, Ella needs to be accepted into a major ballet company and then become the best dancer in that company," I explain, my voice tinged with a mix of hope and realism.

"So, it's like being a general," Diarmuid muses.

"Or a King," I add, my words a bridge between our worlds.

The conversation shifts subtly, Diarmuid's gaze intensifying. "And you? What do you want to do with your life?"

I take a deep breath, my own dream suddenly feeling small and insignificant in the grandeur of this setting. "I want to conquer the Oceans Seven. To be a professional swimmer."

Diarmuid's response is noncommittal, a simple nod that prompts me to push further. "What about you? What's your dream?"

"When you don't really own your life, what use are dreams?" His words are a whisper, heavy with a resignation that surprises me.

The question hangs in the air, unanswered, as the ballet resumes onstage.

The rest of the show unfolds in a shared silence that feels both comfortable and charged with unspoken thoughts. For a moment, the world outside this private box, with its dangers and complexities, fades away. I'm just Niamh, Ella's sister, lost in the beauty of the ballet.

As Ella takes her final bow, something within me ignites. I forget about the formalities, the presence of Diarmuid, the weight of his world pressing in on us. Rising to my feet, I applaud with abandon, my hands coming together in a loud, fervent praise for my sister's performance. In this moment, I am every inch the proud sister, my heart swelling with pride.

As the curtain falls for the last time and the audience begins to filter out, the reality of my surroundings—and my company—settles back in. I'm still standing when I turn to Diarmuid. "Thank you," I say, sincere in my gratitude for the experience, for the view, for the momentary escape from my parents, who I bet haven't even wondered where I am.

He gestures for me to wait, moving with a deliberate calm to close the outer curtains of our box, sealing us away from the departing crowd. The privacy feels suddenly intimate, a world apart from the grand spectacle we've just witnessed.

Then, he turns to me, the intensity in his eyes a stark contrast to the quiet endearment of earlier conversations. "I want to know exactly how you like to be touched, what you want from me," he says, his voice low and earnest. "I've been restless since our last encounter, and I want to make it right."

His words hang between us: a confession, a question, a plea. It's a moment of vulnerability, of honesty, that strips away the layers of his guarded existence. In this secluded box, away from the prying eyes of the world, Diarmuid is not a figure shrouded in mystery and power but a man seeking connection, seeking understanding.

The air shifts around us, filled with a new tension, a new possibility. As I meet his gaze, a thousand thoughts race through my mind, each one a reflection of my own uncertainties, desires, and fears. Yet, beneath it all, there's a flicker of something else—curiosity, perhaps, or even the thrill of stepping into unknown territory.

In this moment, the roles we play—the King and his Bride, the protector and the protected—seem to fade, leaving us simply as Diarmuid and Niamh.

"I'm not sure." My voice rattles.

Diarmuid nods and clears the distance between us. "If I did this, would it be okay?" He takes my face gently in his hands and presses a soft kiss to my lips. I taste mint, it's refreshing, and when he sinks his tongue into my mouth, I press mine into his.

The buzz of the voices of hundreds of people below us sounds distant. The lighting in the private box is dim as Diarmuid breaks the kiss and smiles down at me. Being around Diarmuid before, I've always been shy, but having him alone, changes something in me. I reach up and press my lips to his.

The invitation unleashes a desire in him that surprises me; his kisses are hungry, and his hand warms around my waist, pulling me closer to him. I can feel the full extent of his excitement. My own dampen between my thighs.

A thought assaults me as fast as the strike of lightning. *I like having him to myself.*

I want to be as bold as Amira and as courageous as Selene. I let my hand slip across his wide chest and move lower and lower until I touch him. The outline of his bulge feels huge against my hands. He had almost taken my virginity the other night, but the act wasn't complete. The idea of having him take it with just the two of us present makes me grip him harder and rub his full length. He groans into my mouth, his minty breath filling my own.

He breaks the kiss and rests his forehead against mine. He's still as I continue to rub his full length. His eyes are closed, but he groans in pleasure.

After a moment, he opens his eyes, and his hand leaves my waist and trails down further, where he bunches up the fabric of my dark red dress, gathering it slowly. The cool air touches my bare legs, and anticipation has me frozen for a moment.

"Is this okay?" he asks as his fingers touch my bare skin.

I nod.

He continues until he touches my panties, damp with a need that both shocks me and has me leaning in closer to his hand.

"Can I taste you?"

I glance at the balcony that gave us a view of the stage. It's now blocked with the red, heavy drapes. People still talk below us, but I know no one can see.

I nod again.

Diarmuid sits back down in my chair and pulls up my dress. He's kneeling at my feet, his gaze fixed on my face for a moment before he bends over. Pulling my panties aside, he gives his tongue access to my folds and parts them.

I hiss with pure pleasure. He's done this before, but being alone makes it feel different, more intimate.

He laps at my folds, and my body threatens to release the build-up that's quickly gathering inside me, but Diarmuid's licking turns to kisses that he continues pressing the whole way down my leg. Each kiss feels like he's branding ownership into my flesh. He lifts my leg and slips off my high heel. He watches me as he places kisses along the inside of my foot.

"What do you want, Niamh?" His voice is husky with his own desire.

I know what I want. I'm just not sure I'm brave enough to say it, but as I glance around the dimly lit box, I know it's now or never.

"I want you to take me," I say.

He nods and rises to his feet, holding out his hands. I take them as he pulls me to my feet. He spins me and pulls me into his chest. Kisses are placed along my neck, and the sound of the crowd below turns to a soft buzz as I'm consumed by his roaming hands across my torso and the kisses he continues to place along my neck. He gathers the fabric of my dress again until he's holding it securely around my waist. His lips touch my ear. "Hold onto the chair and bend over," he whispers and kisses my ear again.

I grip the velvet of the back of the chair and arch my ass into him. I take over, holding up my dress. Warmth rushes to my checks as I hear his zipper, belt, and then the shuffle of the material of his trousers.

His fingers prod between my legs, and my eyes flutter closed at the contact. It's only for a moment before he removes them, and a larger, meatier body part is placed at my opening. He pushes himself into me slowly, stretching me, filling me up. It burns slightly, just like the last time. One hand grips my hip while the other runs up and down my spine in a soothing movement, and then he withdraws slightly before pushing into me again. This time, the burn is less. My core squeezes around his cock, and I find myself pushing further into him.

"You are perfect," he whispers. His voice sounds strained, like he's struggling to hold this steady flow of in and out. He's being careful with me, knowing I'm a virgin, and that makes my stomach tighten with appreciation and need.

He moves inside me again and pulls out; he keeps this steady rhythm as he runs his fingers up and down my spine.

I keep my hold on my dress that's gripping the chair, and with my other hand, I reach down and touch my clit. The contact sends a new thrill through me, and I gasp as more sensations seem to override everything else.

I don't know how he senses my need, but he starts to move faster, in and out of me. I'm so close to coming my body screams for release. I turn just enough to meet Diarmuid's gaze. I don't know what he sees, but he starts to move faster, and I turn back, working my clit. The climax is like the final scene of the ballet, and I'm falling, calling out as the lights shatter behind my eyes, and I come fast and hard.

When I open my eyes, I'm panting, a light sprinkle of sweat on my forehead.

"Oh, fuck." I manage to say between dry lips.

A half-snort laugh from Diarmuid, has me apologizing.

"Please don't apologize," he says, removing himself from me and fixes my dress into place. I take a moment to gather some courage to face him, and when I stand straight, Diarmuid is dressed and holding my shoe in his hand.

He smiles happily at me as he kneels down and helps me place my foot into the shoe. It's a real Cinderella moment as my prince rises.

"Did you..." I trail off, I'm wondering if he came.

"No," he confesses and takes my face in his hands. "But we will have plenty of time for that."

I nod, slightly disappointed.

"It was perfect." His kind words chase my worries away.

"I'd better get back to my parents. I'm sure they are wondering where I got to." It's a half-truth; I doubt they even noticed my disappearance.

But Diarmuid places a kiss on my forehead. "Okay."

I'm flustered as I enter the lobby. My parents stand together, with no sign of Ella. She will be with the production crew celebrating tonight.

My mother is the first to spot me. I blush, but she doesn't seem to notice.

"Where were you? I want to get home," she says.

I scramble to find an answer.

But she's so caught up in her own thoughts that she waves me off. "The car is here, let's go."

I follow my parents out of the theater and take one final look at the milling crowd in the lobby, but I don't see Diarmuid. I hope I won't have to wait long to see him again.

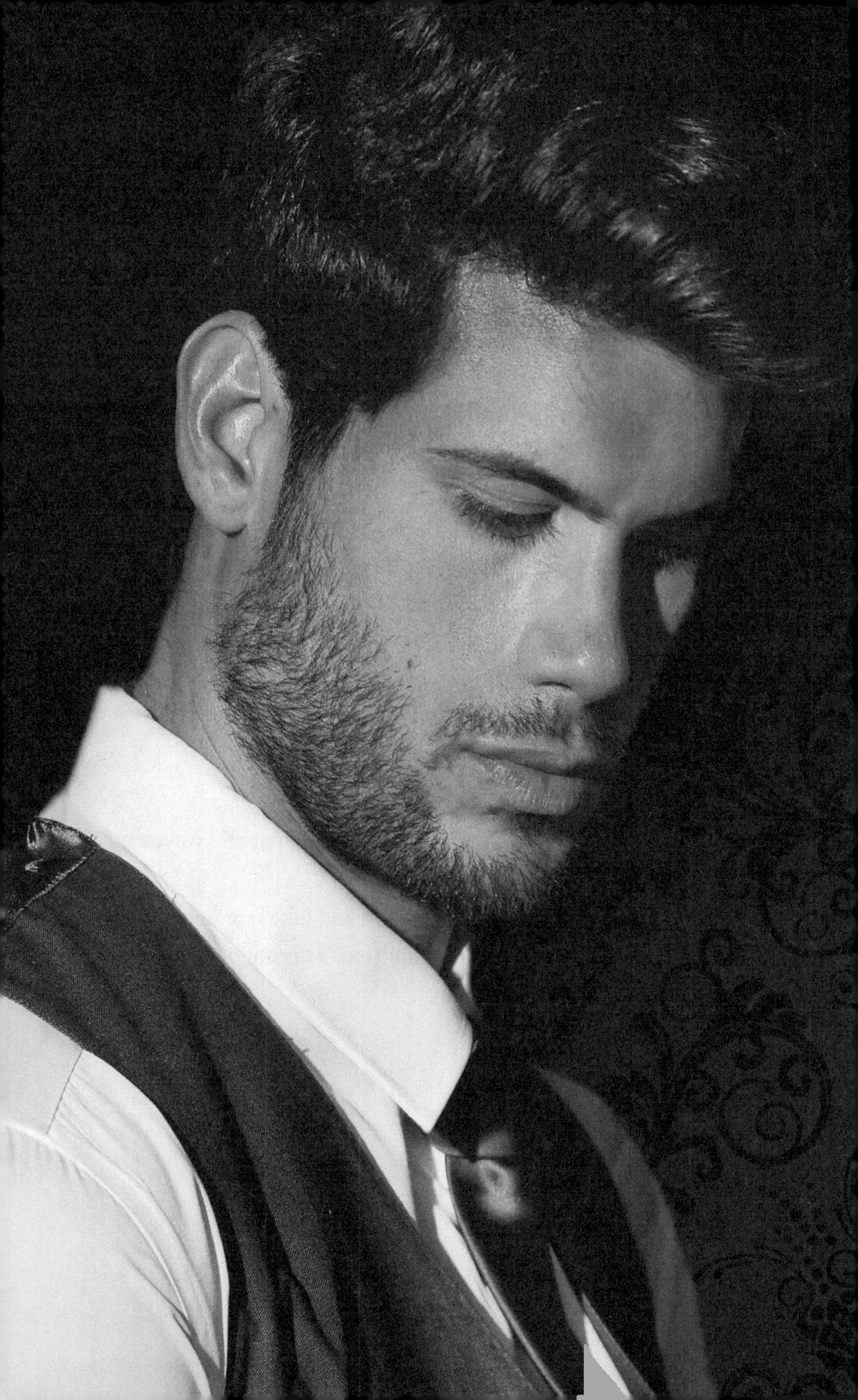

CHAPTER TWENTY-TWO

Diarmuid

I'M DRIVING HOME; the night air is cool against my skin, and my mind is a tumultuous sea. The last remnants of Niamh's taste linger on my lips, and unconsciously, my fingers trace them, seeking more of her, more of that sweet, intoxicating essence. The streets are nearly empty, lit by the occasional streetlamp.

Three women. Each one is so different.

Amira. Her mistake was a crack in the otherwise impeccable facade she presented to the world. But who among us is without fault? My hope by placing her on the driveway the other night was to let her cool down. I need to go check on her and see if she has, in fact, calmed down.

Then there's Selene. Fierce doesn't even begin to cover it. She's a fortress with walls I've been trying to scale since the moment we met. Her resistance only fuels my desire, turning every encounter into a battle of wills I'm determined to win. There's something about the chase, the constant push and pull, that's exhilarating.

And Niamh. Sweet, delicate Niamh. Every moment with her is a tightrope walk between joy and despair. She's fragile, not in body but in spirit, and I fear the world I inhabit will shatter her. She deserves so much more, and I'm caught between wanting to give her everything and fearing that I'll be her undoing.

The ring of my cell phone cuts through my reverie like a knife. It's one of my men. "Boss, we've got a problem. The shipment's delayed."

Right, the business. My life isn't just consumed right now by my Brides; there's the ever-present weight of my empire. Between juggling alliances and unearthing traitors, I've had to lean heavily on my crew. The O'Sullivan arms trade doesn't run itself, after all.

"There's more," he hesitates, and in that pause, a cold shiver runs down my spine. I hear a murmur in the background, a voice that shouldn't be there. A voice I know.

Wolf.

I grip the steering wheel tighter, the leather groaning under my fingers. Wolf's presence is never a harbinger of good news. He's the shadow in my already dark world, a reminder that there are always bigger predators lurking, waiting.

The turn of my wheel toward the Dublin Docklands is a decisive one, pulling me away from any lingering thoughts of checking in on Amira. Priorities constantly shift in my line of work, and the call involving Wolf demands immediate attention. The road stretches before me, leading to a place that blends day-to-day commerce with the undercurrents of a world unseen by most.

The Dublin Docklands, with its bustling activity and scenic views, is a veneer of normalcy and tourism. It's almost laughable how one of the country's vital arteries, handling the lion's share of Ireland's imports and exports, doubles as a stage for criminal enterprises. Tourists flock here, oblivious to the underbelly, drawn by the promise of leisure and the charm of waterside eateries and sporting events. They wander, dine, and celebrate, all under the watchful gaze of cranes that toil in the distance.

When Kings Rise

I pull into a private space beside The Silent Prince Tavern, a pub that's mastered the art of camouflage. Its exterior is a careful construction designed to appeal to tourists with its quaint charm and the allure of an old-world tavern. The sign swaying gently in the night breeze—a young prince, crowned and commanding silence with a finger pressed to his lips—is a fitting emblem for the secrets it guards.

Inside, the atmosphere is rich in Irish culture. The air is alive with the strum of folk music, a melody that's both balm and a blade, cutting through my thoughts. Around me, tourists and locals laugh and chatter. Televisions flicker with the vibrant greens of the Croke Park football field.

Alan, the head bartender, catches my eye from across the room. He signals to one of the bartenders and begins to make his way over. His approach is a casual saunter. No one gives us a second glance as we move together toward the back of the pub, the world of drinks and banter falling away with each step.

The door to the back office closes with a soft click, sealing us away from the lively pulse of the pub.

The office gives way to a hidden world behind a false wall. The sound of hissing kegs filters through as the front continues its normal operation of serving drinks. Inside, the air is cooler, filled with the scent of wood, and a quiet tension seems to fill the room. Organized crates line the walls while a group of men stand huddled around a table in the center.

Without hesitation, I address the matter at hand, my voice cutting through the murmur of conversation. "Where's Wolf?"

"He just left before you arrived," Alan says with a frown, like my line of questioning shouldn't matter.

One of my men steps forward, worry clear in his expression. "The ship was supposed to have our order, but it arrived empty. Clients are waiting, Diarmuid."

I approach the table, poring over the logs laid out before me. Two possibilities unfold in my mind: either our contacts faced unexpected trouble, redirecting our cargo to one of two alternate ships, or our shipment has been intercepted and stolen or discovered. Neither scenario bodes well for us.

"Who's on this?" I ask Alan.

His response, a single name, Fergal, does little to quell the churn of thoughts in my mind. "Is Fergal up to this?" I press.

Alan's nod is firm. "He's earned his stripes, Diarmuid. Been through the fire with us."

"Where was the shipment coming from?" I ask.

"Russia."

The thought of Russia tightens the coil of tension in my gut. International complications are the last thing we need.

As we agree to wait for word from Fergal, Alan pulls me aside, his voice low. "Should we get the others involved?" he whispers.

"The others?" I question.

"Yeah, you know, the others." Alan raises both brows.

I know he's referring to the Hand of Kings. The O'Sullivans have navigated treacherous waters before without needing the Hand of Kings' help. "The O'Sullivans have been handling shit like this for centuries," I assert firmly. "We just need to know what screw needs twisting. Get on that." My voice leaves no room for argument.

As I step back into the lively atmosphere of the bar, the familiar sights and sounds wrap around me like a cloak. In another life, or perhaps just a different chapter of my own, I would have melted into this scene with ease. The counter, the clink of glasses, the hum of conversation—a backdrop against which I'd play out the night's possibilities. The apartments above, silent witnesses to countless nights where I've brought women to fuck. Yet, tonight, that doesn't interest me.

Maybe I am starting to feel something for one of my Brides. With that thought, I find myself at the counter, not to find a woman to fuck, but to seek the simple comfort of whiskey. The bartender places a white napkin in front of me before placing a freshly poured whiskey on top of it. The amber liquid holds a promise of temporary respite, a fleeting escape.

As I lift the glass, the presence of another at my elbow pulls me back to the present. Wolf. His appearance is both unexpected and not. He apparently didn't go too far.

"Why are you here?" I ask and bring the drink to my lips. I take a sip.

"I'm an O'Sullivan, too. There is no reason why I can't be here. It's part of my family's history."

I drink half the glass before turning to Wolf.

"Bullshit," I call out. I wasn't born yesterday.

"Fine." Wolf shrugs. "I was looking for you. I'm going stir crazy since our meeting with Victor." Wolf speaks too loudly but I don't get to scold him as he turns to the bartender and raises two fingers, beckoning him forward.

"You have to sit tight," I say and finish my drink. Staying here with Wolf isn't something I want to do.

"Get me a whiskey," Wolf orders the bartender, who looks at me. I shake my head letting him know I don't want any more.

"How is Amira?" Wolf asks.

The mention of Amira, said so casually from Wolf's lips, ignites a fury within me that's hard to contain. "Why the hell are you interested in her?" My question is a demand.

"Relax, Diarmuid. I've just got something for her." His words, meant to diffuse, only fan the flames. The notion that Amira might need something from him, that there's a connection there I'm unaware of, is intolerable. "Anything she needs can go through me," I assert, a line drawn in the sand.

"I'm afraid that this isn't something that can simply be handed over, Diarmuid," He picks up his drink and takes a swallow.

Who the fuck does he think he is? I take a step toward him, thinking I could break him like I broke his father. Would he scream and plead as loud as his father? I'm sure Wolf would cry and offer up anything or anyone just to save his skin. My hands reach out and grab Wolf by the collar of his jacket. His brows shoot up as if he's surprised that I'm pissed.

Alan appears on the far side of Wolf, and he is the only thing that prevents me from hurting Wolf. A nod from me has Alan stepping back. I can control myself. I release Wolf, but he doesn't appear to be relieved that I didn't hurt him. His eyes tighten in anger.

"You are always protected, watched over." A bitterness enters Wolf's voice as he watches Alan walk away, and when Wolf turns back to me, he gives a laugh. "Of course, I can't touch Diarmuid O'Sullivan. Not in his own place."

Is he fucking mocking me?

"Lorcan is protected by his political circle. Ronan is protected by the thugs he can hire through his business contacts." Angry words roll from his tongue before he comes to the punchline.

"It's obvious Victor is planning to cut me out of my inheritance."

I find myself momentarily at a loss for words. The O'Sullivan family, my family, is indeed a well-oiled machine, each cog turning in unison even amidst the turmoil of our patriarch's decline. Victor's silence on the matter of succession has left us in a state of suspended anticipation, each of us awaiting his signal to align ourselves accordingly.

The right to choose our leader, once held firmly within our grasp, was relinquished the moment we entwined our fate with the Hand of Kings. A decision that, while expanding our reach and solidifying our power, also bound us to their will, making the succession a matter of their interest as much as ours.

Wolf, as a leader, is a thought I can barely entertain. His temper and his darkness make him unsuitable for a role that demands not just strength but restraint. Our world, as unforgiving as it is, requires a leader who can navigate its shadows without being consumed by them. Wolf's love for breaking things wouldn't bode well for him.

"My father was going to give me everything. I was going to take over."

This is a complete surprise to me. I know I should say I'm sorry, but the words would sound as empty as they are.

"Do you remember where you were that night, the night Andrew disappeared?"

The bartender arrives back, and I'm so grateful for the distraction. "I'll have another." I turn to Wolf, but his drink is still full.

"I was probably in the pub, managing my affairs," I respond with a nonchalance I don't quite feel. The question, pointed as it is, dredges up memories better left undisturbed.

"This man was your uncle," he presses, an edge of accusation in his tone. "I know exactly where I was and what I was doing that night. Why can't you remember?"

"I'm a busy man," I counter, the defense sounding feeble even to my own ears.

"We were all at the Church for drinks. Lorcan and Ronan were in town, everyone was there... except you." Wolf has never sounded so sure about anything in his life.

The world around us grows silent, and I know I better get my head straight, quickly.

"If I wasn't there, then I would have been at The Silent Prince," I offer and reach for my drink.

He nods, a gesture heavy with unspoken implications. "Lorcan called Alan that night. Alan said you weren't at the pub, either."

The silence that follows is charged, a tangible thing that stretches between us, laden with questions and accusations unvoiced. I look into Wolf's eyes, seeing not just my cousin but the memories of a shared past. Wolf knows me, perhaps better than anyone—my preferences, my weaknesses, my secrets, well, not all my secrets. A formidable ally, indeed, but in another life, perhaps an even more formidable foe.

Wolf nods, like he got an answer, before he reaches across and picks up my drink.

He drinks the entire glass in one swallow and walks away.

When Kings Rise

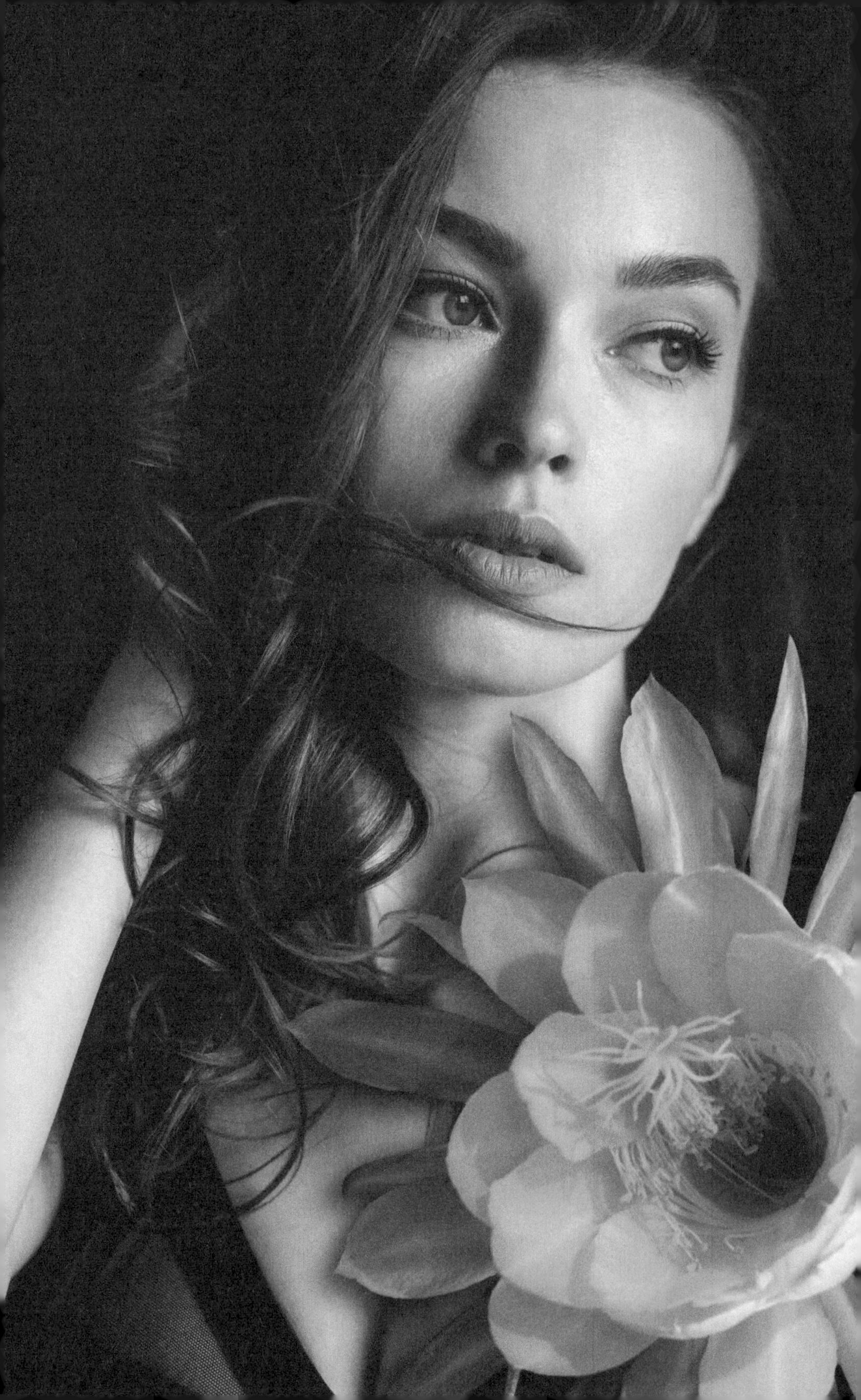

CHAPTER TWENTY-THREE

Selene

THE AIR IS brisk and carries with it the scent of city life—coffee, the faint hint of exhaust, and the promise of rain. Grafton Street buzzes around us, alive and vibrant. I'm walking alongside Niamh, her presence a comforting constant in the pulsating heart of the city. The shops gleam with the allure of luxury, their windows filled with colors and lights, but it's the simple joy of exploration with Niamh that I find myself cherishing the most.

We come to a halt before a statue, its bronze form a tribute to a woman whose story is woven into the fabric of Dublin's history. The statue depicts her with corsets that daringly reveal the top part of her breasts, a silent yet bold testament to her existence—or the lack thereof, depending on who you ask.

"People say she never really existed," I muse aloud, tracing the lines of the statue with my eyes.

Niamh looks at me, her brows furrowing slightly. "That's ridiculous. How can they say Mary Malone was just…made up?"

I lean closer, dropping my voice to a conspiratorial whisper. "There was a mix-up with the record-keeping. Turns out, Mary Malone wasn't exactly who everyone thought she was."

Niamh's gaze drifts to the statue, to the baskets in the wheelbarrow, empty.

"That's probably why she had to sell herself," I say softly with humor.

Niamh laughs softly and links her arm with mine as we resume walking.

"I think she was real." Niamh declares as we walk along the repaved street.

The murmur of conversations envelops us. "What are you craving to eat?" I ask.

Niamh's response is hesitant, tinged with vulnerability. "I'm still getting used to not counting calories or carbohydrates in my food. Being an athlete...it did a number on how I see food."

I squeeze her arm gently, a silent vow forming between us. "We'll fix that. First, we go for Chinese. Then, when we're ready for round two, we'll hit the chocolate store."

Niamh looks at me, surprise etched on her face.

"An entire store dedicated to chocolate. It will be glorious," I grip her hand for a moment with excitement.

As we continue our stroll, a thought bubbles to the surface. "I once dreamed of opening a store here," I confess, the words slipping out before I can weigh their impact.

Niamh's interest is piqued, her eyes lighting up with curiosity. "Oh? What kind of store? A bookstore?"

The guess brings a smile to my face. "You would think, but no. I wanted to open an amezaiku store."

"Amezaiku?" She echoes, her expression a blend of confusion and intrigue.

"It's a very artistic style of Japanese candy making," I explain, the memories of my fascination unfurling like the pages of a well-loved book. "I was obsessed with it for about two years. It was one of those dreams that shine brightly for a moment before fading into the backdrop of reality."

Niamh's interest seems to deepen. "You had other dreams?"

"Too many," I admit with a laugh, a sound that feels both free and a little sad. "My parents let me explore anything that caught my fancy."

"That sounds nice," she muses, a note of wistfulness in her voice.

"It was, in its own way. But in the end, we both ended up in the same place, didn't we?" I say, a subtle acknowledgment of our shared journey, of paths that diverged and converged in the most unexpected of ways.

Our conversation shifts to lighter topics as we secure some Chinese takeout, the warmth of the containers promising comfort and satiation. However, our search for a bench to enjoy our meal proves fruitless; every potential spot is already claimed, a testament to the city's bustle and life.

Undeterred, I lead us to a quiet piece of wall on the side of a building, an improvised spot that offers respite and a view of the street's vibrant dance. Sitting down, I lean my back against the cool brick, feeling its solid presence grounding me. Niamh joins me, her own back finding the wall, and we sit side by side in companionable silence, the city's hum a backdrop to our shared meal.

The rice is divine, with just the right amount of honey sauce poured on, and I find small cuttings of filet beef mixed like prizes amongst the rice and peppers.

"So, what happened at the theater?" I ask. I had tried to ask earlier, but Niamh had avoided my question. She was so obvious; now I need to know what took place.

Niamh's cheeks color with a sudden blush, her eyes darting away, and I can't help but smile at her bashfulness. "Why are you blushing?" I tease, trying to ease her discomfort. "You're a grown woman, and there's nothing to be ashamed of. And, if you're worried about someone overhearing, there's no one close enough to care. Even if they did, no one knows us, anyway."

She hesitates, then, looking at me with a mix of admiration and incredulity, she changes the subject. "How can you think about shopping, food, and... other things when there's so much bad happening in the world?"

Her question strikes a chord deep within me, a reminder of a truth I seldom visit. I recall the moment my parents revealed the nature of my existence to them—not as a daughter cherished and loved for who she is but as a commodity, a pawn in their social maneuverings. I had idolized them, believing in their affection and support, only to discover their warmth was as hollow as the echoes in a deserted hall. The realization had come crashing down on me during a stay at my grandparents' where the absence of genuine love in my upbringing became painfully clear.

Yet, this revelation, this understanding of my place in my family's world, is not something I wish to lay upon Niamh's shoulders. So, I choose simplicity over the weight of my history. "If you worry about something that's going to happen, you suffer twice," I say, hoping to offer a sliver of wisdom amidst the uncertainty.

Niamh pauses, considering my words. "Buddhism?" she ventures, a hint of a smile tugging at her lips.

"Close!" I laugh, shaking my head at Niamh's guess. "Seneca, a Roman philosopher."

Niamh rolls her eyes playfully, and for a moment, the tension between us eases. We're just two friends sharing lunch, not competitors in a bizarre contest for love. But even as we banter, my mind drifts to Diarmuid—charismatic, enigmatic Diarmuid. It's easy to get lost in discussions about his many fine qualities; his charm is undeniable, his smile infectious. But beneath the surface, there's a darkness that nags at me, a shadow I can't ignore.

"He's got this... aura, doesn't he?" Niamh muses, her voice tinged with a mix of admiration and curiosity.

I nod, trying to focus on the conversation, but my thoughts betray me, wandering to that haunting image. Diarmuid, a man capable of killing without hesitation. The rumor of him killing a child whispers in my mind, a sinister lullaby that won't let me rest. I glance at Niamh, wondering if I should share my fears. But what if knowing puts her in danger? What if ignorance is her shield?

We drift to lighter topics, like Diarmuid's peculiar habit of always keeping his shirt on. "Maybe he's hiding a tattoo," Niamh suggests, her eyes sparkling with mischief.

"Or maybe scars from some secret past," I add, trying to match her levity. We weave theories as fantastical as the tales of old, each more absurd than the last. But the laughter doesn't reach my eyes, and I wonder if Niamh notices.

Lunch ends, and we dispose of our garbage, the mundane act grounding me for a moment. As we link arms, heading towards the chocolate store, I can't help but marvel at the strangeness of our situation. Here we are, acting like lifelong friends, yet we're rivals, each hoping to win Diarmuid's heart. I think of Amira and the last time we saw her.

"Have you heard from Amira lately?" I ask, the question slipping out before I can stop it.

Niamh's expression sours slightly. "I try not to talk to her even when we're in the same room," she admits, and there's a bitterness in her voice that surprises me.

As I lean over the counter, watching the chocolatier package our order with an artisanal touch, Niamh's phone shatters the cozy atmosphere of the shop. She steps aside, her expression shifting from casual curiosity to intense focus. I try to distract myself with the array of chocolates, but the undercurrent of our situation tugs at me, pulling me back to a reality I'd rather forget.

When Niamh ends the call, her eyes meet mine, holding a storm within them. "It was Rian," she says, her voice a mix of hope and dread. "He thinks he has the identity of the woman."

A chill runs down my spine, and the delicious aromas around me suddenly don't register. The weight of our investigation crashes into me with renewed force. We're not just playing a game of affection and intrigue; we're knee-deep in a conspiracy, a murder. The realization makes my earlier worries seem naive, a fool's errand of trying to compartmentalize my life into manageable, unthreatening pieces.

I watch as the last of the chocolates are tucked into the box, the ribbon tied with a flourish that now feels grotesquely out of place. Turning to Niamh, I muster a faint smile; the question about the chocolate feels hollow. "Do you... want your chocolate?" My voice is barely above a whisper, laced with a sudden lack of appetite that mirrors my inner turmoil.

Niamh shakes her head, her gaze distant. "Not hungry," she murmurs, and I can see the gears turning behind her eyes, processing the call with Rian, the implications of what he's discovered.

We leave the shop, the box of chocolates in my hand feeling like a leaden weight. Neither of us has the heart to indulge in them, not with the shadow of the murder looming over us. As we walk, the streets seem less vibrant, the laughter and chatter around us a discordant soundtrack to the grim reality we're entangled in. The sweet anticipation of enjoying our treats evaporates, replaced by a cold determination to face whatever comes next.

Rian's apartment feels like a storm's epicenter as we step inside. He dashes back to his cluttered table in a whirlwind of papers without a word of greeting. His excitement is palpable, infectious even, but I'm rooted to the spot, a sense of dread building within me.

"It wasn't easy," Rian starts, his words tumbling out as fast as he moves, shuffling through the documents with frenzied precision. "I've been at this non-stop since you described the composite sketch. And you won't believe where I finally found her identity—a paparazzi blog, of all places."

Niamh and I exchange a look. The absurdity of the situation is not lost on us, yet the gravity of what Rian says anchors us to the moment. He slides a photograph across the table toward us, his fingers trembling slightly with the weight of his discovery.

The woman in the photo turns, her gaze caught by the camera as if she knew this moment was coming. She's beautiful, undeniably so, with an elegance that seems at odds with her fate. Her hair, styled perfectly, frames her face, and the black dress under her peacoat speaks of a night out, perhaps one filled with laughter and life—so starkly different from how her story ended.

Looking at her, alive and vibrant, sends a chill through me, a visceral reaction as the memory of the bruising around her neck flashes in my mind. It's a stark reminder of the brutality she faced, a contrast so jarring against the glamorous image before us. My stomach turns, the injustice of her death a heavy, suffocating blanket.

"She looks..." Niamh starts, her voice trailing off, lost in the same morass of emotions that I'm drowning in.

"Like she didn't deserve what happened to her," I finish for her, my voice barely a whisper. The room feels smaller somehow, the walls closing in as the reality of what we're dealing with settles heavily on my shoulders. We're not just hunting shadows; we're seeking justice for a woman who had her life cruelly snatched away.

"Her name is Sofia Hughes," Rian announces, the solemnity in his tone contrasting sharply with his earlier excitement. My gaze shifts back to the photo, to Sofia's image, and I don't doubt him for a second. The resemblance to the body we saw, to the sketch the coroner had, is undeniable. The article he's referring to paints a picture of a woman caught in a dangerous liaison—a freelance journalist rumored to be entangled with someone high up in the government.

"A cover-up for an affair," I murmur, the words tasting bitter on my tongue. "But murder... It seems so drastic."

Niamh is quiet for a moment; her thoughts are obviously elsewhere. Then, hesitantly, she asks, "Is Sofia's family looking for her?" Her voice is tinged with a personal anguish that doesn't escape me. I know she's thinking of her own sister, the fear of something happening to her lurking in the back of her mind.

Rian doesn't miss a beat, pulling up a social media post on his laptop. It's a plea from Sofia's sister, Nessa, asking about Sofia's whereabouts. The post is accompanied by a photo of Sofia and Nessa together, laughing as they share ice cream. Their joy is palpable, their smiles bright, yet now, knowing Sofia's fate, those smiles haunt me.

"Sofia has a family," I state, the realization hitting me with full force. These are not just names and faces in a case file. These are real people torn apart by tragedy, their lives irrevocably altered. Niamh and I lock eyes, a silent agreement passing between us. We have to find them.

Seeing Sofia's smile and thinking of Nessa's unanswered questions solidifies my resolve. This is more than just solving a crime; it's about bringing peace to a family shattered by loss.

"We'll find them," Niamh says, her determination mirroring my own. "For Sofia."

For a moment, the room is filled with an unspoken vow, a commitment to this cause that goes beyond curiosity or the thrill of the hunt. We're bound by a sense of justice, a need to right a wrong that's all too common in a world that often turns a blind eye to the pain of others. Sofia's story is a tragic reminder of that, but in her memory, we find our mission.

When Kings Rise

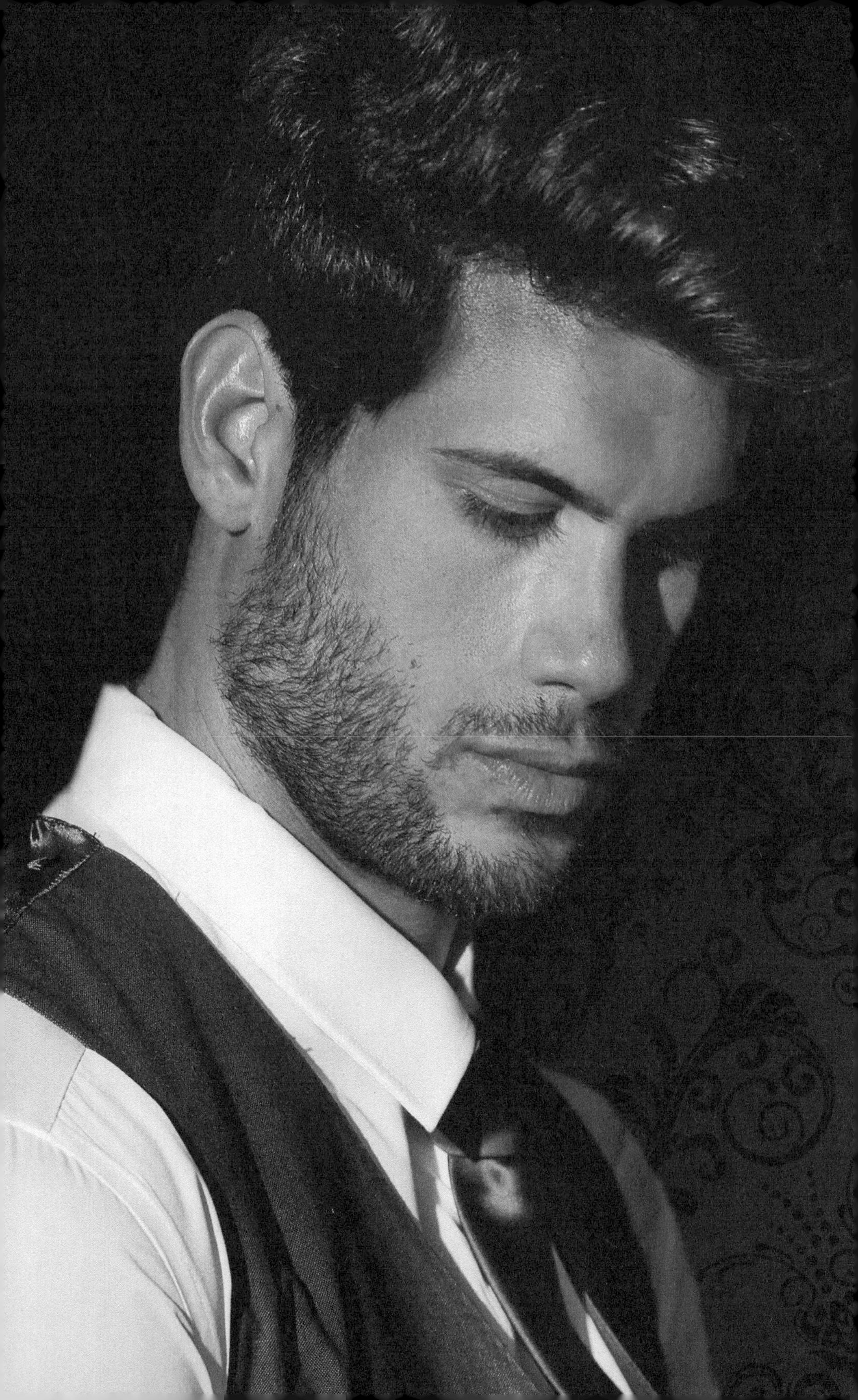

CHAPTER TWENTY-FOUR

Diarmuid

THE WORLD OUTSIDE the car window blurs into a palette of greens and grays as I navigate the now familiar road toward Amira's house. My attempts to reach her have been met with silence, her deliberate avoidance echoing louder than any words she might have hurled my way. It frustrates me, angers me, even. My life is busy enough as it is and chasing after Amira's dramatics is a complication I can ill afford.

Wolf knows—or at least, I think he does. That revelation alone carries the weight of a looming confrontation, a debt between us that's yet to be settled. Part of me had expected him to come at me guns blazing, quite literally, given the chance. But this version of Wolf, one that's cold, calculating, and ominously patient, is unfamiliar. It's unsettling, not knowing what to expect from him now. This isn't the man I once understood, and that unpredictability adds an edge to my already frayed nerves.

As I turn into the driveway, the extensive stretch of it reminds me of Amira's family's involvement with the Hands of the Kings. Out of the three Brides, Amira's family undoubtedly boasts the most expansive property, a fact that traces back to a time when her father, John Reardon, had ambitions aligned with those of the O'Sullivans. He was supposed to initiate the same gun trade I find myself entrenched in now. However, fate had a cruel twist in store when John, in assembling his crew, unwittingly welcomed an undercover Interpol agent into their midst.

The fallout was nearly catastrophic, threatening not just the O'Sullivans but the Hand of Kings itself. In the aftermath, Michael Reardon, the eldest son, became collateral, a hostage in a game of power and retaliation. Andrew O'Sullivan, driven by a thirst for vengeance against John, saw no solution other than death. But Victor, ever the strategist, saw potential where others saw only ruin. Thus, the Reardons were thrust into our world via a transaction of blood for loyalty, but they never received the same standing as my brothers and me.

My hand tightens on the steering wheel, the car's steady hum a contrast to the storm brewing within me.

I park in front of the house; the familiar is now a grim marker of what awaits. The front door, slightly ajar, sends a shiver down my spine. With a cautious push, I open it wider and draw my gun, the weight of it in my hand a cold comfort.

The first thing that hits me is the state of the foyer. It's a mess, a shadow of its former glory. This house, with its sprawling estate close to the shore, was designed to breathe in the fresh sea air, to stand as a testament to the Reardon's wealth and taste. Now, it's suffocating under a layer of neglect. The air is stale, thick with the scent of decay.

As I move silently through the darkened corridors, a sliver of light from the kitchen beckons. What I find there stops me in my tracks. The kitchen is in ruins. Empty bottles clutter the countertops and floor, a silent testimony to despair or madness. The sink overflows, water spilling over the broken tiles, seeping into the grout like open wounds.

The destruction is complete, a cabinet door hanging off its hinges like a final, desperate cry.

With my gun leading the way, I navigate the chaos toward Amira's room. The door is off its hinges, the window wide open, curtains fluttering like ghostly sentinels. The disarray speaks of a hasty departure, or worse. "What the hell happened here?" I whisper to myself, though I'm not sure I want the answer.

A noise from the depths of the house catches my attention. I move towards it, gun ready, my heart racing. Then, he steps into the dim light—John Reardon. But he's not the man I remember. This John is a shell, his suit hanging loosely on his frame, his shirt stained and crumpled as though he's been wearing it for days.

For a moment, we just stare at each other. The standoff is surreal, a moment frozen in time amidst the wreckage of a life once lived here. This man before me, once powerful and feared, is now just another casualty of the life we've chosen. The realization doesn't bring me any comfort. If anything, it's a stark reminder of how easily everything can come crashing down.

"Where's Amira?" The demand slices through the tense air, my voice steady despite the storm raging inside me. John, looking every bit the defeated man, meets my gaze with a mix of resignation and defiance.

"I don't know," he admits, his voice barely above a whisper. "I just got home a few hours ago."

"And where have you been?"

He doesn't answer, but he doesn't have to. Rumors have a way of traveling fast in our circles, and I've heard enough to put the pieces together. "So, while you were out fucking, something happened to my Bride?" I can't keep the edge of accusation from my voice; my patience wearing thin.

He doesn't hold my eye now, and I don't know if it's his first time really looking around him, but a sense of shame washes across his face. I don't give a fuck. I want to know where Amira is.

I cock my gun for extra emphasis. "Where is Amira?"

John's hands spring up instantly. "I don't know. Amira and Tess... they don't have the best relationship," he says. He slowly drops his hands and takes another look around. "From the state I found the kitchen in, I'd say they had another row."

The implication of his words hits me like a physical blow. If there was a fight here, it wasn't just an argument; it was violent, destructive. John's lack of surprise at the violence tells its own story, a grim testament to the norm in this household.

A fury builds within me, a tidal wave of anger at the parade of shitty parents I've encountered in this world. John's mistakes have cost him dearly—Michael, taken as a hostage because of him, and two sons are dead because of his actions. And now, his only daughter is left to fend for herself against an abusive mother while he indulges in another woman. The injustice of it, the sheer negligence, pushes me over the edge.

"You leave your daughter with that woman while you're off playing Casanova?" My voice rises, a mix of disbelief and contempt. "Michael and your other sons paid for your mistakes with their lives. And now Amira... What? She's just another casualty in your long list of failures?"

John's face crumples, the weight of my words hitting home. But my sympathy is long gone, burned away by the sight of the chaos his negligence has wrought. This isn't just about Amira or the Reardons; it's about every child left vulnerable by those who should protect them. My grip on my gun tightens, not with the intent to use it, but as a physical reminder of the control I must maintain. Losing it now won't help Amira, but every fiber of my being screams for some semblance of justice. I want to fucking shoot him, but his dying isn't justice enough.

"She beats Amira?" I say.

He clears his throat and frowns. "You know mothers and daughters." He tries to laugh off his statement.

I put my gun away and grin at him. "No, I don't. Enlighten me."

I take a step closer to him.

"They fight."

I nod again. "She hits Amira?" I ask.

He shrugs. "She drinks a lot. Tess doesn't mean it."

My fist collides with his face, and he hits the ground hard. "Did I mean that?" I ask, kneeling over him. He's shocked for a second before he raises a shaky hand and wipes blood from his face.

I don't give him a moment to recover before I slam my fist into his face again. "Did I mean that, or did my fucking hand slip?" I roar, rage riding high with words that spill from my mouth.

My hands are around his throat, and I squeeze. I squeeze the life out of him. He deserves to die. He claws at my arms, and when his hands fall to the floor as his life slips away, I come to my senses and release him. He doesn't move or gasp for air. I stare at him for a second before he starts to gasp and rolls on his side.

He's alive.

I leave him before I finish what I started.

Fury courses through my veins. My fists clench at my sides as I stare down at him one last time, crumpled and defeated on the floor. My heart races, not with the thrill of victory, but with a ferocious concern for Amira. I need to find her, ensure she's safe from the harm that John's wife has clearly inflicted on her.

Powerful strides carry me to my car, the evening air doing little to cool the heat of my anger. My mind races as I consider where Amira might have sought refuge. The Hand of Kings manse flashes through my thoughts, but I dismiss it just as quickly. No, I had thrown her out, a decision that now twists in my gut like a knife. She wouldn't return there, not after everything.

Where could she be? The city sprawls before me, a labyrinth of possibilities and dead ends. She wouldn't be with Niamh and Selene, as they have no time for her. If it was one of them, I could go to the other for help.

Before I can start the car, my phone pierces the silence. My heart skips, hope and dread mingling in equal measure as I answer. "Do you know where your Brides are?" The voice is unfamiliar, edged with a sinister amusement that sends a shiver down my spine. Selene and Niamh are safe, accounted for, but Amira... This caller knows of her, of the danger she's in.

"I don't know who you think you are," I growl into the phone, my voice a low rumble of barely contained rage, "but you'll regret this call." My mind races, trying to place the voice.

When the caller speaks again, recognition slams into me like a physical blow. Oisin Cormick. The hitman who trained me, the man whose name I've taken as my professional moniker. Him. A flood of memories washes over me, moments of brutal training and begrudging respect. Cormick had been a constant, a harsh mentor but one who had looked out for me in his own twisted way.

All of that feels like a lie now.

"Cormick," I say, the name tasting like betrayal on my tongue. "What have you done with her?"

His laughter is cold, devoid of the warmth I once thought I knew. "Nothing. But I know where two of your Brides are, and it's somewhere they shouldn't be."

I have so many questions, like why is he tracking my Brides? He calls out an address I'm not familiar with.

I type the address into my navigation app. "I am more than twenty minutes away," I say and start up the engine.

"It's a good thing Victor gave you three." The implication of his words sends a fresh wave of urgency coursing through me. The line goes dead, and I grip the steering wheel tighter as I drive toward the address.

The man who had shaped me into the weapon I am today, who had watched over me with a cold, detached kind of care, now plays the most dangerous game with the lives of those under my protection. I have no idea why he is following my Brides, but I need to get to them. I need to protect them from Cormick and from themselves.

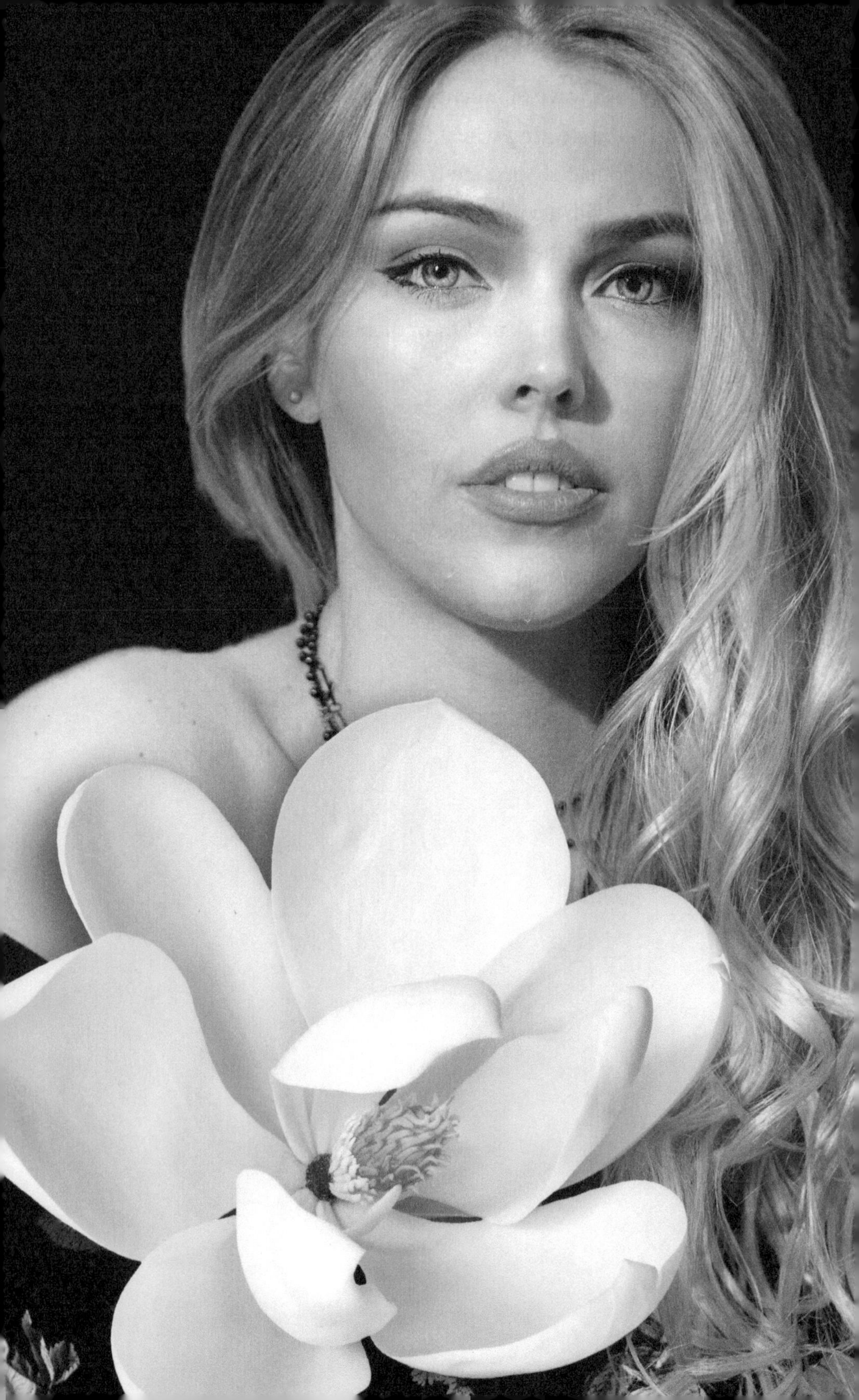

CHAPTER TWENTY-FIVE

Niamh

SOFIA HUGHES WAS more than a name in a file, more than a statistic in the dark underbelly of the city. She was a person, a sister, a part of a family torn asunder by her disappearance. As I flip through the documents Rian has painstakingly gathered, I can't help but admire his diligence. For someone who's not a professional, Rian's work is impressive. The array of Sofia's social media accounts sprawls across the table, a digital footprint frozen in time. The last post dates back two years, a smiling photo that betrays no hint of the darkness to come.

This timeline doesn't add up. Sofia's death was a recent affair, yet her digital life halted long before her last breath—the discrepancy nags at me, a puzzle piece that refuses to fit.

Rian's next revelation is a stack of articles, each penned with the kind of fervor that spoke of Sofia's passion for her work. She was a prolific writer. The majority of names of those she interviewed were politicians. Yet, there's a glaring gap in her professional output. No articles published in the last year of her life. What silenced Sofia Hughes?

In the kitchen, Selene and Rian share a quiet moment over cups of tea, the steam swirling between them. Rian had offered me a cup, but the chaos of his apartment made the very thought unappealing. It's not that Rian's living conditions reflect a lack of cleanliness; rather, it's organized chaos. Everything has its place, though that place makes sense only to him. Selene doesn't seem to mind the mess around us as they speak with ease over their cup of tea.

I can't share their ease. The disorder clashes violently with the world I grew up in, a world of precision and predictability. My comfort zone is a rigid structure, a framework within which I know how to operate.

But it's not the time to dwell on personal discomforts. Sofia's life, her legacy, demands more than that. As I sift through the documents, a plan begins to crystallize. We need to follow the threads Sofia left behind, to trace her last days through the shadows she chased. Her sister deserves answers.

"Rian, how did you come across this?" I ask, motioning towards the stack of articles. My voice cuts through the comfortable silence, a reminder of the work yet undone.

Rian turns toward me. "It wasn't easy. Sofia was meticulous, maybe too much so for her own good. It's like she knew she was onto something big."

Something big. The words hang between us, heavy with implication. Sofia Hughes didn't just vanish from the digital world without reason. She was silenced, but not before she uncovered a truth someone wanted buried.

Selene's question about our next move anchors us back to the task at hand, pulling my attention momentarily away from the chaotic spread of Rian's apartment. Rian's enthusiasm is palpable as he outlines a strategy that includes reaching out to Sofia's sister, contacting secretaries of politicians entangled in Sofia's articles, and possibly even approaching the publications that had purchased her work.

When Kings Rise

As Rian speaks, my gaze drifts, taking in the layers of his obsession that wallpaper the room. Amongst the clutter, a project centered on a 'Lizzie O'Neill' with a "1925" label catches my eye. It's a web of information, a historical puzzle he's piecing together with the patience of a saint. Close by, another collection focuses on Moll McCarthy, and I marvel at Rian's capacity to dive into the past, to resurrect stories long buried.

But it's a familiar symbol that snags my attention, halting the idle wandering of my eyes—a crown cradled in the palm of a hand. I rise, drawn to the wall, where this symbol acts as a nexus for an elaborate network of strings that branch out to maps, photographs, and timelines. It's a conspiracy theorist's dream, connecting dots between organized crime, law enforcement, and even religious figures. There, scrawled in Rian's hand, is the name "Hand of Kings."

The sight of it sends a shiver down my spine. The Hand of Kings, a name whispered in shadows, a name that has brushed against my own life in ways I wish it hadn't. I can't resist the pull of curiosity. "Tell me about this," I urge, my voice tinged with an intensity that mirrors the fixation displayed on the wall.

Rian's excitement is a tangible thing as he turns to the wall, his eyes lighting up with the fire of obsession; he leaves his tea on the counter as he approaches me and pushes his glasses up on his nose. "The Hand of Kings," he begins, unaware that he speaks to someone far more entwined in that world than he could imagine. "I've been tracking this cult since I was a teenager. Most people laugh, call me crazy, but there are too many connections, too many coincidences. They're real, and they have their hands in everything—crime, the law, even the church."

Listening to him, I can't help but feel a twinge of fear mixed with a profound sadness. Here is Rian, a man consumed by a truth too dangerous to pursue.

"I've seen this symbol before," I confess, my voice low, laden with a weight of knowledge I wish I didn't carry. "Your work, your theories... they're not as farfetched as you might think."

Rian's eyes meet mine, a flicker of realization, of validation, passing between us.

Rian's conviction seems to grow with every word, painting a world where hidden clues and shadowy councils pull the strings of global events from behind a veil of secrecy. His theory that there's a vault filled with the world's darkest secrets, with the entrance clue hidden on a grave in Glasnevin Cemetery, sounds like something out of an adventure novel.

"A council?" Selene interjects, skepticism threading through her voice. "There isn't one leader?"

"Not from what I've discovered," Rian responds, his eyes alight with the thrill of sharing his findings.

I find myself drawn into the conversation despite my reservations. "No, wouldn't there be one guy? A Hand to guide the Kings?" I ask, trying to fit Rian's revelations into the framework of what I know.

"The Hand does the bidding of the council. Only he knows their identities," Rian clarifies, his statement echoing through the room like a prophecy. The implication of his words, the existence of someone more powerful than Victor, hangs in the air.

A thump from the hallway outside, a sudden noise that cuts through our discussion, has us all spinning toward the noise. We all freeze, our gazes snapping toward the door as if it were the only thing anchoring us to reality. Selene's whisper slices through the silence, a sharp edge of fear in her voice. "Rian, do you have a weapon?"

Rian's response is almost comical in its naivety. "Why would I have a weapon?" He looks genuinely puzzled, as if the concept of needing physical protection in his own home is a foreign one.

Rian moves toward the door, his determination masking the uncertainty that flickers in his eyes. Selene and I can only watch as he reaches for the handle, the simple act charged with the potential to change everything.

The door swings open to reveal an older man. He introduces himself as someone who works for Diarmuid, claiming concern for Selene and me. His words are smooth, with a practiced ease that belies the tension of our unexpected encounter. Rian's confusion is evident, the name Diarmuid holding no significance for him, a stark reminder of the worlds colliding at his doorstep.

The older man's smile is enigmatic. "The ladies shouldn't be here; they do not have permission." I can sense a veiled threat wrapped in politeness. I watch Rian, trying to gauge his reaction, to see if he senses the danger that's seeped into his home with this stranger's arrival. Rian's body language, open and unconcerned, betrays his inexperience and his inability to see beneath the surface of our visitor's calm demeanor.

Despite the man's unassuming appearance, something about him sets my instincts on edge. His hair, more gray than black, speaks of years and experiences far beyond what any of us can claim. And there, hidden beneath the benign exterior, is the subtle suggestion of a body honed by training.

"Who is Diarmuid?" Rian asks.

I exchange a glance with Selene, a silent communication that speaks volumes.

When Rian glances back at me, I can't find any words. He turns back to the old man. "I think it's best you leave." I don't know what Rian sees on mine and Selene's faces—fear, maybe. Dread. But he pushes the door closed.

A foot wedged in the door stops Rian from closing it. The older man pushes it open with ease and strength. When he reaches for Rian, I rush toward him, but it's a blur of movement. One minute Rian is standing there, the next the old man's arms are around his neck. Rian's death is swift, a chilling demonstration of the older man's lethal skill. The crack of his neck sends my stomach swirling. My heart beats rapidly in my ears. Blood rushes through my body at what just happened in front of us.

He killed Rian.

With a swift movement, he's in the apartment with the door closed behind him and Rian hanging lifeless from his arm. He removes a gun, and Selene screams at the same time as I jump back.

"Be quiet," he warns.

Tears run down my face, and with a shaky hand, I reach up and touch my face. I wasn't even aware that I had started crying.

"Now, get rid of all of this," he orders, his weapon sweeping the room, encompassing the entirety of Rian's life's work in one dismissive gesture.

Selene has her hands raised and nods, backing toward the wall with all its maps and connections.

I swallow bile. I can't look away from Rian's lifeless body.

"Move," I'm ordered, and I find myself with trembling hands reaching out to remove Rian's work. I can't see from the blur of tears.

As we tear it all down, he orders us to place it all in the sink. The man lowers Rian to the ground but still holds the gun toward me and Selene. The flames consume every piece of paper, every note and article, as he lights it all on fire.

Death reduces Rian's body, that was once vibrant with life and curiosity, to an object to be concealed, rolled in a rug as if he were nothing more than refuse.

Guided by the barrel of the man's gun, we move in a daze.

He swings the rug across his shoulder and opens the apartment door. "Outside." He orders.

A strangled cry falls from my lips. A hand takes mine, and I jump for a second until I look into Selene's tear-filled eyes.

The streets are silent witnesses to our grim procession. The alley was a makeshift route to an ending none of us could have predicted. The trunk of the man's car becomes Rian's final resting place, a thought that churns my stomach with a mix of rage and despair.

The man levels his gun once more. "Get in the car." His words are a sentence, mostly a death sentence.

Selene tightens her hand on mine.

"No," she says, keeping me rooted at her side.

I tremble at the cock of the gun. How can this be our end? I glance around the darkened alleyway. Something moves in the shadows. And then Diarmuid is there, a gun in his hand.

It takes me a moment to really allow what I am seeing to sink in. He has blood on his shirt, and he looks like he's been in a battle. In his eyes, there burns a fierce resolve.

The old man turns to Diarmuid. "You got here quicker than I thought." His gun is now pointed at Diarmuid.

"I was only ten minutes away. I lied," Diarmuid confesses and takes a step closer. "You will let them go, Cormick."

Cormick's smile is a flash in the darkness. "And why would I do that?"

There is a shift in the air as Diarmuid puts his gun away and steps even closer to Cormick. "You don't need a gun."

Cormick laughs. "It's a quicker way."

"If you fire that gun, people will hear." Diarmuid holds up his hands and takes another step closer to Cormick; he hasn't looked at Selene and me who are holding hands.

I want to tell Diarmuid that he killed Rian, but it's like there is no air in the alleyway.

Cormick lowers his gun, his smile no longer visible. "You want to dance? Let's dance." He whips out a knife, and my stomach sinks to my feet.

With a growl, Diarmuid moves with precision, a knife in his own hand I hadn't even seen him extract. His movements are both horrific and mesmerizing. He swipes a large arch, and the knife nicks Cormick's arm. Selene drags me back until our backs hit the wall.

Cormick retaliates instantly, his own knife drawing blood across Diarmuid's chest. I scream when I see the injury, but Diarmuid's onslaught is relentless, driven by a primal need to protect, to avenge. He swipes quickly at Cormick again, this time cutting the man's torso. I'm waiting for his guts to spill out across the asphalt, but crimson red soaks his shirt. He hasn't a moment to recover when Diarmuid swipes out and takes his legs out from under him with one quick movement.

When Cormick hits the ground, Diarmuid is on top of him, one knee crushing the man's hand that held the knife. With a pressure that has Cormick releasing his knife, the clang of the weapon echoes. But relief that Diarmuid has the upper hand has me stepping away from the wall, thinking it's over.

"I trained you well," Cormick says before spitting to his left. Diarmuid's fist slams into Cormick's face with a viciousness that takes the remaining air from my lungs. He raises the knife and pierces one of Cormick's eyes. The sound is too much, and I want Diarmuid to stop.

"Please stop." It's all too much.

I don't think Diarmuid can hear me as he removes his knife with an eyeball hanging on the end.

"Never touch what is mine," Diarmuid says before he drives the knife deeper until only the handle sticks out of Cormic's eye, and his body goes still.

I pivot just in time as bile claws its way up my throat, and I empty the meager contents of my stomach along the alleyway wall.

Arms circle me, and I expect Selene, but it's Diarmuid. "It's okay. You are safe." He pulls me into his chest, blood soaking into my clothes, and I can't stop the sobs that take over.

"Are you hurt?" I find myself saying.

He brushes it off. "We need to leave." He doesn't look like he could walk two feet without collapsing, but he pulls me and Selene away from the gruesome scene.

"Rian, he killed Rian," I say. "We need to get him out of the trunk."

"No," Diarmuid orders as he drags us away from the bodies.

"We can't just leave him here," I object.

"It's just the way with our world," he states, a harsh truth spoken with a finality. His refusal is not cruel but a necessity born from years within a world that devours its weak and sentimental.

We leave the alleyway and Rian behind.

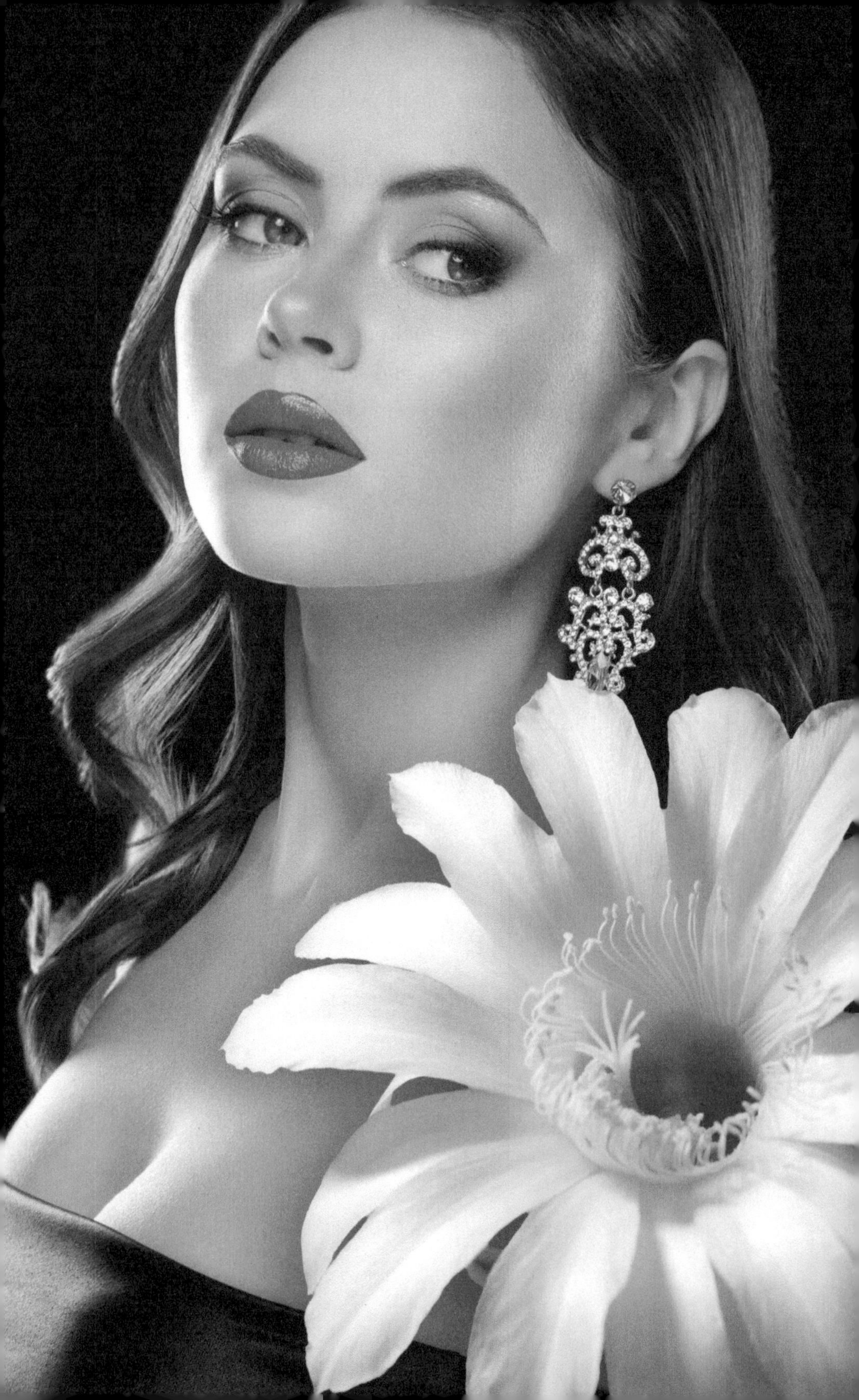

CHAPTER TWENTY-SIX

Amira

I'VE NEVER KNOWN emptiness like this. It's a void, an abyss that swallows every thought, every feeling, leaving behind a shell of who I used to be. My brothers, Dominic and Kevin, were my world, and now they're gone. Dominic's death hit like a storm, sudden and devastating. He had been the rock, the one who took on all of the responsibility after Michael was taken away. The pressure to keep our family afloat led him down paths he shouldn't have taken, risks that cost him everything. Then, there was Kevin. Watching him spiral into the abyss of drugs, trying to escape a reality too harsh to bear, was a slow torture. I knew, deep down, that it was only a matter of time for him.

Rain droplets begin to pummel the canvas above me. This is not my first night in Dublin's tent cities, and it won't be my last. But moving is life and staying in one place too long is dangerous. The thought of taking a plane out of the country is laughable—I don't even have a passport, and my finances are a joke. Sneaking onto one of the ferries has crossed my mind, but the risk of getting caught without knowing how long I'll need to stretch my meager funds is too great. This tent, my makeshift home, was a find, a bargain from a secondhand store that's become my biggest purchase in weeks.

I zip down the front of the tent and quickly zip it back up as raindrops try to enter as if they can find warmth inside my tent. They won't. It's freezing, and the angry gray sky doesn't look like it's going to give up pouring any time soon.

Hostels might have been an option, a chance for a warm bed and a shower, but they'd want identification, something I cannot give. The last thing I need is for anyone to know where I am. The fear of being found, of being dragged back into a life that's already taken everything from me, is a constant shadow.

As these thoughts swirl through my mind, I find myself mechanically opening a can of premade pasta and sauce. I don't even bother to heat it; realistically I have no way to do so.

Between mouthfuls, I shove my hands deep into the pockets of my worn jacket, seeking solace in the fleeting warmth. The cold has a way of creeping into your bones here, uninvited and relentless. Occasionally, I pause to take a sip from my water bottle. How anyone survives here in Tent City is a mystery. Two nights I've been here, and already I can sense death tapping away along with the rain on the tent.

The rain outside intensifies, its rhythm a constant backdrop to my rummaging through the clothes I picked up from the secondhand store. I'm searching for anything else I can layer. . My fingers brush against a slightly thicker sweater, and a small, triumphant smile crosses my lips. It's a minor victory in the grand scheme of things, but it's mine.

But as I layer the sweater on, the weight of my situation presses down on me with renewed vigor. Winter is approaching, and if I don't figure out what to do, it won't be the lack of a home or the constant running that will kill me—it'll be the cold. The thought sends a shiver down my spine that has nothing to do with the temperature.

As the rain ceases, I finish my pasta and sit in the tent, knowing staying here will kill me. I need to walk around and get some warmth back into my body. I'm wary, all too aware of the little I possess and how easily it could be lost or stolen. I take what's left of my money, hiding it in various parts of my clothing, securing it against my body. It's not much, but it's all I have.

Stepping out of the tent, the cold hits me anew, as if the very air sharpens its teeth against my skin. My breath forms a mist before me, a ghostly apparition in the early morning light. I pull my jacket tighter around me, huddling against the biting wind as I navigate through the camp. Around me, life stirs in muted tones, and a few faces lift in greeting, but I can't afford the luxury of acknowledgment. I can't let myself become familiar, become a part of something. The more I blend into the shadows, the safer I am.

Because of the cult that's probably still searching for my mother and me, I don't know where safety lies anymore.

As I make my way through the camp, a sudden commotion breaks out around me. People are scrambling, fear etched into their movements as they grab their belongings and flee. Through the sparse trees, the unmistakable shape of a gardai car becomes visible.

"Oh, God," I whisper to myself, a prayer to no one. The last thing I need is to be caught up in whatever is happening. With the Garda here, the camp will no longer be safe.

Not that it ever truly was.

The presence of Garda officers is overwhelming, their figures materializing from between the tents like specters of my worst fears come to life. Panic sets in, a wild, thrashing thing inside me as I try to find an escape route through the chaos. Screams pierce the early morning calm. My heart races, each beat a loud drum in my ears as I dodge and weave, desperate to remain unseen. But luck isn't on my side. After a frantic, brief chase, a firm hand closes around my arm, and I'm caught, the reality of my situation crashing down on me with the weight of a thousand bricks.

"Don't struggle." The deep male voice has me going still, and I look around as I'm led to a Garda car. I'm very aware that I'm the only one being arrested.

Sitting in the back of the Garda car, a mix of emotions courses through me. There's a twisted relief in the warmth of the car. The thought of going to jail brings with it the possibility of a warm meal. But the relief is fleeting, smothered by the realization that being booked could lead to my discovery. If my name makes it into the papers, they will find me.

The officer's phone call pierces the haze of my thoughts, a simple phrase that chills me to the bone. "I have her." The finality in his tone, the implication that I am known to them, sends a wave of dread crashing over me. This isn't just an arrest; it's a capture. They know who I am, and this changes everything. He starts the car, and I want to argue, but I already know that there is no getting away from this.

The car pulls up to a building that strikes a familiar chord of fear within me—Wolf's building. The realization hits me like a physical blow. Diarmuid has rejected me, discarded me to the whims of fate without so much as a backward glance. I am to be passed straight to the bottom, no bargaining, no chance of mercy. The officer's demeanor is icy, his eyes never meeting mine, as if I'm already condemned, already nothing.

This is it—the end of the line. All my running, all my hiding, has led to this moment—the fear, the cold, the loneliness. Diarmuid's rejection is a sentence worse than any jail could impose, a fate I'd been desperate to avoid.

As I'm led into the building, each step feels heavier than the last. The warmth of the car is a distant memory, replaced by the cold reality of my situation. I'm entering the lion's den, a place where mercy is a foreign concept and survival is a game played by rules I no longer understand.

The sensation of not having to fight anymore washes over me with an almost surreal relief. For so long, I've been running, dodging, hiding—survival was my only goal. Now, as I'm led into the lobby by the officer, the fight seems to drain out of me, leaving behind a weary acceptance of whatever fate awaits.

Wolf is there.

"Thank you, officer," he says with a fresh smile.

The officer releases me and nods at Wolf. His departure is swift, leaving me alone with Wolf, the architect of my current predicament. He gestures for me to follow, and I do, my body moving of its own accord, my mind numb to the implications of his command.

As we walk, a heavy dread settles in my stomach, a weight so profound it threatens to drag me down. I can imagine the horrors that might await me, each scenario more terrifying than the last. A voice in the back of my mind whispers that death might be preferable to what lies ahead.

Wolf's voice cuts through my dark reverie. "I have a present for you." The word sends a chill through me. As we approach a door, the sound of screaming filters through, a warning of the nightmare to come. He opens the door, and the scene that unfolds is something out of a twisted fantasy.

My mother, the woman who gave me life, is tied to a post in the center of the room, a spectacle of despair. Her eyes are wild, filled with an animalistic fear as she screams and twists, more monster than human. The room is lined with tables that hold an array of implements: weapons, syringes, bottles, blindfolds, and even a stereo. It's a tableau of torture, a display of human depravity.

"I went to your home to find you, but I found your mother instead." Perverse satisfaction laces Wolf's voice.

Like a circus presenter, he holds out his arms towards my mother. "A present for you."

The sight of her, so broken and lost, ignites something within me—a rage, a sorrow, a desperate urge to protect, even now.

As my mother's gaze locks onto mine, the air shifts. The pitiable creature I saw moments ago vanishes, replaced by the all-too-familiar specter of anger and venom that I grew up with. Her curses fill the room, each word a barb that finds its mark with practiced ease. The years of verbal abuse, the emotional scars—they all come flooding back as she lunges at me, restrained only by her bonds.

In a moment of raw, unfiltered emotion, I clutch my jaw, stepping into the room with a determination I hadn't known I possessed. My eyes fixate on one of the tables, and without thinking, I grab a bat. My grip is tight, knuckles white as I advance, every step fueled by years of suppressed anger and hurt. But before I can swing, my arm is caught in an iron grasp. Wolf's smile is chilling, a silent reminder of the power he holds in this moment.

"When was the first time she hurt you?" he asks, his voice calm.

The memory surfaces unbidden—Dominic's funeral, six years ago, a day when grief was met with cruelty instead of comfort.

"Six years ago. At Dominic's funeral." I say, still looking at my mother. She's still struggling, still screaming at me, telling me I'm a whore. I'm useless.

"If you hurt her now, it will be over in a moment. A lifetime of hurt deserves a lifetime of pain." He releases my arm as if allowing me to make my choice.

He steps behind me and speaks in my left ear.

"LSD is coursing through her veins. You could create nightmares for her." He steps to my left.

"You could be the architect of her terror." He smiles at me.

"Why? Why do this for me?" I ask a question that feels too small for the gravity of this moment.

He exhales and glances at my mother, who's still struggling. "I feel a kinship with you. The world hasn't been kind to either of us."

The revelation is unsettling. To be seen, understood, and aided by someone like Wolf—a man capable of such cruelty yet ready to offer solace in revenge—is to stand at the edge of an abyss. The choice before me is stark: to embrace the darkness offered as salvation or to reject it.

I have no one else. No one who sees me like Wolf does. Diarmuid has cast me aside, and in the space of a moment, I make my mind up.

I hand the bat to Wolf. "The world has been fucking cruel," I snarl and he smiles.

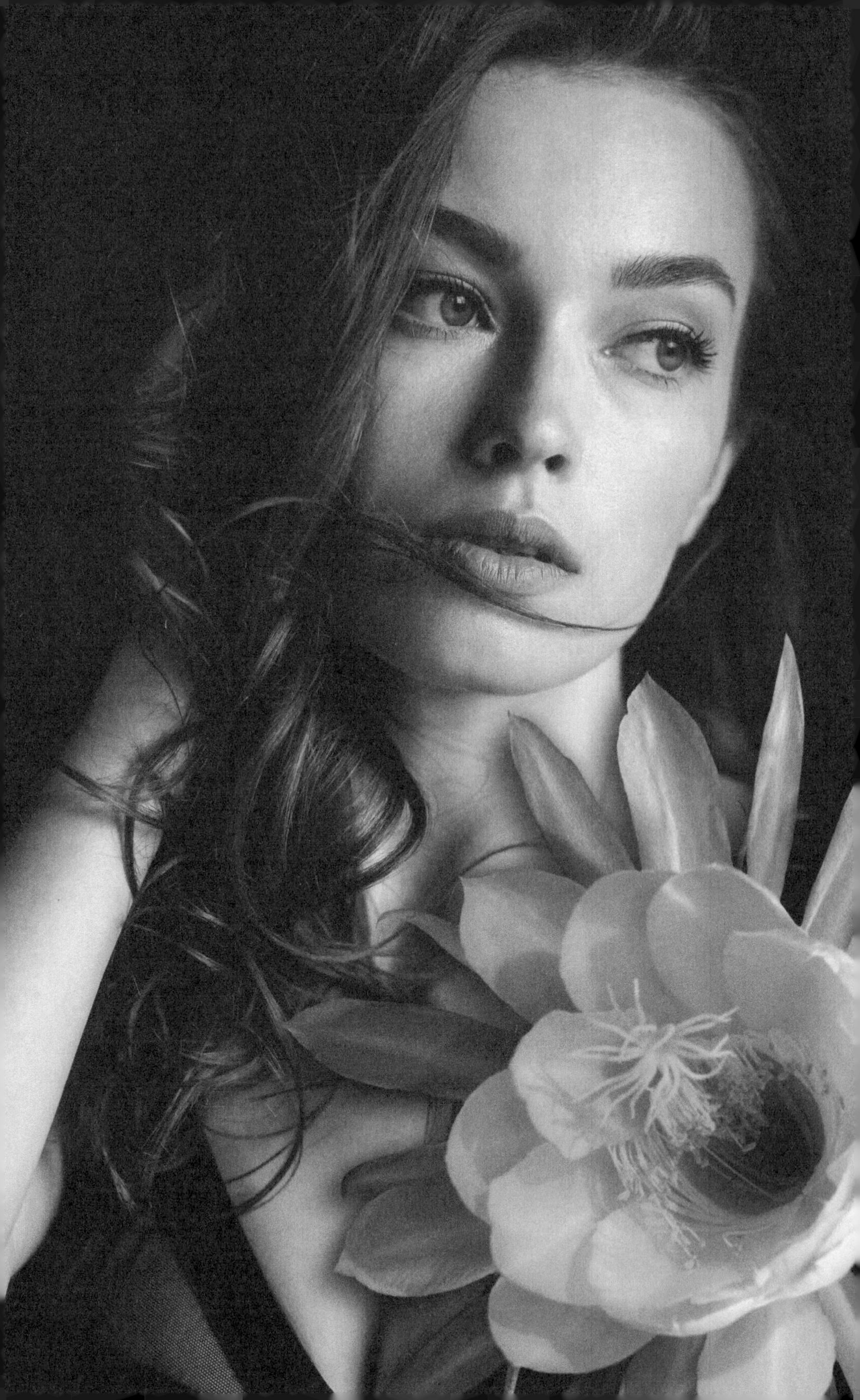

CHAPTER TWENTY-SEVEN

Selene

THE INTERIOR OF Diarmuid's townhouse is bathed in the soft glow of evening light. But tonight, the elegance of my surroundings is marred by the scene unfolding within its walls. Diarmuid stands at the center, a figure of defiance and pain, his shirt clinging to his skin with blood. Cuts and bruises litter his skin, and one of his knuckles looks to be broken.

I can't think about what happened to Rian. I can't think about the violence that I watched pour from Diarmuid. A man always so finely dressed, well presented… I never would have known such darkness lived inside him. But he protected me and Niamh.

"Sit down," I say to him, He looks ready to fall down. He does, slumping into a white plush couch. Niamh has already gone to the kitchen, and I hear water running. I think her mind is as fumbled as mine.

I stand over Diarmuid, trying to figure out where to start. When Niamh returns with water and a washcloth, we work in tandem. She hands me one, and I start to mop up some of the blood on Diarmuid's arm, each stroke showing me the blood isn't his. He had arrived with blood on him…who else had he killed?

Diarmuid pushes our hands away. His phone is in his hand, and he keeps hitting a call button. The noise of the engaged tone has his worry growing.

"Have you seen Amira?" he asks.

I glance at Niamh and shake my head.

"No," Niamh says.

He hits the call button again. I had always thought Amira to be overdramatic, her tendency to find trouble a constant source of irritation. But tonight, the fear in Diarmuid's eyes mirrors my own. Despite everything, despite the harsh words and harsher feelings between Amira and myself, the thought of her in danger, potentially facing death, is a cold wake-up call. She may be a bitch, but she doesn't deserve to die.

I start to open the buttons of his shirt, and he continues to call Amira. Each time I hear the dead line, my stomach drops a little further again. There is so much blood on his chest, and this time it's his own. The largest cut was from Cormick.

"You need stitches," I say as I hold the cloth to his chest.

"I was supposed to keep you all safe, protect you from harm." His voice rises. He doesn't seem to be aware of the damage that was inflicted on him. Niamh washes his free hand, and he hisses as she rubs across his damaged knuckles.

"Sorry." Her voice is a whisper.

Diarmuid seems to come out of the fog he is under and looks at Niamh. "I'm sorry," he says, the gentleness in his voice bringing tears to Niamh's eyes.

"My sister, Ella. I need to get her. She could be in danger."

Diarmuid nods. "I will find her and protect her."

Niamh sniffles. "Thank you."

"I will take the four of you somewhere safe." His hopes for Amira are still high.

When Diarmuid looks at me, fire lights up in his eyes.

"Why were you at Rian's apartment?" I can hear the jealousy in his voice.

The intensity of his gaze is unnerving.

"We heard there was a body placed on top of Andrew O'Sullivan's, and we wanted to know who she was."

He's glaring at me. His reaction is a mixture of concern and barely contained fury.

"How could you be so stupid?" His voice is grave, and I'm stunned at the level of venom in his words.

"I needed to know more about you. I needed to learn about what I might be marrying into." My words send my heart racing. I did this to learn about him.

"Jesus, Selene. This ends now. Do you understand how dangerous this is? Someone wants that murder covered up, and they will stop at nothing."

I nod because the moment I watched Rian die, I knew we had gone too far.

"How do you not want to know who killed your uncle?" Niamh asks. I know her mind is still on Ella. If Ella had died in such a brutal way, she would want to know who killed her.

"I already know who killed Andrew O Sullivan," Diarmuid says. Something shifts in his gaze, and he glances from me to Niamh. "I did."

His confession seems to draw the very air from the room. The murder of Andrew O'Sullivan wasn't just another headline in the news; it was a deed done by Diarmuid's own hand.

I want to ask why, but I can't bring the words to my lips.

Niamh quietly excuses herself, retreating to the sanctuary of the guest bedroom. Her departure is a silent echo of the turmoil that Diarmuid's confession has stirred within us.

At Diarmuid's words, I dropped the washcloth, and blood found its way onto the white sofa. I watch it soak in.

"I'm going for a shower." Diarmuid rises, and I give him some space as I try to process what he just told us. He killed Andrew O'Sullivan.

He killed Cormick, too, to save us. Maybe he had his reasons for killing Andrew. My stomach continues to curl, and I find myself following him. I never thanked him for saving us. Our fate would have been in the hands of Cormick, and most likely, right now, we would be rotting in a shallow grave.

I open the door to thank him, my words poised on the tip of my tongue, but they die away at the sight that greets me. Reflected in the mirror, Diarmuid's back is a tapestry of scars, each mark a story of pain endured, of survival against odds that would break lesser men. The scars, numerous and brutal, speak of hundreds of strokes of the whip, each one a testament to his torment.

My heart clenches at the sight, a mix of horror, sorrow, and an indescribable urge to reach out, to somehow ease the pain that each scar represents. "What happened to you?" The question is out before I can think, a demand for understanding, for the story behind the scars that mar the skin of a man who has become an enigma. Did Andrew do this? Is that why he killed him? But these marks are years in the making.

Diarmuid's shoulders tense, and he turns to me. His chest is marred with fresh cuts. Jesus, he has endured so much.

"Andrew and Victor had a very unique way of punishing me." His words bring tears to my eyes. No one should suffer like this.

"That's why you killed him?" I ask and swallow the sorrow that threatens to consume me.

"I will kill Victor, too." His words should terrify me, should have me running from him.

"Good," I say as tears make a pathway down my cheeks.

The man before me is not just a protector. He is a survivor, carrying the weight of his past with every step, every decision marked by the trials he has endured. The revelation does not weaken my perception of him; rather, it deepens my respect and my understanding of the battles he has fought, those both visible and those hidden beneath the surface.

"Rian spoke of a council, one that's even higher than Victor."

"I know it's deeper than they allow me to see. But I will find every player."

I step deeper into the bathroom. "I'll help you."

He immediately shakes his head. "I won't let you get hurt."

"I won't give up." I raise my head. Seeing the marks on his body has solidified my decision.

Diarmuid steps closer to me and takes my face in his hands. "Troublemaker," he says with a soft smile.

"But we must find out what happened to Sofia Hughes. I can't let it go, Diarmuid," I state.

He rests his forehead against mine. "Okay, Troublemaker."

I lean back so I can look up at him. "Now, get in the shower." He needs to wash off the blood; we need to see the true extent of his injuries.

He slowly takes off his trousers, his back to me again, and my heart squeezes. I want to ask when the abuse started, but instead, I peel off my clothes. He glances at me for a brief moment but finishes getting undressed. The water pours from the showerhead, and he steps in under the spray. I grab a washcloth and enter the shower with him.

He's facing me, and all of a sudden, I'm nervous and heartbroken, but I swallow the emotions and start to wash him down.

He hisses now and then but closes his eyes. "I have sent men to find Amira and bring Ella here."

I nod but remember he can't see me. "Okay," I say.

He glances down at me, and his large hand covers mine. He loosens the cloth from between my fingers and starts washing me. I allow it. I need some form of kindness now after all the violence. My parents never showed me any affection, and I lean into this moment with Diarmuid. My hands rise up, and I touch his broad shoulders, allowing them to run down his arms. This is why he never took his shirt off. He was hiding the brutality that was inflicted on him.

The water pours down his chest, the stream of blood turning from red to pink. He still needs stitches, but the cut isn't as ugly as I thought it would be.

"Cormick trained you?" I ask as my hands trail down to his elbow before running along his forearms.

"Yes, as an assassin."

My heart races, and I look into his steel eyes. I need to know.

"Did you kill a child?" I ask.

He tilts his head, the cloth he was using to wash me forgotten.

"No."

I can't help the relief that floods my body. "Was each death for the greater good?" I ask.

This time, his gaze shadows over. "No," he answers honestly.

I nod.

"I just do as I'm instructed." He sounds so tired.

He tilts my chin up so I'm looking at him. "But, I won't for much longer. I'll take you all somewhere safe."

I want to ask him how he really thinks that will play out, but time will tell.

His lips brush mine, and they are warm and gentle as he kisses me, like he's sealing our fates deeper together.

Stepping into the shower with Diarmuid was a choice. My one move in this game of kings and pawns...perhaps the only one worth making.

But mine, nonetheless.

My choice right now is to kiss him back, our mouths moving

together assuredly with passion and conviction. I am no longer a pawn in this game.

I'm now a player.

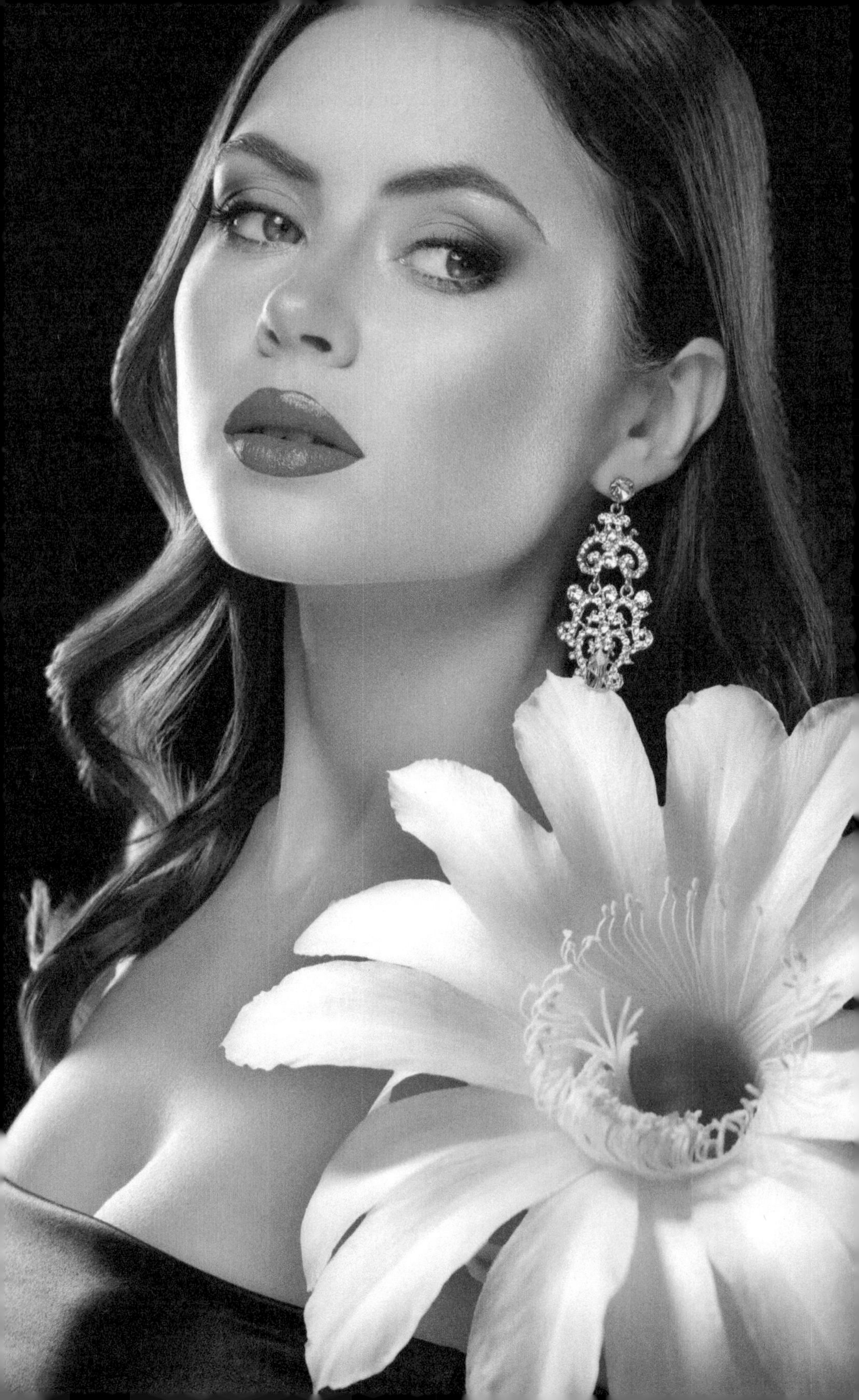

CHAPTER TWENTY-EIGHT

Amira

AS WOLF LEADS me down to the basement, my heart races. I wonder if my mother is still alive or has the constant pumping of drugs into her system taken her away. The stairway creaks under our weight, each step sending a shiver through me. At the bottom, a single bulb casts long shadows on the walls.

I spot my mother huddled in the corner of the cage. Her hands cover her head, and her knees are buried in her face. Does she look thinner? I'm not sure. Wolf's fingers graze mine, and I glance up at him.

"I want to show you something." A slow smile creeps along her lips as he guides me to a small, cold room at the far end. There, I see her—the maid who had struck me, now confined within a rusty iron cage. My stomach churns at the sight, a mix of fear and vindication swirling inside me. Her wrists are covered in white bandages, her hands gone. Diarmuid really had them cut off.

The idea sends butterflies bursting through my belly.

The maid is standing with her back to me, her handless arms hanging at her sides.

Wolf pulls a chair close to the cage and motions for me to sit. As I do, he perches on an old crate across from me, his eyes never leaving my face. "Amira, I want to tell you a story," he begins, his voice steady but filled with intensity.

He reaches into his pocket and takes out a small gold box. He opens it, and inside, I see white powder.

I know what this is: cocaine. After slicing the cocaine into fine, thin lines, he uses a note and rolls it tight. With one end at his nostril and the other at the start of the line, he inhales the powder before standing and walking toward me. "I think it's best you have some before I tell you the story."

I take the note and repeat what he just did. The rush is immediate and Wolf laughs when I sneeze and scrunch up my nose.

"First time?" he asks. The gold box disappears back into his pocket, and he sits back on the crate.

I blink several times. Everything seems more focused, more defined. The maid still has her back to me in a show of pure disrespect.

"Amira!" My voice reaches my ear in a sing-song note.

I turn to Wolf, who grins. "I want to tell you a story."

I sit up. "Tell me."

"There once was a princess who was loved deeply by her three brothers, surrounded by warmth and affection. But tragedy struck, and two brothers died, leaving her vulnerable."

I snort. I wasn't a princess, but when Wolf narrows his eyes at me, I keep my sarcasm to myself.

"Her mother, consumed by grief and madness, turned against her, the once kind touches now twisted into pain. The princess was eventually handed off to a man named Diarmuid, who promised protection but soon abandoned her as well."

That part has me shifting in my seat. He had dumped me outside without a second glance. Left me to fend for myself on the street with no one. I bet he hasn't even looked for me.

"And then," Wolf's voice grows softer, "a prince came. He offered her a life where she'd never be alone again, a life where she could exact revenge on those who wronged her."

My hands tremble in my lap as he speaks, the story weaving into the fabric of my own life, mirroring the abandonment and pain I've known.

He reaches under his seat and produces an axe, the metal glinting ominously in the dim light. He offers it to me with a solemn nod. "I will give you a world of cages that will hold your enemies, Amira. A world where you can be the queen of your own justice."

As I take the axe, feeling its weight in my hands, a dark satisfaction fills me. Images of Niamh and Selene, the sources of so much of my misery, locked away as this maid is now, dance through my mind. I can't help but smile—a cold, hard smile that doesn't quite reach my eyes.

"Now, let them know they have messed with the wrong person," Wolf says, his voice a low growl as he strides to the cage and swings the door open. The maid's screams pierce the silence, her terror a stark contrast to the calm that settles over me.

With the axe in hand, I stand, my resolve hardening. Today marks the end of my victimhood and the beginning of something new—something fierce. As Wolf steps back, watching me with an approving gaze, I know I've crossed into a realm from which there is no return. My heart, once fragile, now beats a violent rhythm, ready to claim what is mine by right of pain and survival.

FIND OUT WHAT HAPPENS NEXT IN
WHEN KINGS BEND:

When Kings Rise

OTHER BOOKS BY VI CARTER

THE CELLS OF KALASHOV
THE COLLECTOR #1
THE SIXTH #2
THE HANDLER #3

MURPHY'S MAFIA MADE MEN
SINNER'S VOW #1
SAVAGE MARRIAGE #2
SCANDALOUS PLEDGE #3

SONS OF THE MAFIA
SINS OF THE MAFIA #0.5
VENGEANCE IN BLOOD #1

YOUNG IRISH REBELS SERIES
MAFIA PRINCE #1
MAFIA KING #2
MAFIA GAMES #3
MAFIA BOSS #4
MAFIA SECRETS #5

WILD IRISH SERIES
FATHER (PREQUEL)
VICIOUS #1
RECKLESS #2
RUTHLESS #3
FEARLESS #4
HEARTLESS #5

THE BOYNE CLUB
DARK #1
DARKER # 2
DARKEST #3
PITCH BLACK #4

THE OBSESSED DUET
A DEADLY OBSESSION #1
A CRUEL CONFESSION #2

BROKEN PEOPLE DUET
DECIEVE ME #1
SAVE ME #2

ABOUT THE AUTHOR

Vi Carter - the queen of **DARK ROMANCE**, the mistress of suspense, and the high priestess of *PLOT TWISTS!*

When she's not busy crafting tales of the **MAFIA** that'll leave you on the edge of your seat, you can find her baking up a storm, exploring the gorgeous Irish countryside, or spending time with her three little girls.

Vi's Young Irish Rebels series has been praised by readers and can be found in English, Dutch, German, Audible and soon will be available in French.

And let's not forget her two greatest loves: ***coffee and chocolate.*** If you ever need to bribe her, just offer up a mug of coffee and a slab of chocolate, and she'll be putty in your hands.

So, if you're ready to join Vi on a wild journey with the mafia, sign up for her newsletter and score a free book! Just be warned - her stories are so **ADDICTIVE**, you might not be able to put them down.

WHAT READERS ARE SAYING

EDITORIAL REVIEWS

"Vi Carter has once again blown my mind with another outstanding story. She never fails to create a masterpiece with memorable characters that leap off the page. This book is complete perfection."- USA Today Bestselling Author Khardine Gray

Vi is one of those authors who never disappoints. She weaves LOVE & DANGER effortlessly. *5 stars!*

I definitely recommend this book. It is SUSPENSEFUL and exciting. I enjoy reading Vi Carter's book. *5 stars!*

HOW TO KEEP IN TOUCH WITH VI CARTER

Visit Vi's website: https://author-vicarter.com/.
Join the newsletter: t.ly/yZWbX
Or scan the code below:

On Facebook, Instagram, TikTok and YouTube @darkauthorvicarter
and on Twitter @authorvicarter
Or scan the code below:

ACKNOWLEDGEMENTS

I'm very lucky to have such amazing readers and Beta Readers. I want to thank the following people who worked with me on this book.

Developmental Editor: Lori Wray White
Proofreader: Sherry Schafer
Blurb was written by: S.R. Frederick
Formatter: Elise Hoffman
Co-plotter: S. R. Frederick (https://pen-nibblers.com/about/)

Beta Readers:
Lucy Korth
Tami Thomason